Symbols for writing numbers: the **digits** or **numerals**
The **ten basic digits** for writing numbers (decimal system) are 0, 1, 2, 3, 4, 5, 6, 7, 8, a...

Natural Numbers = {1, 2, 3, 4, 5, 6, 7, 8, 9, 10, 11, 12, 13,...}
The natural numbers are also known as the counting numbers or the positive integers.

Whole numbers = {0, 1, 2, 3, 4, 5, 6, 7, 8, 9, 10, 11, 12, 13,...}
Integers = {...,-7, -6, -5. -4, -3, -2, -1, 0, 1, 2, 3, 4, 5, 6, 7,...}

Rational number : A number which can be written as the ratio of two integers.

Irrational number : A number which **cannot** be written as the ratio of two integers.

Example of a number showing the **place-value** relationships:

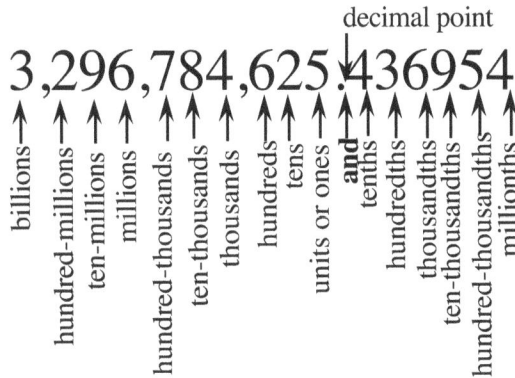

decimal point

$$3,296,784,625.436954$$

billions
hundred-millions
ten-millions
millions
hundred-thousands
ten-thousands
thousands
hundreds
tens
units or ones
and
tenths
hundredths
thousandths
ten-thousandths
hundred-thousandths
millionths

The above number is read " three billion, two hundred ninety-six million, seven hundred eighty-four thousand, six hundred twenty-five **and** four hundred thirty-six thousand, nine hundred fifty-four millionths".

Signed number operations: **1**. -7 - 9 = -16 (add); **2**. -7 - (-9) = -7 + 9 = 2 (subtract); **3**. -7(-9) = +63 (multiply).

Slope, m, of the line passing through the points (x_1, y_1) and (x_2, y_2) is given by $m = \dfrac{y_2 - y_1}{x_2 - x_1}$

Slope-intercept form of the **equation** of a line with slope m and y-intercept b, is $y = mx + b$.

An **equation** of the line with y-intercept b, and x-intercept a, is given by $y = -\dfrac{b}{a}x + b$.

Factoring: $a^2 - b^2 = (a + b)(a - b)$; **2**. $3a^2 - 12 = 3(a^2 - 4) = 3(a + 2)(a - 2)$.

Quadratic equations: **1**. If $(x - 2)(x + 3) = 0$, then $x = 2$ or $x = -3$; **2.** If $x^2 = 4$, then $x = \pm 2$.

Radicals: **1**. $\sqrt{12} = \sqrt{4}\sqrt{3} = 2\sqrt{3}$; **2**. $\sqrt{18} = \sqrt{9}\sqrt{2} = 3\sqrt{2}$.

Scientific notation: **1**. $436.7 = 4.367 \times 10^2$; **2**. $.00721 = 7.21 \times 10^{-3}$

Ratio: The ratio $a:b$ is the fraction $\dfrac{a}{b}$

Proportion: If a is to b as c is to d, then $\dfrac{a}{b} = \dfrac{c}{d}$

Exponents: **1**. $x^2 \cdot x^3 = x^5$; **2**. $(x^2)^3 = x^6$; **3**. $\dfrac{x^6}{x^2} = x^4$; **4**. $(xy)^3 = x^3 y^3$.

Factoring: 3. $a^3 + b^3 = (a + b)(a^2 - ab + b^2)$. **4.** $a^3 - b^3 = (a - b)(a^2 + ab + b^2)$.

Area & Perimeter of a Circle: Area, A, is given by $A = \pi r^2$; Perimeter, P, is given by $P = 2\pi r$

Distance Formula: $d = \sqrt{(x_2 - x_1)^2 + (y_2 - y_1)^2}$

Midpoint of a Line

The **midpoint** of the line, $P_1 P_2$, connecting the points $P_1(x_1, y_1)$ and $P_2(x_2, y_2)$ has the coordinates given by the following formulas:

x-coordinate, x_m, of the midpoint is given by $x_m = \dfrac{x_1 + x_2}{2}$

y-coordinate, y_m of the midpoint is given by $y_m = \dfrac{y_1 + y_2}{2}$

1	2	3	4	5	6
$1 \times 1 = 1$	$2 \times 1 = 2$	$3 \times 1 = 3$	$4 \times 1 = 4$	$5 \times 1 = 5$	$6 \times 1 = 6$
$1 \times 2 = 2$	$2 \times 2 = 4$	$3 \times 2 = 6$	$4 \times 2 = 8$	$5 \times 2 = 10$	$6 \times 2 = 12$
$1 \times 3 = 3$	$2 \times 3 = 6$	$3 \times 3 = 9$	$4 \times 3 = 12$	$5 \times 3 = 15$	$6 \times 3 = 18$
$1 \times 4 = 4$	$2 \times 4 = 8$	$3 \times 4 = 12$	$4 \times 4 = 16$	$5 \times 4 = 20$	$6 \times 4 = 24$
$1 \times 5 = 5$	$2 \times 5 = 10$	$3 \times 5 = 15$	$4 \times 5 = 20$	$5 \times 5 = 25$	$6 \times 5 = 30$
$1 \times 6 = 6$	$2 \times 6 = 12$	$3 \times 6 = 18$	$4 \times 6 = 24$	$5 \times 6 = 30$	$6 \times 6 = 36$
$1 \times 7 = 7$	$2 \times 7 = 14$	$3 \times 7 = 21$	$4 \times 7 = 28$	$5 \times 7 = 35$	$6 \times 7 = 42$
$1 \times 8 = 8$	$2 \times 8 = 16$	$3 \times 8 = 24$	$4 \times 8 = 32$	$5 \times 8 = 40$	$6 \times 8 = 48$
$1 \times 9 = 9$	$2 \times 9 = 18$	$3 \times 9 = 27$	$4 \times 9 = 36$	$5 \times 9 = 45$	$6 \times 9 = 54$
$1 \times 10 = 10$	$2 \times 10 = 20$	$3 \times 10 = 30$	$4 \times 10 = 40$	$5 \times 10 = 50$	$6 \times 10 = 60$
$1 \times 11 = 11$	$2 \times 11 = 22$	$3 \times 11 = 33$	$4 \times 11 = 44$	$5 \times 11 = 55$	$6 \times 11 = 66$
$1 \times 12 = 12$	$2 \times 12 = 24$	$3 \times 12 = 36$	$4 \times 12 = 48$	$5 \times 12 = 60$	$6 \times 12 = 72$
$1 \times 13 = 13$	$2 \times 13 = 26$	$3 \times 13 = 39$	$4 \times 13 = 52$	$5 \times 13 = 65$	$6 \times 13 = 78$
$1 \times 14 = 14$	$2 \times 14 = 28$	$3 \times 14 = 42$	$4 \times 14 = 56$	$5 \times 14 = 70$	$6 \times 14 = 84$
$1 \times 15 = 15$	$2 \times 15 = 30$	$3 \times 15 = 45$	$4 \times 15 = 60$	$5 \times 15 = 75$	$6 \times 15 = 90$
$1 \times 16 = 16$	$2 \times 16 = 32$	$3 \times 16 = 48$	$4 \times 16 = 64$	$5 \times 16 = 80$	$6 \times 16 = 96$
$1 \times 17 = 17$	$2 \times 17 = 34$	$3 \times 17 = 51$	$4 \times 17 = 68$	$5 \times 17 = 85$	$6 \times 17 = 102$
$1 \times 18 = 18$	$2 \times 18 = 36$	$3 \times 18 = 54$	$4 \times 18 = 72$	$5 \times 18 = 90$	$6 \times 18 = 108$
$1 \times 19 = 19$	$2 \times 19 = 38$	$3 \times 19 = 57$	$4 \times 19 = 76$	$5 \times 19 = 95$	$6 \times 19 = 114$
$1 \times 20 = 20$	$2 \times 20 = 40$	$3 \times 20 = 60$	$4 \times 20 = 80$	$5 \times 20 = 100$	$6 \times 20 = 120$

7	8	9	10	11	12
$7 \times 1 = 7$	$8 \times 1 = 8$	$9 \times 1 = 9$	$10 \times 1 = 10$	$11 \times 1 = 11$	$12 \times 1 = 12$
$7 \times 2 = 14$	$8 \times 2 = 16$	$9 \times 2 = 18$	$10 \times 2 = 20$	$11 \times 2 = 22$	$12 \times 2 = 24$
$7 \times 3 = 21$	$8 \times 3 = 24$	$9 \times 3 = 27$	$10 \times 3 = 30$	$11 \times 3 = 33$	$12 \times 3 = 36$
$7 \times 4 = 28$	$8 \times 4 = 32$	$9 \times 4 = 36$	$10 \times 4 = 40$	$11 \times 4 = 44$	$12 \times 4 = 48$
$7 \times 5 = 35$	$8 \times 5 = 40$	$9 \times 5 = 45$	$10 \times 5 = 50$	$11 \times 5 = 55$	$12 \times 5 = 60$
$7 \times 6 = 42$	$8 \times 6 = 48$	$9 \times 6 = 54$	$10 \times 6 = 60$	$11 \times 6 = 66$	$12 \times 6 = 72$
$7 \times 7 = 49$	$8 \times 7 = 56$	$9 \times 7 = 63$	$10 \times 7 = 70$	$11 \times 7 = 77$	$12 \times 7 = 84$
$7 \times 8 = 56$	$8 \times 8 = 64$	$9 \times 8 = 72$	$10 \times 8 = 80$	$11 \times 8 = 88$	$12 \times 8 = 96$
$7 \times 9 = 63$	$8 \times 9 = 72$	$9 \times 9 = 81$	$10 \times 9 = 90$	$11 \times 9 = 99$	$12 \times 9 = 108$
$7 \times 10 = 70$	$8 \times 10 = 80$	$9 \times 10 = 90$	$10 \times 10 = 100$	$11 \times 10 = 110$	$12 \times 10 = 120$
$7 \times 11 = 77$	$8 \times 11 = 88$	$9 \times 11 = 99$	$10 \times 11 = 110$	$11 \times 11 = 121$	$12 \times 11 = 132$
$7 \times 12 = 84$	$8 \times 12 = 96$	$9 \times 12 = 108$	$10 \times 12 = 120$	$11 \times 12 = 132$	$12 \times 12 = 144$
$7 \times 13 = 91$	$8 \times 13 = 104$	$9 \times 13 = 117$	$10 \times 13 = 130$	$11 \times 13 = 143$	$12 \times 13 = 156$
$7 \times 14 = 98$	$8 \times 14 = 112$	$9 \times 14 = 126$	$10 \times 14 = 140$	$11 \times 14 = 154$	$12 \times 14 = 168$
$7 \times 15 = 105$	$8 \times 15 = 120$	$9 \times 15 = 135$	$10 \times 15 = 150$	$11 \times 15 = 165$	$12 \times 15 = 180$
$7 \times 16 = 112$	$8 \times 16 = 128$	$9 \times 16 = 144$	$10 \times 16 = 160$	$11 \times 16 = 176$	$12 \times 16 = 192$
$7 \times 17 = 119$	$8 \times 17 = 136$	$9 \times 17 = 153$	$10 \times 17 = 170$	$11 \times 17 = 187$	$12 \times 17 = 204$
$7 \times 18 = 126$	$8 \times 18 = 144$	$9 \times 18 = 162$	$10 \times 18 = 180$	$11 \times 18 = 198$	$12 \times 18 = 216$
$7 \times 19 = 133$	$8 \times 19 = 152$	$9 \times 19 = 171$	$10 \times 19 = 190$	$11 \times 19 = 209$	$12 \times 19 = 228$
$7 \times 20 = 140$	$8 \times 20 = 160$	$9 \times 20 = 180$	$10 \times 20 = 200$	$11 \times 20 = 220$	$12 \times 20 = 240$

13	14	15	16	17	18
$13 \times 1 = 13$	$14 \times 1 = 14$	$15 \times 1 = 15$	$16 \times 1 = 16$	$17 \times 1 = 17$	$18 \times 1 = 18$
$13 \times 2 = 26$	$14 \times 2 = 28$	$15 \times 2 = 30$	$16 \times 2 = 32$	$17 \times 2 = 34$	$18 \times 2 = 36$
$13 \times 3 = 39$	$14 \times 3 = 42$	$15 \times 3 = 45$	$16 \times 3 = 48$	$17 \times 3 = 51$	$18 \times 3 = 54$
$13 \times 4 = 52$	$14 \times 4 = 56$	$15 \times 4 = 60$	$16 \times 4 = 64$	$17 \times 4 = 68$	$18 \times 4 = 72$
$13 \times 5 = 65$	$14 \times 5 = 70$	$15 \times 5 = 75$	$16 \times 5 = 80$	$17 \times 5 = 85$	$18 \times 5 = 90$
$13 \times 6 = 78$	$14 \times 6 = 84$	$15 \times 6 = 90$	$16 \times 6 = 96$	$17 \times 6 = 102$	$18 \times 6 = 108$
$13 \times 7 = 91$	$14 \times 7 = 98$	$15 \times 7 = 105$	$16 \times 7 = 112$	$17 \times 7 = 119$	$18 \times 7 = 126$
$13 \times 8 = 104$	$14 \times 8 = 112$	$15 \times 8 = 120$	$16 \times 8 = 128$	$17 \times 8 = 136$	$18 \times 8 = 144$
$13 \times 9 = 117$	$14 \times 9 = 126$	$15 \times 9 = 135$	$16 \times 9 = 144$	$17 \times 9 = 153$	$18 \times 9 = 162$
$13 \times 10 = 130$	$14 \times 10 = 140$	$15 \times 10 = 150$	$16 \times 10 = 160$	$17 \times 10 = 170$	$18 \times 10 = 180$
$13 \times 11 = 143$	$14 \times 11 = 154$	$15 \times 11 = 165$	$16 \times 11 = 176$	$17 \times 11 = 187$	$18 \times 11 = 198$
$13 \times 12 = 156$	$14 \times 12 = 168$	$15 \times 12 = 180$	$16 \times 12 = 192$	$17 \times 12 = 204$	$18 \times 12 = 216$
$13 \times 13 = 169$	$14 \times 13 = 182$	$15 \times 13 = 195$	$16 \times 13 = 208$	$17 \times 13 = 221$	$18 \times 13 = 234$
$13 \times 14 = 182$	$14 \times 14 = 196$	$15 \times 14 = 210$	$16 \times 14 = 224$	$17 \times 14 = 238$	$18 \times 14 = 252$
$13 \times 15 = 195$	$14 \times 15 = 210$	$15 \times 15 = 225$	$16 \times 15 = 240$	$17 \times 15 = 255$	$18 \times 15 = 270$
$13 \times 16 = 208$	$14 \times 16 = 224$	$15 \times 16 = 240$	$16 \times 16 = 256$	$17 \times 16 = 272$	$18 \times 16 = 288$
$13 \times 17 = 221$	$14 \times 17 = 238$	$15 \times 17 = 255$	$16 \times 17 = 272$	$17 \times 17 = 289$	$18 \times 17 = 306$
$13 \times 18 = 234$	$14 \times 18 = 252$	$15 \times 18 = 270$	$16 \times 18 = 288$	$17 \times 18 = 306$	$18 \times 18 = 324$
$13 \times 19 = 247$	$14 \times 19 = 266$	$15 \times 19 = 285$	$16 \times 19 = 304$	$17 \times 19 = 323$	$18 \times 19 = 342$
$13 \times 20 = 260$	$14 \times 20 = 280$	$15 \times 20 = 300$	$16 \times 20 = 320$	$17 \times 20 = 340$	$18 \times 20 = 360$

19	20	21	22	23	24
$19 \times 1 = 19$	$20 \times 1 = 20$	$21 \times 1 = 21$	$22 \times 1 = 22$	$23 \times 1 = 23$	$24 \times 1 = 24$
$19 \times 2 = 38$	$20 \times 2 = 40$	$21 \times 2 = 42$	$22 \times 2 = 44$	$23 \times 2 = 46$	$24 \times 2 = 48$
$19 \times 3 = 57$	$20 \times 3 = 60$	$21 \times 3 = 63$	$22 \times 3 = 66$	$23 \times 3 = 69$	$24 \times 3 = 72$
$19 \times 4 = 76$	$20 \times 4 = 80$	$21 \times 4 = 84$	$22 \times 4 = 88$	$23 \times 4 = 92$	$24 \times 4 = 96$
$19 \times 5 = 95$	$20 \times 5 = 100$	$21 \times 5 = 105$	$22 \times 5 = 110$	$23 \times 5 = 115$	$24 \times 5 = 120$
$19 \times 6 = 114$	$20 \times 6 = 120$	$21 \times 6 = 126$	$22 \times 6 = 132$	$23 \times 6 = 138$	$24 \times 6 = 144$
$19 \times 7 = 133$	$20 \times 7 = 140$	$21 \times 7 = 147$	$22 \times 7 = 154$	$23 \times 7 = 161$	$24 \times 7 = 168$
$19 \times 8 = 152$	$20 \times 8 = 160$	$21 \times 8 = 168$	$22 \times 8 = 176$	$23 \times 8 = 184$	$24 \times 8 = 192$
$19 \times 9 = 171$	$20 \times 9 = 180$	$21 \times 9 = 189$	$22 \times 9 = 198$	$23 \times 9 = 207$	$24 \times 9 = 216$
$19 \times 10 = 190$	$20 \times 10 = 200$	$21 \times 10 = 210$	$22 \times 10 = 220$	$23 \times 10 = 230$	$24 \times 10 = 240$
$19 \times 11 = 209$	$20 \times 11 = 220$	$21 \times 11 = 231$	$22 \times 11 = 242$	$23 \times 11 = 253$	$24 \times 11 = 264$
$19 \times 12 = 228$	$20 \times 12 = 240$	$21 \times 12 = 252$	$22 \times 12 = 264$	$23 \times 12 = 276$	$24 \times 12 = 288$
$19 \times 13 = 247$	$20 \times 13 = 260$	$21 \times 13 = 273$	$22 \times 13 = 286$	$23 \times 13 = 299$	$24 \times 13 = 312$
$19 \times 14 = 266$	$20 \times 14 = 280$	$21 \times 14 = 294$	$22 \times 14 = 308$	$23 \times 14 = 322$	$24 \times 14 = 336$
$19 \times 15 = 285$	$20 \times 15 = 300$	$21 \times 15 = 315$	$22 \times 15 = 330$	$23 \times 15 = 345$	$24 \times 15 = 360$
$19 \times 16 = 304$	$20 \times 16 = 320$	$21 \times 16 = 336$	$22 \times 16 = 352$	$23 \times 16 = 368$	$24 \times 16 = 384$
$19 \times 17 = 323$	$20 \times 17 = 340$	$21 \times 17 = 357$	$22 \times 17 = 374$	$23 \times 17 = 391$	$24 \times 17 = 408$
$19 \times 18 = 342$	$20 \times 18 = 360$	$21 \times 18 = 378$	$22 \times 18 = 396$	$23 \times 18 = 414$	$24 \times 18 = 432$
$19 \times 19 = 361$	$20 \times 19 = 380$	$21 \times 19 = 399$	$22 \times 19 = 418$	$23 \times 19 = 437$	$24 \times 19 = 456$
$19 \times 20 = 380$	$20 \times 20 = 400$	$21 \times 20 = 420$	$22 \times 20 = 440$	$23 \times 20 = 460$	$24 \times 20 = 480$

If you can remember a needed information, you make decisions faster, you work faster, and you are more productive.

Yes, you can memorize them.
Squares of Natural Numbers

$1 \times 1 = 1$	$26 \times 26 = 676$
$2 \times 2 = 4$	$27 \times 27 = 729$
$3 \times 3 = 9$	$28 \times 28 = 784$
$4 \times 4 = 16$	$29 \times 29 = 841$
$5 \times 5 = 25$	$30 \times 30 = 900$
$6 \times 6 = 36$	$31 \times 31 = 961$
$7 \times 7 = 49$	$32 \times 32 = 1024$
$8 \times 8 = 64$	$33 \times 33 = 1089$
$9 \times 9 = 81$	$34 \times 34 = 1156$
$10 \times 10 = 100$	$35 \times 35 = 1225$
$11 \times 11 = 121$	$36 \times 36 = 1296$
$12 \times 12 = 144$	$37 \times 37 = 1369$
$13 \times 13 = 169$	$38 \times 38 = 1444$
$14 \times 14 = 196$	$39 \times 39 = 1521$
$15 \times 15 = 225$	$40 \times 40 = 1600$
$16 \times 16 = 256$	$41 \times 41 = 1681$
$17 \times 17 = 289$	$42 \times 42 = 1764$
$18 \times 18 = 324$	$43 \times 43 = 1849$
$19 \times 19 = 361$	$44 \times 44 = 1936$
$20 \times 20 = 400$	$45 \times 45 = 2025$
$21 \times 21 = 441$	$46 \times 46 = 2116$
$22 \times 22 = 484$	$47 \times 47 = 2209$
$23 \times 23 = 529$	$48 \times 48 = 2304$
$24 \times 24 = 576$	$49 \times 49 = 2401$
$25 \times 25 = 625$	$50 \times 50 = 2500$

FREMPONG'S STEP-BY-STEP SERIES IN MATHEMATICS

Elementary Algebra

Excellent for Exam Preparation

SIXTH EDITION

A.A. FREMPONG

Elementary Algebra

ISBN 978-1-946485-16-8

Sixth Edition

Printed in the United States of America

In Memory of My Parents

Mom:
She was a devoted mother, sharing, kind, kinder to strangers and generous to a fault. She never cursed, she never hated; she never cheated, and she never envied. She never lied, and she never got angry. Once, she nursed an almost dying stranger renting a room in her house back to good health to the extent that the relatives of this renter later travelled one hundred miles just to thank mom. She was always peaceloving and forever forgiving.
An angel once lived on this earth to serve others.

Dad:
A great dad, kind, generous and forgiving. He emphasized and was an example of both formal education and self-education. A veterinarian, a bacteriologist, an Associate of the Institute of Medical Laboratory Technology (UK), a Fellow of the Royal Society of Health (UK); an incorruptible civil servant; his book on ticks has always inspired me to write whenever the need arises.

NOTE TO THE STUDENT

This book was written with you in mind at all times. You may use this book as the course textbook or as a review book since the book gets to the point quickly on all relevant topics and yet covers these topics in detail.

Begin to master the definitions and the solutions of the sample problems thoroughly. (You have mastered a sample problem if you can solve the sample problem and similar problems without any reference to this book or any other source. For some problems, two or more methods are presented. Read the various methods and decide which methods you would like to remember; but always be aware of the existence of the other methods, in case the need arises. After having mastered the sample problems, try the exercise problems. The answers to these problems are presented immediately after the problems. You may cover the answers with paper before you attempt these problems, if the answers are too obvious. You may refer back and forth to the solved problems when you do not remember how to proceed.

You may also attempt some of the sample problems first, if you have been exposed to the topics previously, before reading the solution methods, and in this approach, the sample problems become more practice problems for you.

As a reminder, in any book, do not dwell on the few inadvertent errors you may find, but rather concentrate on what is useful to you.

For this book to be useful both as the course textbook, as well as review for exams, it is **important** to **Understand, Remember, Apply**, and Remember the material covered.

Wishing you Good Luck on all the exams
A.A.Frempong

Books in the series by the author: Integrated Arithmetic; **Elementary Algebra**; Intermediate Algebra, Elementary Mathematics; Intermediate Mathematics; Elementary & Intermediate Mathematics (combined); College Algebra; College Trigonometry; College Algebra & Trigonometry and Calculus 1 & 2

PREFACE TO THE SIXTH EDITION

This **volume** retains the spirit of the previous editions. The topics have been expanded and covered quite extensively, especially those topics students find difficult to understand or sort out. For such topics, I have endeavored to organize the coverage in such a way as to minimize student confusion. One such topic whose coverage is well-organized is "finding an equation of a straight line" (page 161-174).

   In writing this edition, I was encouraged by comments from students who had used the previous editions. One student remarked: "Your book gets to the point. It does not beat about the bush."

   In addition to the regular programs, this book can also be used for short term programs such as mini-sessions and immersion programs, and distance learning programs, with the instructor, perhaps, supplementing with hand-out homework problems, when necessary.

I gratefully acknowledge the help and encouragement of students, colleagues, and friends.

A. A. Frempong
New York,

CONTENTS

Page

CHAPTER 1
1

Signed Numbers and Real Number Operations

Lesson 1: **Addition and Subtraction of Signed Numbers** 1

Lesson 2: **Multiplication and Division of Signed Numbers** 6

Lesson 3: **Operations Involving Zero, Powers and Roots of Numbers** 9

CHAPTER 2
14

Order of Operations and Evaluation of Algebraic Expressions

Lesson 4: **Order of Operations in Performing Algebraic operations** 14

Lesson 5: **Evaluation of Algebraic Expressions Containing Variables** 17

CHAPTER 3
21

Exponents

Lesson 6: **Rules of Exponents and Applications** 21

Lesson 7: **Scientific Notation** 25

CHAPTER 4
27

Algebraic Expressions and Polynomials

Lesson 8: **Basic Definitions; Addition of Like terms** 27

Lesson 9: **Removing Grouping Symbols; Addition,** Subtraction of Polynomials 30

Lesson 10: **Multiplication and Division of Polynomials** 33

CHAPTER 5
45

First Degree Equations Containing One Variable
(Simple Linear Equations)

Lesson 11: **Axioms for Solving Equations** 45

Lesson 12: **Solving First Degree Equations** 47

Lesson 13: **Solving Decimal and Literal Equations** 54

CHAPTER 6
59

WORD PROBLEMS

Lesson 14: **Words and Phrases for Translating Word Problems into Equations** 59

Lesson 15: **Solving Number Word Problems** 61

Lesson 16: **Perimeter Problems; Work-Time Problems** 63

Lesson 17: **Motion Problems** (Distance-Time-Speed Problems) 66

Lesson 18: **Consecutive Integer Problems; Age Problems; Percent Problems** 78

Lesson 19: **Mixture and Mixture-Like Problems** 85

CHAPTER 7
RATIO
92 ᵡ

Lesson 20: **Definition; Reduction of Ratios to Lowest Terms** 92
Lesson: 21 **Using Ratios to Compare Quantities** 94
Lesson: 22 **Using Ratios to Divide a Quantity into Parts** 96

CHAPTER 8
PROPORTION (VARIATION)
98

Lesson 23**: Introduction to Direct and Indirect Proportion** 98
Lesson 24: **Methods for Solving Direct Proportion Problems** 103
Lesson 25: **Methods for Solving Inverse Proportion Problems** 119
Lesson 26: **Compound Proportion (Compound Variation) Problems** 124

CHAPTER 9A
Factoring Polynomials
128

Lesson 27: **Greatest Common Factor and Common Monomial Factoring** 128
Lesson 28: **Difference between two Squares; Factoring by Grouping** 131
Lesson 29: **Factoring Monic Quadratic Trinomials** (x^2-term coefficient is 1) 133
Lesson 30: **Non-monic Quadratic Trinomials** (x^2-term coefficient is **not** 1) 138
Lesson 31**: Factoring trinomials quadratic in form (Substitution method)** 142
Lesson 32: **Solving Quadratic Equations by Factoring and Applications** 143

CHAPTER 9B
More Factoring Polynomials
147

Lesson 33: **Factoring the Sum and Difference between two Cubes** 147

CHAPTER 10A
149

Lesson 34A: **Rectangular Coordinate System** 149
Lesson 34B: **Drawing the Graph of a Straight Line** 152
Lesson 35: **Points on a Line; Slopes of lines; Intercepts** 156
Lesson 36: **Equations of Straight Lines** 161

CHAPTER 10B
171

Lesson 37A: **Equations of Parallel lines and Perpendicular Lines** 171
Lesson 37B: **Midpoint of a Line; Distance between Points;** Distance Formula 174
Lesson 37C: **Applications: 1.** Distance from a point to a line whose equation is given; 178
2. Perpendicular Bisector of a Line

CHAPTER 11
184
Solving Systems of Linear Equations (Simultaneous Equations)

Lesson 38: **Addition Method; Substitution Method; Equality Method** 184
Lesson 39: **Graphical Method** 189

CHAPTER 12 190
Applications of Solutions of Systems of Linear Equations
Lesson 40: **Coin Problems and Coin-Like Problems** 190
Lesson 41: **Distance-Time-Speed Problems: Travel by Water or by Air** 195

CHAPTER 13A 201
Exponents and Radicals
Lesson 42: **Review of Exponents and Radicals** 201
Lesson 43: **Definitions and Simplification of Radicals** 204
Lesson 44: **Addition and Multiplication of Radicals** 208

CHAPTER 13B 212
More Radicals
Lesson 45: **Division of Radicals; Rationalization of Denominators** 212
Lesson 46: **Fractional Exponents and Reduction of Indices** 216

CHAPTER 14 220
Lesson 47. **Right Triangle, Pythagorean Theorem and Applications** 220

CHAPTER 15 224
Areas and Perimeters
Lesson 48: **Area and Perimeter of Triangles, Rectangles, and Trapezoids** 224
Lesson 49: **Area and Perimeter of a Circle; Arc Length and Sector of a Circle Area and Perimeter of Composite Figures** 229

CHAPTER 16 237
Algebraic Fractions
Lesson 50: **Reduction, Multiplication, Division of Algebraic Fractions** 237
Lesson 51: **Addition of Fractions; Simplification of Complex Fraction** 241

CHAPTER 17 249
Inequalities
Lesson 52: **Set Notation, Set Operations, Interval Notation** 249
Lesson 53: **Linear Inequalities; Compound Inequalities; Word Problems** 252

Extra Chapters
The following chapters may be included in some programs.
Consult the course syllabus.

CHAPTER 18: Quadratic Equations p.259

CHAPTER 19: Functions, Relations, Functional Notation p.271

CHAPTER 20: Sketching Parabolas p.278

CHAPTER 21: Solving Fractional (Rational Equations) p.285

CHAPTER 22: Solving Radical Equations p.289

CHAPTER 23: Basic Review For Geometry p.292

APPENDIX A

EXTRA: **Factoring Perfect Square Trinomials** p.305
Conversion Factors for Measurements p.317

APPENDIX B
About Measurements p..307
Standard Unit, Error, Rounding-off Numbers, Significant Digits, Scientific Notation

INDEX p. 319

CHAPTER 1

Signed Numbers and Real Number Operations

Lesson 1: **Addition and Subtraction of Signed Numbers**

Lesson 2: **Multiplication and Division of Signed Numbers**

Lesson 3: **Operations Involving Zero, Powers and Roots of Numbers**

Lesson 1
Addition and Subtraction of Signed Numbers

Signed Numbers

A signed number is a number with either a plus sign " +" or a minus sign "-" preceding it (in front of it). If there is no sign in front of a number , we will assume that the number has a plus sign. We call a number with a plus sign a positive number, and we call a number with a minus sign a negative number. For a positive number, we sometimes do not write the plus sign, but for a negative number, we **must** always write the minus sign.

The Real Number Line and Real numbers

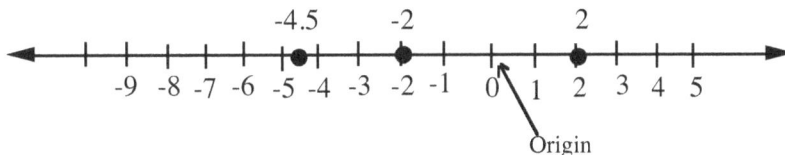

Figure 1

The real number line is a horizontal straight line with equally spaced intervals as in Figure 1 above. We label a point called the origin, 0 (zero). Points to the right of the origin are labeled positive and points to the left of the origin are labeled negative. The numbers increase as one moves from the left to the right on the real number line. Roughly speaking, a real number is a number that can be represented by a point on the real number line. The real numbers consists of the integers, fractions, mixed numbers, decimals, and radicals. In Figure 1, if the real numbers, -4.5, -2, and 2 are of interest, we can represent them by the dots shown. Every point on this line is associated with a real number; and every real number is associated with a point on this line. We can also say that the set of real numbers consists of the signed numbers and zero.

Absolute Value

The absolute value of a signed number may be defined as the number without its sign.

Examples: **1**. The absolute value of -4 , symbolized $|-4| = 4$.

   2. The absolute value of 6, symbolized $|6| = 6$.

   3.   The absolute value of $+9$ is 9.

   4. The absolute value of 0 is 0.

Note: The absolute value of a signed number is also its distance from zero on the number line.

Addition of Two Signed Numbers

To add any two signed numbers, first determine which of the following two cases is involved.

Case 1: The two numbers have the same sign
>   Step 1: Add the absolute values of the numbers (the unsigned parts)
>   Step 2: Write the common sign in front of the sum from Step 1.

Examples   **1** . $(+2) + (+3) = +5$ or 5

2 . $(-3) + (-4) = -7$

3. $(-.5) + (-.25) = (-.75)$

Case 2: The two numbers have different signs

Step 1 : Subtract the smaller absolute value from the larger absolute value
>   (This subtraction is exactly like the subtraction in arithmetic, where we always subtract a smaller number from a larger number)

Step 2. Write the sign of the number with the larger absolute value in front of the difference (result) from Step 1.

Examples   **1** . $(+5) + (-7) = -2$

2 . $(-6) + (+9) = +3$ or 3

3. $(-8) + (+6) = -2$

Note that   **1** . $-8 + 6$ implies $(-8) + (+6) = -2$
        2 . $-6 + 9$ implies $(-6) + (9) = +3$ or 3
        3 . $-8 + 9 - 7$ implies that we are to add the numbers, with each number carrying the sign in front of it, and thus
            $-8 + 9 - 7 = (-8) + (+9) + (-7) = -6$

Absolute value defined more formally:

The absolute value of a real number x is x if x is a positive number or zero, but it is $-x$ if x is a negative number .   (i.e. the negative of a negative number).

Subtraction of Signed Numbers

Procedure:

Change the sign of the number being subtracted, **and add** the changed number to the number from which you are subtracting (i.e., add the negative of the subtrahend to the minuend).

Examples   **1.**   (+5) - (+7)   (subtract)   (Here, +5 is the minuend and +7 is the subtrahend)

Step 1: We change +7 to -7

Step 2: We add +5 and -7   (-7 is the changed number; it is the negative of +7)

(+5) + (-7)   (add)

= - 2

2.   (-6) - (-9)   (subtract)

Step 1: We change -9 to +9

Step 2: We add -6 and +9   (+9 is the changed number)

(-6) + (+9)   (add)

= +3

3.   -7 - (-10)   (subtract)

Step 1: We change -10 to + 10

Step 2: We add -7 and + 10   (+10 is the changed number)

-7 + (+10)   (add)

= +3

4.   8 - (12)   (subtract)

Step 1: We change 12 to -12

Step 2: We add 8 and -12   (-12 is the changed number)

8 + (-12)   (add)

= -4

Note the following about the plus "+" and minus "-" signs.

The plus sign "+" may imply either the sign of a positive number, for example, +2, positive 2; or it may imply the operation of addition, for example, (+2) + (-8).

⎯⎯ Add

The minus sign "-" may imply the sign of a negative number, for example, -6, negative 6, or the operation of subtraction, for example, (-6) - (+7).

⎯⎯ Subtract

Note also that whenever we are to "add" we must use the rules for the addition of signed numbers and to subtract we must use the subtraction rule(s) for signed numbers.

Subtraction defined more formally:

To subtract a signed number y from another signed number x means add the negative of y to x.

that is, $x - y = x + (-y)$.

Lesson 1 Exercises

A. Add: **1.** (-3) + (-9);   **2.** (+5) + (+2);   **3.** (-7) + (-6)

Answers **1.** -12;   **2.** +7 or 7 ; **3.** -13

B. Add: **1.** (-8) + (3);   **2.** (5) + (-9); **3.** (7) + (-4)

Answers   **1.** -5; **2.** -4 ;   **3.** +3 or 3

C. Simplify: **1.** 5 - 7 ;   **2.** - 4 - 7;   **3.** -8 + 3 - 2;   **4.** -3 + 8; **5.** -2 + 9 + 3;

Answers **1.** -2 ; **2.** -11 ; **3.** -7;   **4.** 5 ;   **5.** 10.

D. Perform the indicated operations:

1. (+7) + (+8) ;   **2.** $-3\frac{2}{5} + 4$;   **3.** (-6) + (-8) ;   **4.** 12 + (-15)

5 . (+9) + (7) ;   **6.** (-3) + (4) ;   **7.** 9 + (8) ;   **8.** -3 + 1

9. $(-2\frac{1}{2}) + (\frac{1}{4})$;   **10.** - 5 + (8) ;   **11.** (4.3) + (-.21);   **12.** 9 - 12

13. (-6) + (-5);   **14.** - 6 + 5 ;   **15.** 8 - 0 ;   **16.** (-3) + (-2)

17. - 4 + 8;   **18.** -1 - 1;   **19.** $-\frac{11}{2} + 3$;   **20.** 5.32 - 7.45

Also find the following:

21. | -2 |=   **22.** | +3 | =   **23.** $| -1\frac{1}{2} |=$ **24.** | -5 | =

Answers: **1.** 15 ; **2.** $\frac{3}{5}$; **3.** -14 ; **4.** -3 ; **5.** 16; **6.** 1; **7.** 17 ; **8.** -2; **9.** - $2\frac{1}{4}$; **10.** 3; **11.** 4.09 ; **12.** -3 ;

13. -11 ; **14.** -1 ; **15.** 8 ; **16.** -5 ; **17.** 4 ; **18.** -2 ; **19.** $-\frac{5}{2}$ or $-2\frac{1}{2}$; **20.** -2.13 ; **21.** 2 ; **22.** 3 ;

23 $1\frac{1}{2}$; **24.** 5

E.

Simplify the following:

1. (-5) - (-7) =   **2.** (-6) - (+7) =   **3.** -6 - (2)   **4.** 6 - (9)

5. -1 - (-1) =   **6.** $-2\frac{1}{2} - (2\frac{3}{4})$ **7.** (8) - (-10)

8. Subtract 6 from 9   **9.** Subtract 9 from 6.   **10.** From 4, subtract 6

11. $\frac{3}{4} - (-\frac{1}{2}))$ **12.** (-4) - (7)   **13.** (-6) - (-8)   **14.** - 8 - (6)

Answers: **1.** 2 ;   **2.** -13;   **3.** -8 ;   **4.** -3 ;   **5.** 0 ; **6.** $-5\frac{1}{4}$; **7.** 18 ;   **8.** 3;   **9.** -3;   **10.** -2; **11.** $1\frac{1}{4}$;

 12. -11 ; **13.** 2 ; **14.** -14

Lesson 2

Multiplication and Division of Signed Numbers

Multiplication of Two Signed Numbers

Implication of Multiplication Example: $2 \times 3 = 2(3) = (2)(3) = (2) \cdot (3) = (2)3 = 2 \cdot 3 = 6$

Step 1: Multiply the absolute values.

Step 2: The product is positive (that is, use a plus sign or no sign) if the two numbers have the same sign; but the product is negative (that is, use a minus sign) if the two numbers have different signs.

Examples. The two numbers having the same sign

>   **1**. $(+2)(+3) = +6$ or 6

>   **2**. $(-2)(-3) = +6$ or 6

>   **3**. $-4(-2) = + 8$

The two numbers having different signs

>   **1**. $(-4)(+2) = -8$

>   **2**. $(+5)(-3) = - 15$

>   **3**. $-6(4) = -24$

>   **4**. $(3)(-7) = - 21$

Note the following distinctions:

>   **1**. $-7 - 8 = (-7) + (-8) = -15$ (Addition)

>   **2**. $-7 - (-8) = -7 + (+8) = 1$ (Subtraction)

>   **3**. $-7(-8) = 56$ (Multiplication)

Note also that $(-3)(-2)(-4) = (+6)(-4) = -24$ (The multiplication rules are for **two** numbers at a time.)

Division of Signed Numbers

The rules of signs for division are the same as those for multiplication.

Procedure:

Step 1: Divide the absolute values of the numbers.

Step 2: The quotient is positive (that is, use a plus sign or no sign) if the two numbers have the same sign; but the quotient is negative (that is, use a minus sign) if the two numbers have different signs.

Examples: The two numbers have the same sign

1. $\frac{+8}{+4} = +2$ or 2; 2. $\frac{-8}{-4} = +2$ or 2; 3. $\frac{-12}{-9} = +\frac{4}{3})$ or $\frac{4}{3}$

The two numbers have different signs

1. $\frac{-6}{+2} = -3$; 2. $\frac{8}{-2} = -4$; 3. $\frac{-13}{3} = -\frac{13}{3}$

Reciprocals: The reciprocal of a real number A is $\frac{1}{A}$. Example: The reciprocal of $\frac{2}{3}$ is $\frac{3}{2}$ .Thus, to find the reciprocal of a number, invert the number (or interchange the numerator and the denominator).

The reciprocal of number is also known as the multiplicative inverse of that number.

The product of a number and its reciprocal is 1. Example $\frac{1}{4} \times \frac{4}{1} = 1$

Application: In $\frac{3}{7} \div \frac{9}{28}$ ($\frac{3}{7}$ is the dividend; $\frac{9}{28}$ is the divisor)

Procedure : Multiply the dividend by the **reciprocal** of the divisor. (Same as invert the divisor and multiply)

$$\frac{3}{7} \div \frac{9}{28} = \frac{3}{7} \times \frac{28}{9} = \frac{4}{3}$$

Lesson 2 Exercises

A. Simplify: **1.** (-5)(-4);   **2.** -8(-9) ;   **3.** 6(7) ;   **4.** +5(4);   **5.** (-6)(+7);   **6.** -4(9)

       7. -6(+3);   **8.** 2(-8)

Answers **1.** 20;   **2.** 72;   **3.** 42;   **4.** 20 ;   **5.** -42;   **6.** -36;   **7.** -18;   **8.** -16

B. Perform the indicated operations:
     1. -8 - 8;   **2.** -8(-8)**;**   **3.** -6 - (8);   **4.** -6 - (-8)**;**   **5.** 9 - 7 + 3 - 8

Answers **1.** -16 ; **2.** 64; **3.** -14 **4.** +2;   **5.** -3

C. Perform the indicated operations:

1. (-4)(-5) =   **2.** (-8)(6) =   **3.** 8(-7) =

4. -8(-7) =   **5.** -1(-5) =   **6.** +6(+4) =

7. 5(6) =   **8.** $4\frac{1}{3}\left(\frac{1}{2}\right)) =$ **9.** (2.3)(.5) =

10. (-4)(+2) =   **11.** (-4)(2) =   **12.** (-3)(-2)(-4) =
Review

13. - 3 - 4 =   **14.** 7 - (-10) =   **15.** -8 - 9 =

16. - 4 + (3) =   **17.** 3 - (-8) =   **18.** 4 - (9) =

19. $5\frac{3}{4} - 8\frac{1}{4}$ =   **20.** (-10) - (-12) =

Answers: **1.** 20 ;   **2.** -48 ;   **3.** -56 ;   **4.** 56 ;   **5.** 5 ;   **6.** 24 ;   **7.** 30;   **8.** $2\frac{1}{6}$;   **9.** 1.15 ;   **10.** -8 ;
     11. -8 ;   **12.** -24 ;   **13** -7 ;   **14.** 17;   **15.** -17 ;   **16.** -1 ;   **17.** 11 ;   **18.** -5 ;   **19.** $-2\frac{1}{2}$; **20.** 2

D. Simplify: **1.** (-6) ÷ (-3);   **2.** $\frac{-12}{-4}$;   **3.** $\frac{-4}{-6}$;   **4.** (-10) ÷ (2);   **5.** $\frac{24}{-6}$;   **6.** $\frac{-15}{3}$

Answers   **1.** 2 ;   **2.** 3 ;   **3.** $\frac{2}{3}$;   **4.** -5 ;   **5.** -4;   **6.** -5

E. Carry out the indicated operations:

1. (+12) ÷ (+2);   **2.** (-6) ÷ (-2);   **3.** $\frac{14}{-7}$;   **4.** (6) ÷ (-2);   **5.** $-\frac{1}{2} \div \frac{1}{4}$;

6. Divide -9 by 2;   **7.** Divide $2\frac{1}{4}$ by $-\frac{1}{4}$;   **8.** Simplify: $\frac{4\frac{1}{5}}{2\frac{1}{5}}$;   **9.** Simplify: $\frac{-7.5}{.3}$;

   10. Simplify: $\frac{4.28}{-100}$;   **11.** $\frac{+12}{-16}$ **12.** $\frac{-16}{12}$;   **13.** Find the reciprocal of $\frac{4}{5}$.

Review: **14.** -4 - 2 =   **15.** 8 - 3 =   **16.** 3 - 8 =

       17. -1 - 1 =   **18.** 3 - (-4) =   **19.** 4 - (12) =

Answers: **1.** 6 ; **2.** 3 ; **3.** -2 ; **4.** -3 ; **5.** -2 ; **6.** $-\frac{9}{2}$; **7.** -9; **8.** $1\frac{10}{11}$;   **9.** -25 ; **10.** -.0428; **11.** $-\frac{3}{4}$;
     12. $-\frac{4}{3}$ or $-1\frac{1}{3}$;   **13.** $\frac{5}{4}$;**14.** -6 ; **15.** 5 ; **16.** -5 ; **17.** -2 ; **18.** 7 ; **19.** -8.

Lesson 3
Operations Involving Zero, Powers and Roots of Numbers

Operations Involving Zero

Addition
Examples **1.** $6 + 0 = 6$;   **2.** $0 + 4 = 4$;   **3.** $-7 + 0 = -7$;   **4.** $0 - 5 = -5$;   **5.** $b + 0 = b$

Multiplication

Examples **1.** $5(0) = 0$;   **2.** $(7)(9)(0)(14) = 0$
   Note: **3.** $0 \times 0 = 0$
Division
Examples **1.** $\frac{0}{5} = 0$ (because $5 \times 0 = 0$)

   2. $\frac{0}{-6} = 0$ (because $-6 \times 0 = 0$

But **3.** $\frac{5}{0}$ is undefined.(There is **no** number such that $0 \times$ "that number" $= 5$. Do not divide by 0)

   4. $\frac{0}{0}$ is   indeterminate.   (Any number will do since $0 \times$ "any number" $= 0$)

Square root: $\sqrt{0} = 0$

Powers of Signed Numbers

Examples **1.** $2^5 = (2)(2)(2)(2)(2) = 32$

   2^5 is read as 2 to the fifth power. (or 2 raised to the fifth power)
   The exponent "5" indicates how many times the base "2" is being used as a factor.

   2. $(-2)^3 = (-2)(-2)(-2)$
         $= -8$
   3. $(-3)^2 = (-3)(-3)$
         $= 9$

Zero as an Exponent
   Any (nonzero) number raised the to power zero is 1.

Examples 1. $b^0 = 1$;   **2.** $4^0 = 1$;   **3.** $(7x^2 dz)^0 = 1$

Note that 0^0 is indeterminate. Example: $0^0 = \frac{0^2}{0^2} = \frac{0}{0}$ which is indeterminate (see above)

Roots of Signed Numbers

Root finding and Power finding are inverse operations (in much the same way as multiplication and division are inverse operations).

Square Root: Symbol " $\sqrt{}$ " or " $\sqrt[2]{}$ "

Finding the square roots of perfect squares by inspection or by factoring

1. We may cheaply define the square root (principal square root) of a nonzero number as one of the two equal **positive** factors of that number. We exclude negative roots.

2. The square root of 0 is 0.

Examples

1. **The** square root of 9 is symbolized $\sqrt{9}$, and

$$\sqrt{9} = 3$$

because $3^2 = 9$

Note that , according to the above definition,

$$\sqrt{9} = \sqrt{(3)(3)} = 3$$

2. The square root of 64 is written $\sqrt{64}$, and

$\sqrt{64} = 8$, because $8^2 = 64$.

Also, $\sqrt{64} = \sqrt{(8)(8)} = 8$

3. **Note:** $\sqrt{0} = 0$, because $0^2 = 0$

If the given number is not a perfect square, we will use tables or a calculator to find the square root.

The **Cube Root** : Symbol" $\sqrt[3]{}$ "

From the above square root definition, we can similarly define the cube root of a number as one of the three equal factors of that number. (In this definition, we do **not** specify **positive root,** since the cube root may be positive or negative, depending on whether the given number is positive or negative.)

Examples

1. $\sqrt[3]{8} = 2$; because $2^3 = 8$.   Note that $\sqrt[3]{8} = \sqrt[3]{(2)(2)(2)} = 2$

2. $\sqrt[3]{-8} = -2$; because $(-2)^3 = -8$.   Note that $\sqrt[3]{(-2)(-2)(-2)} = -2$

More formal definition of square root

1. $\sqrt{A} = r$ if $r^2 = A$, where A, and r are both real and positive.

2. The square root of zero is zero (i.e. $\sqrt{0} = 0$,).

Lesson 3 Exercises

A.. Simplify: **1.** 9 + 0;   **2.** 3 - 0;   **3.** -6 - 0 ;   **4.** (6)(0);   **5.** (0)(-7);

6. $\frac{0}{4}$;   **7.** $\frac{0}{-5}$;   **8.** $\frac{9}{0}$;   **9.** 0(0) ;   **10.** (78)(5)(0)(32);   **11.** $\frac{0}{+6}$;   **12.** 0(-8)

Answers **1.** 9; **2.** 3 ; **3.** -6 ; **4.** 0; **5.** 0 ; **6.** 0; **7.** 0 ; **8.** undefined; **9.** 0 ; **10.** 0 ; **11.** 0 ; **12.** 0

B. Evaluate the following:

1. 3^3

2. 2^4

3. 8^2

4. $(-5)^3$

5. $(-1)^3$

6. $(-1)^4$

7. $(-1)^5$

8. $(-\frac{1}{2})^3$

9. 4^0

10. 0^0

11. 0^3

12. $(\frac{3}{2})^2$

13. $(-5)^0$

14. $(-2)^3$

15. $(-2)^4$

Review:

16. $-6-6$;

17. $-7(-5)$;

18. $9-(-8)$;

19. $9-8$

20. $-9-8$

21. $8-9$;

22. $\frac{-16}{4}$

23. $\frac{0}{-3}$;

24. $\frac{6}{0}$;

25. $(.6)(-.7)$

Answers: **1.** 27; **2.** 16 ; **3.** 64 ; **4.** -125 ; **5.** -1 ; **6.** 1 ; **7.** -1 ; **8** $-\frac{1}{8}$; **9.** 1 ; **10.** indeterminate; **11.** 0 ; **12.** $\frac{9}{4}$; **13.** 1 ; **14.** -8 ; **15.** 16; **16.** -12; **17.** 35 ; **18.** 17 ; **19.** 1 ; **20.** -17 ; **21.** -1 ; **22.** -4 ; **23.** 0 ; **24.** undefined ; **25.** -.42

C. Evaluate the following : **1.** 4^3; **2.** $(-3)^4$; **3.** $\sqrt{64}$; **4.** $(-5)^3$; **5.** $\sqrt[3]{-64}$.

Answers **1.** 64 ; **2.** 81 ; **3.** 8 ; **4.** -125 ; **5.** -4

D.

Simplify the following:

1. $\sqrt{25}$

2. $\sqrt{81}$

3. $\sqrt{36}$

4. $\sqrt{256}$

5. $\sqrt{1024}$

6. $\sqrt[3]{64}$

7. $\sqrt[4]{81}$

8. $\sqrt{121}$

9. $\sqrt{\frac{25}{49}}$

10. $\sqrt[3]{-8}$

Review

11. $-2-2$

12. $-3(-2)$

13. $-4-(-7)$

14. $-2 \div 6$

15. $3(0)$

16. 3^4

17. $\sqrt{100}$

18. $6-(-14)$

Answers: **1.** 5 ; **2.** 9 ; **3.** 6 ; **4.** 16 ; **5.** 32 ; **6.** 4 ; **7.** 3 ; **8.** 11 ; **9.** $\frac{5}{7}$; **10.** -2; **11** -4; **12.** 6 ; **13.** 3 ; **14.** $-\frac{1}{3}$; **15.** 0 ; **16.** 81 ; **17.** 10 ; **18.** 20.

Test # 1 - Student's Self-Test (**Always**, **Test yourself before you are tested**)

Attempt all questions on clean sheets of paper, **Do not write in the book** Show all necessary work.

---

1. (a) $(-4) + (-7) =$

(b) $(-6) + (2) =$

2. $-6\frac{2}{5} + 4 =$

---

3. (a) $(4.3) + (-.27) =$
(b) $-6 - 8 =$

4. (a) $(-6) - (-7) =$

(b) $5 - 20 =$

---

5. (a) Subtract 4 from 7.

(b) $-5 - 12 =$

6. (a) $- 4 - (7) =$

(b) $-6(-6) =$

---

7. (a) $(-7)(-3) =$
(b) $(-2)(+6) =$

8. (a) $\left|-4\frac{1}{2}\right| =$
(b) $- 7 - (-3) =$

---

9. (a) $8 - 5 + 7 - 9 =$

(b) $(-20) \div (5) =$

10. (a) $4\frac{1}{3}\left(\frac{1}{2}\right) =$
(b) $\frac{-14}{21} =$

---

11. (a) $6\frac{3}{4} - 8\frac{1}{4} =$
(b) $-5(-7) =$

12. (a) $\frac{0}{-7} =$

(b) $- 1 - 1 - 1 =$

---

13. (a) $\frac{8}{0} =$
(b) $(-5) + (-4) =$

14. (a) $8^0 =$

(b) $-3(-5) =$

15. (a) $\dfrac{0}{-4} =$

(b) $-1 - 5 =$

16. (a) $\sqrt{49} =$

(b) $- 4 - (2) =$

17. (a) $\sqrt{\dfrac{16}{25}} =$

(b) $4(0) =$

18. (a) $\sqrt{512} =$

(b) $7 - 3 - 6 =$

19. (a) $- 3 - 1 + 9 - 8 =$

(b) $\dfrac{-8}{12} =$

20. (a) $8 - 8 =$

(b) $\dfrac{40}{-16} =$

21. (a) $(-12) \div (4) =$

(b) $(- 7) + (-8) =$

22. (a) $- 1 - 5 + 6 - 8 =$

(b) $-9 - 2 =$

23. (a) $8 - (-5) =$

(b) $(-6) + (-9) =$

24. (a) $-2(-6)(-2) =$

(b) $-2(-6)(-2)(-3) =$

25. (a) $- 6 - (-8) =$

(b) $(+3) + (-7) =$

Bonus: (a) $8 - 7 - 5 =$

(b) $\dfrac{0}{0} =$

Answers: **1.** (a) -11; (b) -4 ; **2.** $-2\frac{2}{5}$; **3.** (a) 4.03; (b) -14 ; **4.** (a) 1; (b) -15 ; **5.** (a) 3; (b) -17; **6.** (a) -11; (b) 36 ; **7.** (a) 21; (b) -12 ; **8.** (a) $4\frac{1}{2}$; (b) -4 ; **9.** (a) 1; (b) -4 ; **10.** (a) $2\frac{1}{6}$; (b) $-\frac{2}{3}$; **11.** (a) $-1\frac{1}{2}$; (b) 35; **12.** (a) 0; (b) -3 ; **13.** (a) undefined; (b) -9 ; **14.** (a) 1; (b) 15 ; **15.** (a) 0; (b) -6 ; **16.** (a) 7; (b) -6 ; **17.** (a) $\frac{4}{5}$; (b) 0; **18.** (a) $16\sqrt{2}$; (b) -2; **19** (a) -3; (b) $-\frac{2}{3}$; **20.** (a) 0; (b) $-\frac{5}{2}$; **21.** (a) -3; (b) -15; **22.** (a) -8; (b) -11; **23.** (a) 13; (b) -15 ; **24** (a) -24; (b) 72 ; **25.** (a) 2; (b) -4; **Bonus**: (a) -4; (b) indeterminate.

CHAPTER 2

Order of Operations and Evaluation of Algebraic Expressions

Lesson 4: **Order of Operations in Performing Algebraic operations**

Lesson 5: **Evaluation of Algebraic Expressions Containing Variables**

Lesson 4

Order of Operations in Performing Algebraic Operations

Perform the operations according to the following order
1. Grouping Symbols: If there are grouping symbols, evaluate within the grouping symbols first;
2. Then, evaluate the powers and the roots in any order ; followed by
3. Division and Multiplication from left to right (or invert the divisors and multiply in any order); and then
4. Addition and Subtraction from left to right.

Note: If grouping symbols occur within other grouping symbols, evaluate within the innermost grouping symbols first, followed by the next innermost symbols and so on.

Example 1

$5 + 2(4 - 7) - 9$
$= 5 + 2(-3) - 9$ Evaluating within the parentheses first
$= 5 - 6 - 9$ Multiplying next
$= - 1 - 9$ Adding from left to right
$= -10$

Example 2

$2^3 - 9(4 + 3) + 8$ Evaluating within the parentheses first

$= 2^3 - 9(7) + 8$

$= 8 - 9(7) + 8$ **Scrapwork:** $2^3 = (2)(2)(2) = 8$
$= 8 - 63 + 8$
$= - 47$

Example 3

$18 \div 6 \times 3$ From left to right, we encounter division first and
$= 3 \times 3$ therefore, we divide first, according to
$= 9$ the order of operations.

Note in Example 3 above that, if we had multiplied first, we would **not** obtain the correct answer of 9.
In multiplying first, $18 \div 6 \times 3 = 18 \div 18 = 1$ **<-----This is the wrong solution.**

Example 4

$6 \times 3 \div 18$
$= 18 \div 18$ (From left to right, we encounter multiplication
$= 1$ first, therefore, we multiply first and then divide)
Note however that, if we divide first, we will still get the correct answer.

$$\text{Thus,} 6 \times 3 \div 18 = \frac{6}{1} \times \frac{3}{18} = \frac{6}{1} \times \frac{1}{6} = 1$$

We conclude, from Examples 3 and 4, that if we always divide first (whether or not division comes first) , we will obtain the correct answer; but if multiplication does not come first in going from left to right, then we should **not** multiply first. Thus if we always perform division first, we will obtain the correct answer.
Note that Example 3 and Example 4 are different problems.

Lesson 4: Order of Operations in Performing Algebraic operations

EXTRA. Simplify: $20 \div \frac{1}{2}$ of 10 (Hint: Perform the "of" operation first before dividing.)

Solution $20 \div \frac{1}{2}$ of 10

$= 20 \div (\frac{1}{2} \times 10)$ (The notion of changing "of" to multiplication and performing the operations from
$= 20 \div 5 = 4$ left to right will lead to the **wrong result** in this problem: "of" precedes $\times$ and $\div$.)

Example 5 $10 - [7 - 3(5 - 1)]$
$= 10 - [7 - 3(4)]$
$= 10 - [7 - 12]$
$= 10 - [- 5]$
$= 10 + [+5]$
$= 15$

(Add the terms in the numerator, then

Example 6 $\dfrac{-18 - 6}{9 - 3} = \dfrac{-24}{6}$ add the terms in the denominator before dividing)

$= -4$

Example 7   Evaluate.

$0 - 4^2$ $= 0 - (4)(4)$
$= 0 - 16$
$= - 16$

Thus, $-4^2 = -16$

Example 8   Evaluate. $0 + (-4)^2$

Solution: $0 + (-4)^2 = 0 + (-4)(-4)$
$= 0 + 16$
$= 16$

Thus, $(-4)^2 = 16$; but note that $- 4^2 = - 16$ (From Examples 7 and 8)

Example 9   Evaluate $(-3)^2 + 7(-3)$

Procedure: Apply the order of operations.
$(-3)^2 + 7(-3)$

                    Scrapwork: 1. $(-3)^2 = (-3)(-3) = 9$
$= 9 - 21$ 2. $9 - 21 = (+9) + (-21) = -12$
$= -12$

Example 10

$(-2)^3 - 9(-3)$ **Scrapwork:**   1. $(-2)^3 = (-2)(-2)(-2) = - 8$

$= - 8 + 27$ 2. $-9(-3) = +27$
$= 19$

Example 11

$3\sqrt{16} - 2(-6) \div 3$
$= 3(4) + 12 \div 3$
$= 12 + 4$
$= 16$

Lesson 4 Exercises

A.

Evaluate the following: **1.** $7 - 5(3 + 4)$; **2.** $8 + 3(5 + 1)^2$; **3.** $6\sqrt{81} - 5(4)(-2)$

4. $-3[9 - (- 4 - 5) + 8]$; **5.** $15 \div 3 \times 5$ **6.** $12 \times 8 \div 4$

7. $\dfrac{4 - 7}{-8 + 6}$; **8.** $(-6)^2$; **9.** -6^2 ; **10.** $5(-2)^3 - 7(-4)$

Answers: **1.** -28 ; **2.** 116; **3.** 94 ; **4.** -78 ; **5.** 25 ; **6.** 24 ; **7.** $\dfrac{3}{2}$; **8.** 36 ; **9.** -36 ; **10.** -12

B.

Simplify the following:

1. $6 + 2(3 - 8)$ **2.** $4^2 - 3(2 + 5) - 9$ **3.** $12 \div 6 \times 2$

4. $6 \times 2 \div 12$ **5.** $12 - 4[3 - (8 - 2) - 5] + 3$ **6.** $4 -1[5 + 3]^2 - 8$

7. $0 - 3^2$ **8.** $0 - (-3)^2$ **9.** -3^2 **10.** $(-3)^2$

11. $-2(3) - (-4)$ **12.** $5(-4) + (-3)(2)$ **13.** $-7(-6) - 9(-4)$

14. $(-6)2(0)(8)$ **15.** $3 - 4\{ - (6)(-2) + 2[4 - 7]\}$ **16** $\dfrac{- 6 - 3}{-2 - 3}$

17. $-3(2 - 8) - (-3)$ **18.** $\sqrt{4^2 + 3^2}$ **19.** $(\tfrac{2}{3})^3 + \dfrac{1}{9}$

Review

20. $-8^2 + 3(0)$ **21.** $(-4)^2 - (-3)^2$ **22.** $\dfrac{(-8) - (-2)}{-2}$

Answers: **1.** -4 ; **2.** -14 ; **3.** 4 ; **4.** 1 ; **5.** 47 ; **6.** -68 ; **7.** -9 ; **8.** -9 ; **9.** -9 ; **10.** 9 ; **11.** -2 ; **12.** -26 ; **13.** 78 ; **14.** 0; **15.** -21 ; **16.** $\dfrac{9}{5}$ or $1\tfrac{4}{5}$; **17.** 21 ; **18.** 5 ; **19.** $\dfrac{11}{27}$; **20.** -64 ; **21.** 7 ; **22.** 3.

C.

1. At 7 AM. today, the temperature was 10 degrees below zero. By 3 PM, today, the temperature   had increased by 12 degrees. What was the temperature at 3 PM.

2. If an original temperature of 30 degrees above zero decreases by 50 degrees, what is the new temperature?

3. Maria's score on a math test was 4 points above average, while John's score was 3 points below   average. What is the difference between Maria's score and John's score?

Answers: 1. The temperature is 2 degrees above zero; **2.** The temperature is 20 degrees below zero.
3. The difference is 7 points.

Lesson 5

Evaluation of Algebraic Expressions Containing Variables and Numbers, Given or Knowing the Numerical Values of the Variables

Procedure:
Step 1: Replace each letter by its given numerical value in the expression, (enclosing each value in parentheses during the replacement).
Step 2: Evaluate the resulting expressions using the order of operations learned previously (see p.14).

Example 1   Given that $x = -3$, evaluate $x^2 + 7x + 2$

$$\begin{aligned}
\textbf{Solution:} \quad & (-3)^2 + 7(-3) + 2 \\
&= 9 - 21 + 2 \\
&= -12 + 2 \\
&= -10
\end{aligned}$$

Scrapwork: $(-3)^2 = (-3)(-3)$
$= 9$

Example 2   Find the value of $5 - 4[3x - 1]$, if $x = 2$.

Solution   Substituting 2 for x, we obtain

$$\begin{aligned}
& 5 - 4[3(2) - 1] \\
&= 5 - 4[6 - 1] \\
&= 5 - 4[5] \\
&= 5 - 20 \\
&= -15
\end{aligned}$$

Example 3   (a) Evaluate: $200\,(2^n)$ when $n = 3$

Procedure: Substituting n = 3 in $200(2^n)$ and applying the order of operations, then we obtain:

$$\begin{aligned}
& 200\,(2^3) \\
&= 200\,(8) \\
&= 1600
\end{aligned}$$

Scrapwork
$2^3 = (2)(2)(2) = 8$

(b) Evaluate: $1000\,(4)^n$ if $n = 2$

Procedure: Substitute $n = 2$ in $1000\,(4)^n$

$$\begin{aligned}
& 1000\,(4)^2 \\
&= 1000\,(16) \\
&= 16000
\end{aligned}$$

Scrapwork
$(4)^2 = (4)(4) = 16$

Lesson 5 Exercises

A. 1. Find the value of $b^2 - 6b + 7$ when $b = -3$

2. Given that $A = \dfrac{1}{2}bh$, find the value of A if $b = 6, h = 4$

3. Find the value of F if $C = -60$, given that $F = \dfrac{9}{5}C + 32$

4. Given that $A = L \times W$, find the value of A when $L = 8, W = 6$

Answers   1. 34 ;   **2.** 12 ;   **3.** -76;   **4.**   48

B. Evaluate the following if $a = 2,\ \ b = 1,\ \ c = 3,\ \ x = -5,\ \ y = -1$

1. $4x + 5y$ 　　　　　　**2.** $a^2 - b^2$ 　　　　　　**3.** $abcxy$

4. $2ab - xy$ 　　　　　　**5.** $-ax - by + 10$ 　　　　**6.** $8 - a^2 + bc$

7. $(-5)^2 - by$ 　　　　**8.** $\dfrac{3a - 2}{ax + 2}$

9. Find the value of F if $C = -20$, given that $F = \dfrac{9}{5}C + 32$

10. Given that $A = \dfrac{1}{2}bh$, find the value of A when $b = 10$ and $h = 4$.

Answers: 1. -25 ; **2.** 3 ; **3.** 30 ; **4.** -1 ; **5.** 21 ; **6.** 7 ; **7.** 26 ; **8.** $-\dfrac{1}{2}$; **9.** -4 ; **10.** 20

Lesson 5: Evaluation of Algebraic Expressions Containing Variables

Test # 2 - Student's Self-Test (**Always**, **Test yourself before you are tested**)

Attempt all questions on clean sheets of paper, **Do not write in the book** Show all necessary work.

---

1. (a) $(-2)^4 =$

(b) $(-3)^3 =$

2. (a) $\sqrt{81}$

(b) $\sqrt[3]{-8}$

---

3. Evaluate (a) $7 - 2(3 + 4)$

(b) $16 \times 4 \div 4$

4. (a) $5\sqrt{49} - 3(4)(-2)$

(b) $-3[8 - (-4 - 3) + 8]$

---

5. $\dfrac{5 - 3}{-10 + 6}$

6. (a) $5 + 3(4 + 1)^2$

(b) $2(-3)^3 - 7(-5)$

---

7. (a) $(-5)^2 =$

(b) $-5^2 =$

8. (a) $6 \times 12 \div 2$

---

9. (a) $2 - 4\{ - (4)(-2) + 2[4 - 7]$

10. $\dfrac{-6 - 3}{-2 - 3}$

---

11. (a) $0 - 4^2$
(b) $0 - (-4)^2$
(c) $0 + (-4)^2$

12. (a) $\sqrt{4^2 + 3^2}$

(b) $\left(\dfrac{2}{3}\right)^3 + \dfrac{1}{9}$

---

13. (a) $(-4)2(0)(9) =$

(b) $-7^2 + 2(0) =$

14. (a) $-3(2 - 7) - (-3)$

(b) $(-5)^2 - (-3)^2$

---

15. (a) $6 - 2[4 + 3\}^2 - 8$

16. (a) Find the value of $b^2 - 6b + 7$ when $b = -2$

17. Given that $A = \frac{1}{2}bh$ find the value of A if $b = 10, h = 3$

18. Find the value of F if $C = -20$, given that $F = \frac{9}{5}C + 32$

19. Given that $A = L \times W$, find the value of A when $L = 7, W = 4$.

20. $\frac{(-4)-(-1)}{-1}$

Questions 21-24
Evaluate if $a = -2, b = 1, c = 3, x = -5, \ y = -1$
21. $6x + 5y$

22. $abcxy$

23. $\frac{4a-2}{ax+2}$

24. $a^2 - b^2$

25 Find the value of F if $C = -40$, given that $F = \frac{9}{5}C + 32$

.

Bonus:

Answers: **1.** (a) 16; (b) -27 ; **2.** (a) 9; (b) -2 ; **3.** (a) -7; (b) 16 ; **4.** (a) 59; (b) -69 ; **5.** $-\frac{1}{2}$; **6.** (a) 80; (b) -19 ; **7.** (a) 25; (b) -25 ; **8.** 36 ; **9.** -6 ; **10.** $\frac{9}{5}$; **11.** (a) -16; (b) -16; (c) 16 ; **12.** (a) 5 ; (b) $\frac{11}{27}$; **13.** (a) 0; (b) -49 ; **14.** (a) 18; (b) 16 ; **15.** -100 ; **16.** 23 ; **17.** 15 ; **18.** -4 ; **19.** 28 ; **20.** 3; **21.** -35 ; **22.** -30 ; **23.** $-\frac{5}{6}$; **24.** 3 ; **25.** -40.

CHAPTER 3
EXPONENTS

Lesson 6: **Rules of Exponents and Applications**
Lesson 7: Scientific Notation

Lesson 6
Rules of Exponents and Applications

Example 1. $2^5 = 2 \cdot 2 \cdot 2 \cdot 2 \cdot 2 = 32$

2^5 is read as 2 raised to the fifth power.(or 2 to the fifth power)
The exponent "5" indicates how many times the base "2" is being used as a factor.

Example 2. $x^6 = x \cdot x \cdot x \cdot x \cdot x \cdot x$

Mnemonic Examples for the Rules of Exponents
(Match the following examples with the rules of exponents which are presented later, below.)

1. $x^2 x^3 = x^5$

2. $\frac{x^7}{x^4} = x^{7-4} = x^3$

3. $(x^4)^2 = x^8$

4. $(xy)^5 = x^5 y^5$

5. $(\frac{x}{y})^6 = \frac{x^6}{y^6}$

6. $x^{-2} = \frac{1}{x^2}$

7. $x^0 = 1$

8. $\frac{x^4}{x^7} = x^{4-7} = x^{-3} = \frac{1}{x^3}$

Rules of Exponents

1. $x^a x^b = x^{a+b}$ (Product of powers of the same base rule) Example $x^2 x^3 = x^5$

2. $(x^a)^b = x^{ab}$ (Power to a power rule) Example $(x^4)^2 = x^8$

3. $x^{-a} = \frac{1}{x^a}$; also $\frac{1}{x^{-a}} = x^a$ (where a is positive) Negative exponent rule .Ex.1: $x^{-2} = \frac{1}{x^2}$; Ex.2: $\frac{1}{x^{-2}} = x^2$

4. $\frac{x^a}{x^b} = x^{a-b}$ (Quotient of powers of the same base rule) Example $\frac{x^7}{x^4} = x^{7-4} = x^3$

5. $(xy)^n = x^n y^n$ (Product to a power rule) Example $(xy)^3 = x^3 y^3$

6.. $(\frac{x}{y})^n = \frac{x^n}{y^n}$ ($y \neq 0$) (Quotient to a power rule) Example $(\frac{x}{y})^5 = \frac{x^5}{y^5}$

Zero as an exponent: $x^0 = 1$ ($x \neq 0$) Example: $5^0 = 1$

Applications of the rules of exponents

Example 1 Simplify: $(3x^2 y)^2$

$= 3^2 x^4 y^2$ (Using the rules of exponents)

$= 9x^4 y^2$ (**Note**: also that $(3x^2 y)^2 = (3x^2 y)(3x^2 y) = 9x^4 y^2$)

Example 2 Simplify: $(2x^4y^6)^4$

$= 2^4x^{16}y^{24}$

$= 16x^{16}y^{24}$

Scrapwork: $2^4 = (2)(2)(2)(2)$

$= 16$

Lesson 6 Exercises

A. Simplify the following: **1.** x^6x^3 ;   **2.** $(y^7)^3$;   **3.** $(bc^4)^6$; **4.** $\dfrac{x^9}{x^5}$;

     5. $\dfrac{(y^3)^5x^9}{(y^7)^2x^4}$;   **6.** $(x^3y^6)^4$;   **7.** $(-\frac{2}{3})^3$

Answers: **1.** x^9 ;   **2.** y^{21} ; **3.** b^6c^{24} ;   **4.** x^4 ; **5.** x^5y ;   **6.** $x^{12}y^{24}$;   **7.** $-\dfrac{8}{27}$

B. Simplify the following: **1.** $(3x^5y)^3$;   **2.** $(-2x^2y^7)^5$;   **3.** Multiply: $\left(\dfrac{-8b^2c}{c}\right)\left(\dfrac{9d}{3bcd}\right)\left(\dfrac{-15}{27c}\right)$;

     4. $(-2x^2y^7)^4$

Answers: **1.** $27x^{15}y^3$;   **2.** $-32x^{10}y^{35}$;   **3.** $\dfrac{40b}{3c^2}$;   **4.** $16x^8y^{28}$

C. Simplify, leaving answers with positive exponents.

1. $x^8 \cdot x^2$ **2.** $x^{-4} \cdot x^6$ **3.** $x(x^2)$;   **4.** $\dfrac{x^9}{x^6}$

5. $\dfrac{x^6}{x^9}$ **6.** $(a^2b^3)^5$ **7.** $(xy)^0$;   **8.** $(x^{-2})^4$

9. $10^{-2} \cdot 10^6$ **10.** $\dfrac{m^2}{n^{-4}}$ **11.** $\dfrac{10^{-3} \cdot 10^8}{10^3}$;   **12.** $(x^2y^{-3}z)^4$

13. $x^5 x^7$ **14.** $(x^8)^4$ **15.** $(-3x)^3$;   **16.** $(2x^2y^3)^4$

17. $(-3ab^2c^3)^3$ **18.** $\dfrac{x^2y^4}{xy}$ **19.** $\dfrac{xy}{x^2y^5}$;   **20.** $2x(x)$

21. $\dfrac{(x^3y^4)^2}{x^2y}$ **22.** $(a^3)^{12}(a^2)^5$ **23.** $\dfrac{b^5(b^2)^3}{b^8(b^3)^4}$;   **24.** $\dfrac{(x^3)^5(x^4)^2}{(x^5)^2x^9}$

Answers: **1.** x^{10} ;   **2.** x^2 ;   **3.** x^3 ;   **4** x^3 ;   **5.** $\dfrac{1}{x^3}$;   **6.** $a^{10}b^{15}$;   **7.** 1 ;   **8.** $\dfrac{1}{x^8}$;   **9.** 10^4 ; **10.** m^2n^4 ;

11. 10^2 or 100;   **12.** $\dfrac{x^8z^4}{y^{12}}$;   **13.** x^{12} ;   **14.** x^{32} ;   **15.** $-27x^3$; **16.** $16x^8y^{12}$;   **17.** $-27a^3b^6c^9$;   **18.** xy^3;

19. $\dfrac{1}{xy^4}$;   **20.** $2x^2$;   **21.** x^4y^7 ;   **22.** a^{46} ;   **23.** $\dfrac{1}{b^9}$;   **24.** x^4

Test # 3 - Student's Self-Test (Always, Test yourself before you are tested)

23

Attempt all questions on clean sheets of paper, **Do not write in the book** Show all necessary work.

Simplify the following and leave answers with positive exponents

1. (a) $x^4 x^5$

   (b) $(y^5)^4$

2. (a) $(bc^4)^6$

   (b) $\dfrac{x^8}{x^5}$

3. (a) $\left(-\dfrac{2}{3}\right)^3$

   (b) $\dfrac{(y^3)^5 x^9}{(y^7)^2 x^4}$

4. (a) $(x^9 y^6)^4$

   (b) $(-3x^2 y^7)^3$

5. $(2x^5 y)^3$

6. Multiply: $\left(\dfrac{-8b^2 c}{c}\right)\left(\dfrac{9d}{3bcd}\right)\left(\dfrac{-15}{27c}\right)$

7. (a) $x^8 \cdot x^2$

   (b) $x^{-4} \cdot x^6$

8. (a) $x(x^2)$

   (b) $\dfrac{x^9}{x^6}$

9. (a) $\dfrac{x^6}{x^9}$

   (b) $(a^2 b^3)^5$

10. $(-3x)^5$

11. (a) $x^5 x^7$

    (b) $(x^8)^4$

12. $\dfrac{10^{-4} \cdot 10^9}{10^3}$

13. $10^{-2} \cdot 10^6$

14. (a) $(xy)^0$

    (b) $(x^{-2})^4$

Lesson 6: Rules of Exponents and Applications

24

15. $\dfrac{(x^3y^4)^2}{x^2y}$

16. $(a^3)^{12}(a^2)^5$

17. $\dfrac{x^2y^4}{xy}$

18. $\dfrac{m^2}{n^{-4}}$

19. $(-3ab^2c^3)^3$

20. (a) $\dfrac{xy}{x^2y^5}$

(b) $2x(x)$

21. $(2x^2y^3)^4$

22. $(-2x^2y^7)^5$

23. $\dfrac{(x^3)^4(x^4)^2}{(x^5)^2x^9}$

24. $\dfrac{b^5(b^2)^3}{b^8(b^3)^4}$

25. $(x^2y^{-3}z)^4$

Bonus: (a) What is your last name?
(b) What is the fourth power of the number of letters in your last name?

Answers: **1.** (a) x^9; (b) y^{20}; **2.** (a) b^6c^{24}; (b) x^3; **3.** (a) $-\dfrac{8}{27}$; (b) x^5y ;

4. (a) $x^{36}y^{24}$; (b) $-27x^6y^{21}$; **5.** $8x^{15}y^3$; **6.** $\dfrac{40b}{3c^2}$; **7.** (a) x^{10}; (b) x^2 ; **8.** (a) x^3; (b) x^3;

9. (a) $\dfrac{1}{x^3}$; (b) $a^{10}b^{15}$ **10.** $-243x^5$; **11.** (a) x^{12} ; (b) x^{32} ; **12.** 10^2 or 100 ; **13.** 10^4 or 10,000;

14. (a) 1; (b) $\dfrac{1}{x^8}$; **15.** x^4y^7 ; **16.** a^{46}; **17.** xy^3 ; **18.** m^2n^4; **19.** $-27a^3b^6c^9$;

20. (a) $\dfrac{1}{xy^4}$; (b) $2x^2$; **21.** $16x^8y^{12}$; **22.** $-32x^{10}y^{35}$; **23.** x ; **24.** $\dfrac{1}{b^9}$; **25.** $\dfrac{x^8z^4}{y^{12}}$

Bonus: Varies

Lesson 7
Scientific Notation

Preliminaries:

Powers of 10

Examples:
$$\left.\begin{array}{l} 10^1 = 10 \\ 10^2 = 100 \\ 10^3 = 1000 \\ 10^4 = 10,000 \end{array}\right\} \text{positive integral powers of 10.}$$

Also,

$$\left.\begin{array}{l} 10^{-1} = \frac{1}{10} \\ 10^{-2} = \frac{1}{100} \\ 10^{-3} = \frac{1}{1000} \\ 10^{-4} = \frac{1}{10000} \end{array}\right\} \text{negative integral powers of 10.}$$

The inverse of multiplying by 1000 is dividing by 1000 (or multiplying by $\frac{1}{1000}$)

The inverse of multiplying by 10^3 is multiplying by 10^{-3} (Since $10^3 = 1000$ and $10^{-3} = \frac{1}{1000}$)

Definition: A number is in scientific notation if it is written as the product of a number between 1 and 10 and an integral (integer) power of 10.

Examples The following are in scientific notation:

(a) 2.45×10^4;   (b) 5.6×10^{-3} ;   (c) 6.023×10^{23}

Procedure for writing a given number in scientific notation:

To write a number in scientific notation, place the decimal point immediately after the first non-zero digit, and multiply by an appropriate power of 10 so that the value of the original number remains unchanged.

Example   Write the following in scientific notation:
(a)   6434 ;   (b) .27898   (c) 830,000 ;   (d) .004039

Solution

(a) $6434. = \boxed{6.434 \times 10^3}$ (Move the decimal point three places to the left (i.e., $\times 10^{-3}$) and on the right-hand side of the equation, do the opposite of $\times 10^{-3}$, which is $\times 10^3$.)
$\times 10^{-3}$

Check: $6.434 \times 10^3 = 6.434 \times 1000 = 6434$ (It is good practice to always check the result.)

(b) $.27898 = \boxed{2.7898 \times 10^{-1}}$ (Move the decimal point one place to the right (i.e., $\times 10^1$) and on the right-hand side of the equation, do the opposite of $\times 10^1$, which is $\times 10^{-1}$.)
$\times 10^1$

(c) $830000. = 8.3 \times 10^5$ (Move the decimal point five places to the left (i.e., $\times 10^{-5}$) and on the right-hand side of the equation, do the opposite of $\times 10^{-5}$, which is $\times 10^5$.)
$\times 10^{-5}$

(d) $.004039 = 4.039 \times 10^{-3}$ (Move the decimal point three places to the right (i.e., $\times 10^{3}$) , and on the
right-hand side of the equation, do the opposite of $\times 10^{3}$, which is $\times 10^{-3}$.)

$\times 10^{3}$

Note above, for example, that: in (d) .004039

$$= .004039 \times 10^{3} \times 10^{-3} \qquad \text{(Note: } 10^{3} \times 10^{-3} = 10^{0} = 1\text{)}$$
$$= (.004039 \times 10^{3}) \times 10^{-3}$$
$$= 4.039 \times 10^{-3}$$

Lesson 7 Exercises

A. Write in scientific notation: (*a*) 7856.3; (*b*) 890400; (*c*) .000562; (*d*) 45.35; (*e*) 0.0046

Answers: (*a*) 7.8563×10^{3}; (*b*) 8.904×10^{5}; (*c*) 5.62×10^{-4}; (*d*) 4.535×10; (*e*) 4.6×10^{-3}

B. Simplify and leave answer in scientific notation:

1. $[9.3 \times 10^{5}] \times [4.0 \times 10^{-8}]$ **2.** $\dfrac{[8.1 \times 10^{4}] \times [5.1 \times 10^{-8}]}{[1.7 \times 10^{-7}] \times [9.0 \times 10^{2}]}$

Answers: 1. 3.72×10^{-2};   **2.** 2.7×10.

CHAPTER 4

Algebraic Expressions and Polynomials

Lesson 8: **Basic Definitions; Addition of Like terms**
Lesson 9: **Removing Grouping Symbols; Addition,** Subtraction of Polynomials
Lesson 10: **Multiplication and Division of Polynomials**

Lesson 8

Basic Definitions; Addition of Like terms

Definitions

Constant: a symbol representing a quantity that does not change in value in a certain discussion or operation. We usually represent a constant by a number.

Example: In $4x^2 - 3xy$, the constants are 4, 2 and -3.

We may also represent a constant by a letter, such as one of the first few letters of the alphabet. In a discussion in which letters are used to represent constants, it is good practice to state which letters represent constants.

Example: A quadratic trinomial is of the form $ax^2 + bx + c$, where a, b, and c are **constants**.

Variable: a variable is a symbol representing a quantity that changes in value. We usually represent a variable by a letter.

Example: In $4x^2 - 3xy$, the variables are x and y.

Algebraic Expression

An algebraic expression is any variable (letter) or constant (number) or any combination of variables and constants related by signs of operation (such as +, -, ×, etc.) and grouping symbols.

Examples 1. $5x^2 + 8x - 9$.

2. $-2x(x - y) + 8$.

The plus and the minus signs separate an algebraic expression into parts. Each part is called a **term**. $5x^2 + 8x - 9$ has three terms : The first term is $5x^2$; the second term is $+ 8x$, and the third term is - 9. $-2x(x - y) + 8$ has two terms : The first term is $-2x(x - y)$ and the second term is +8.

Monomial: A monomial in x is a one-term algebraic expression of the form ax^n, where a is a real number and n is a whole number. Examples: **1.** $4x^3$ **2.** $5x$;   **3.** $6x^3$;   **4.** 8

A monomial in x and y is a one-term algebraic expression of the form $ax^m y^n$, where a is a real number, and m and n are whole numbers.

Polynomial: A polynomial is an algebraic expression consisting of two or (a finite number of) more unlike monomials.

Examples: **1.** $7x^4 - 5x^3 - 4x + 7$ **2.** $4x + 8$;   **4.** $\frac{2}{3}x^5 + 7x + 4$;   **5.** $3x^5 y^2 + 7x^3 y$.

An example of an expression which is **not** a polynomial: $3x^2 + 6x^4 + \frac{2}{x} - 9$ is **not** a polynomial because of the x in the denominator of the third term. Note also that $\frac{2}{x} = 2x^{-1}$ (has a negative exponent).

Another definition

A polynomial in x is an algebraic expression consisting of a finite number of terms with each term being of the form ax^n, where a is a real number and n is a whole number. (In a polynomial, there are **no** variables in the denominator or **no** negative exponents; no variables with fractional exponents or the radical equivalents.)

Types of polynomials

Binomial : a polynomial of exactly two unlike monomials. Examples: **1.** $5x^2 - 3x^4$;   **2.** $a + b$

Trinomial: a polynomial of exactly three unlike monomials. Examples: **1.** $2x^3 - 5xy - 8$; **2.** $x^2 + 7x + 12$.

Note that a monomial is a special type of term. All monomials are terms but not all terms are monomials.

A polynomial may also be in terms of two or more variables. It may also be in terms of any letter(s).
Definition: A polynomial in x and y is an algebraic expression in which all the terms are of the form $ax^m y^n$, where a is any real number and m and n are whole numbers.

The **degree** of a **term** is the sum of the exponents of the variables of that term.

Example: Find the degree of the following: **1.** $3x^5 y^2$, **2.** $7x^4$, **3.** x, **4.** 8
Solution:
1. degree = 5 + 2 =7, **2.** degree = 4; **3.** degree = 1 ; **4.** degree = 0 (since $8 = 8x^0$ and $x^0 = 1$)

The **degree** of a **polynomial** is the degree of the term with the highest degree.

Example: Find the degree of the following polynomials

1. $7x^4 - 5x^3 - 4x + 7$ **, 2.** $4x + 8$ **, 3.** $\frac{2}{3}x^5 + 7x + 4$ **, 4.** $3x^5 y^2 + 7x^3 y$

Solution   1. Degree = 4 (degree of $7x^4$);   2. degree = 1 (degree of $4x$)
3. degree =5 (degree of $\frac{2}{3}x^5$) ; **4.** degree = 7 (degree of $3x^5 y^2$).

Like Terms

Like terms have exactly the same literal (letter) factors, with each letter raised to same power.
Examples $3x$ and $-7x$ are like terms.
          $5x^2 y$ and $6x^2 y$ are like terms.
          $2bc$ and $-5cb$ are like terms.
However, $3x^2 y^2$ and $-8x^2 y$ are **unlike** terms.

Note: Like terms **can** be added or subtracted,   Unlike terms **cannot** be added or subtracted.

Numerical Coefficient of a Term

Examples   The numerical coefficient of $8x$ is 8
          The numerical coefficient of $-8x$ is -8
          The numerical coefficient of $- 7xy$ is -7
          The coefficient of $6x^2$ is 6.

The Distributive Property | $a(b + c) = ab + ac$ |

Also, $a(b + c + d) = ab + ac + ad$
     $a(b - c - d) = ab - ac - ad$
     $a (b +c) = (b + c)a = ba + ca = ab + ac$

Examples   **1.** $5(2x + y)$ **2.** $3(a - b)$
          $= 10x + 5y$ $= 3a - 3b$

Addition of Like Terms

To add like terms, add the numerical coefficients and keep the literal parts.

		You may skip writing this step
Example 1	$4x + 2y + 6x + 3y$	**Scrapwork**
	$= (4 + 6)x + (2 + 3)y$ <---------	For the x-terms: $4 + 6 = 10$
	$= 10x + 5y$	For the y-terms: $2 + 3 = 5$
Example 2	$-8x + 5x + 2y$	
	$= (-8 + 5)x + 2y$ <---------------------------You may skip this step	
	$= -3x + 2y$	For the x-terms: $-8 + 5 = -3$

Lesson 8 Exercises

A. Add the like terms: **1**. $-7x + 5y + 4x - 3y$;   **2**. $7a + 3b - 8b - 9a$;   **3**. $6y - 6x + 3y - 9z$

Answers   **1**. $-3x + 2y$;   **2**. $-2a - 5b$;   **3**. $-6x + 9y - 9z$

B. Add the like terms:

1. $3x - 6y + 8x - 4y$ **2**. $3a - 4b - 2a - 5b$ **3**. $-5x + 6y - 3x + 4$

4. $2a + 3b - 5a + 4b - 9$ **5**. $-4a^2b + 5ab^2 + 3a^2b + ab^2$ **6**. $-2x^2y - 5xy + 2x^2y + y^2$

7. $6ab - 8ba + 3ab + 6b$ **8**. $7a^2 - 6ab^2 - 4a^2 + 9$ **9**. $5x - 7y + 3y - 4x$

Answers: **1**. $11x - 10y$;   **2**. $a - 9b$;   **3**. $-8x + 6y + 4$;   **4**. $-3a + 7b - 9$;   **5**. $-a^2b + 6ab^2$;
     6. $-5xy + y^2$;   **7**. $ab + 6b$;   **8**. $3a^2 - 6ab^2 + 9$;   **9**. $x - 4y$.

Lesson 9

Grouping Symbols; Addition and Subtraction of Polynomials

Grouping Symbols

1. The parentheses : (   ) , example: $4(x + y)$

2. The brackets : [   ] , example : $4[x + y]$

3. The braces : {   }, example : $4\{x + y\}$

* **4.** The bar : " ——— " example: $6\sqrt{9 + 16}$ (also see note below.)

**Note that the grouping symbols can be used interchangeably, and thus
$3(4 + 2) = 3[4 + 2] = 3\{ 4 + 2\}$

*Note also that $\dfrac{2 + 4}{3 + 5} = \dfrac{(2 + 4)}{(3 + 5)}$

How to Remove Grouping Symbols

We will consider four cases.

Case 1: If there is a plus sign or no sign preceding the grouping symbols, delete the plus sign, delete the grouping symbols, and copy the expression within the grouping symbols.

Example 1: $+ (2x + 4y - 6)$
            $= 2x + 4y - 6$

Example 2 $(2x + 4y - 6)$
            $= 2x + 4y - 6$

Case 2: If a minus sign precedes the grouping symbols, delete the minus sign, delete the grouping symbols, and change the sign of every term within the grouping symbols.

Example: $- (3x + 4y - 7)$ (Note that changing the sign of every term within the parentheses is
        $= - 3x - 4y + 7$ equivalent to multiplying every term within the parentheses by -1)

Case 3: If a factor precedes the grouping symbols, multiply each term within the grouping symbols by this factor. This case is the application of the distributive property:
        $a(b + c) = ab + ac$

Example 1 $2(5x - 6y + 3)$
        $= 10x - 12y + 6$

Example 2 $- 3(4a + 2b - 5c)$
        $= -12a - 6b + 15c$

Case 4: If there are grouping symbols within other grouping symbols, remove the innermost grouping symbols first, followed by the next innermost, and so on.

Example: Remove the grouping symbols.
        $6[- 3(4x - 1) + 11] - 12x$
    $= 6 [- 12x + 3 + 11] - 12x$ (Removing the parentheses first)
    $= - 72x + 18 + 66 - 12x$
    $= - 84x + 84$ (Adding the like terms)

 Note that this interchangeability applies only to the present topics. As you learn more mathematics, you will cover topics such as ordered pairs and sets where these symbols **cannot** be used interchangeably.

Addition of Polynomials

Procedure: To add polynomials, add the like terms.

Example 1 Find the sum of $-2b + 3c$ and $5b - 9c$

Procedure: Add the like terms.
$-2b + 3c + 5b - 9c$
$= 3b - 6c$

Scrapwork: For the b- terms:
$-2 + 5 = 3$
For the c-terms:
$3 - 9 = -6$

Example 2 Add $b^2 + 4c$ and $-3b^2 - 9c$

Procedure: Add the like terms: $b^2 + 4c - 3b^2 - 9c$

$$= -2b^2 - 5c$$

Subtraction of Polynomials

Example Subtract $5x^2 - 3$ from $2x - 4$

Procedure: Write the "from" expression first, followed by the subtracted expression in parentheses with a minus sign in front of the parentheses.
$2x - 4 - (5x^2 - 3)$
Now simplify by removing the parentheses and adding any like terms.

or

Change the signs of all the terms of the subtracted expression and add to the expression from which you are subtracting.

$2x - 4 - (5x^2 - 3)$
$= 2x - 4 - 5x^2 + 3$
$= -5x^2 + 2x - 1$

Note: The subtracted expression (the subtrahend) follows the word subtract and the "from" expression (the minuend) follows the word "from".

Lesson 9 Exercises

A. Remove the grouping symbols: **1.** $-(5x - 8y - 6)$; **2.** $-4(3x - 2y - 9)$;

3. $-(3x + 4y - 7)$; **4.** $7[-5(3y - 2) + 9] - 8x$; **5.** $-6(2x - 3) - (4 - 5x) - 2x$

Answers: **1.** $-5x + 8y + 6$; **2.** $-12x + 8y + 36$; **3.** $-3x - 4y + 7$; **4.** $-8x - 105y + 133$; **5.** $-- 9x + 14$

B. Add the following:

1. $3x - 6y - 5$ and $2y - 4x + 2$

2. $4x^2y - 9xy^2$ and $5x^2y + 3xy^2$

3. $2a - 7b - 8$ and $a - 36$

4. $-a^2b + 3ab^2$ and $-a^2b + 5a^2b - 4a^2b$

5. Simplify: $\frac{1}{2}a^2b + 9 + \frac{1}{4}a^2b$

Answers:**1.** $-x - 4y - 3$; **2.** $9x^2y - 6xy^2$; **3.** $3a - 7b - 44$; **4.** $-a^2b + 3ab^2$; **5.** $\frac{3}{4}a^2b + 9$

Lesson 9: Removing Grouping Symbols; Addition, Subtraction of Polynomials

32

C.

1. Subtract $-6x^3 + 4x - 5$ from $7x - 2$;   **2**. From $7x - 2$, subtract $-6x^3 + 4x - 5$

3. Simplify : $4x^2y^3 - (x^2y^3 + x^2y) + xy^2 - 5x^2y$;   **4. Simplify**: $5a^4b^5 + 3ab - (7a^4b^5 + 3ab)$

Answers: **1**. $6x^3 + 3x + 3$;   **2**. $6x^3 + 3x + 3$;   **3**. $3x^2y^3 - 6x^2y + xy^2$; ;   **4**. $-2a^4b^5$

D.

1. Subtract 6 from 2.   **2**. Subtract 2 from 6

3. Subtract $2y - 4x + 2$ from $3x - 6y - 5$ **4**. Subtract $2a - 7b - 8$ from $a - 36$

5. Simplify: $2a^2 - 4ab + b^2 - (5a^2 + 2ab - 3b^2)$ **6**. Subtract $a - 36$ from $2a - 7b - 8$

7. Simplify: $(6x - 8y + 7) - (-3x + 5y - 1)$

Answers: **1**. -4 ; **2**. 4 ;   **3**. $7x - 8y - 7$;   **4**. $-a + 7b - 28$;   **5**. $-3a^2 - 6ab + 4b^2$; **6**. $a - 7b + 28$;
     7. $9x - 13y + 8$.

Lesson 10
Multiplication and Division of Polynomials

Multiplication of Polynomials

Example 1   Find the product of $4xy^2$ and $-3y^2$

Multiply: $(4xy^2)(-3y^2) = -12\,xy^4$

Example 2 $(-2x^3y)(3xy^5)$
$$= (-2 \cdot 3)x^3 \cdot x \cdot y \cdot y^5$$
$$= -6x^4 y^6$$

Example 3 $2x(x + 4)$
$$= 2x^2 + 8x$$

Example 4 $(x + 2)(x + 5)$ <----------- Multiplication of two binomials
$x(x + 5) + 2(x + 5)$ (Distribute x over $(x + 5)$, followed by distribution of $+2$ over $(x + 5)$.
                You may skip writing this step, and write the next step.
$$= x^2 + 5x + 2x + 10$$
$$= x^2 + 7x + 10$$

Example 5 $(3x + 2)(4x - 3)$
$$= 12x^2 - 9x + 8x - 6$$
$$= 12x^2 - x - 6$$

Special Products: **1.** $(x + y)^2 = (x + y)(x + y)$
$$= x^2 + xy + xy + y^2 \quad (\text{Always read the letters alphabetically})$$
$$= x^2 + 2xy + y^2$$

2. $(x - y)^2 = (x - y)(x - y)$
$$= x^2 - xy - xy + y^2$$
$$= x^2 - 2xy + y^2$$

3. $(a + b)(a - b) = a^2 - ab + ab - b^2$
$$\mathbf{= a^2 - b^2} \qquad\qquad (- ab + ab = 0)$$

The expression $a^2 - b^2$ is known as the difference between two squares. We will meet this expression again when we factor polynomials in the future (Lesson 47).

What is meant by to simplify an algebraic expression?

The meanings given below will be sufficient for the topics in this course.

1. It may mean carry out the indicated operations, which may be addition , subtraction, multiplication, division, root extraction, or power finding; or

2. It may mean remove the grouping symbols and add any like terms, if there are any like terms; or

3. It may mean reduce the fractions to lowest terms.

Examples   Simplify each of the following:

1. $3b + 4(b + c)$
$= 3b + 4b + 4c$
$= 7b + 4c$

2. $5(2x - 3y) - 6(2x - 4y)$
$= 10x - 15y - 12x + 24y$ (Removing the parentheses)
$= -2x + 9y$ **Scrapwork:** For the x-terms
$10 - 12 = -2$
For the y-terms
$-15 + 24 = 9$

3. $11x - 12y + 6x + 4$ **Scrapwork**: For the x-terms
$= 17x - 12y + 4$ $11 + 6 = 17$

4. $- (5b + 8c) - 3(4b - 20 + c) - 8$
$= - 5b - 8c - 12b + 60 - 3c - 8$ (Removing the parentheses)
$= -17b - 11c + 52$ (Adding the like terms)

5. $3xy(2x^2y - 5y) - 2xy^2$

Procedure: Remove the parentheses and add any like terms.

$3xy(2x^2y - 5y) - 2xy^2$

$= 3xy(2x^2y) + 3xy(-5y) - 2xy^2$ <------You may skip this step

Scrapwork
$- 15 - 2 = -17$

$= 6x^3y^2 - 15xy^2 - 2xy^2$

$= 6x^3y^2 - 17xy^2$

6.   Simplify: $6 + 8x^2y - 4x(xy^2 - 3xy)$
$= 6 + 8x^2y - 4x^2y^2 + 12x^2y$

$= 6 + 20x^2y - 4x^2y^2$

$= 20x^2y - 4x^2y^2 + 6$ (Rewriting)

Division of Polynomials

Division of a monomial by a monomial

Example 1 $\dfrac{8x^3y}{2xy}$

$$= \dfrac{8}{2} \cdot \dfrac{x^3}{x} \cdot \dfrac{y}{y}$$

$$= \dfrac{4}{1} \cdot \dfrac{x^2}{1} \cdot 1$$

$$= 4x^2$$

Example 2 $\dfrac{-6x^3y^4}{-2x^8y}$

$$= \dfrac{-6}{-2} \cdot \dfrac{x^3}{x^8} \cdot \dfrac{y^4}{y}$$

$$= \dfrac{3}{1} \cdot \dfrac{1}{x^5} \cdot \dfrac{y^3}{1}$$

$$= \dfrac{3y^3}{x^5}$$ (Leaving answer with positive exponents only)

Division of a polynomial by a monomial

Example 1   Simplify: $\dfrac{12x^2 - 8x}{4x}$

Step 1: $\dfrac{12x^2}{4x} + \dfrac{-8x}{4x}$

Step 2: $3x + (-2)$

Step 3: $3x - 2$

Example 2 Simplify: $\dfrac{15y^3 + 20y^2}{-5y}$

Step 1: $\dfrac{15y^3}{-5y} + \dfrac{20y^2}{-5y}$

Step 2: $-3y^2 + (-4y)$

Step 3: $-3y^2 - 4y$

Note that in Examples 1 and 2 (above), you may skip writing Steps 1 and 2 if you **make sure** that you divide **every term** (taking into account its sign) in the **numerator** by the denominator. The author suggests skipping Steps 1 and 2, provided attention is paid to the above **note**.

Division of a polynomial by a polynomial (The long division method)

We will consider four cases.

Case 1: The divisor divides the dividend exactly (i.e. without a remainder).

Example: Divide $7x^2 + x^3 + 6 + 13x$ by $x + 2$

Note : $7x^2 + x^3 + 6 + 13x$ is the dividend; $x + 2$ is the divisor.

We will use a long division method which is similar to the long division in arithmetic.

Step 1: Check the dividend and the divisor to see if the terms of each expression are arranged in descending powers of the variable, x. Descending powers of x means the term with the highest power of x is written first, followed in decreasing orders by terms with lower powers of x.
The divisor $x + 2$ is already arranged in descending powers of x.
The dividend $7x^2 + x^3 + 6 + 13x$ is not arranged in descending powers of x. In descending powers of x, the dividend becomes $x^3 + 7x^2 + 13x + 6$.

Step 2: Arrange the divisor and the dividend as done in arithmetic long division, as shown below.

$$x + 2 \,\overline{\smash{\big)}\, x^3 + 7x^2 + 13x + 6.}$$

Step 3: Divide the first term (leading term) of the dividend by the first term (leading term) of the divisor, and write quotient x^2 above the x^3 (as we do in arithmetic).

$$x + 2 \,\overline{\smash{\big)}\, \overset{\displaystyle x^2}{x^3 + 7\,x^2 + 13\,x + 6}} \qquad \text{Scrapwork: } \frac{x^3}{x} = x^2$$

Step 4: As done in arithmetic, multiply the divisor, $x + 2$ by the x^2 , and write the product under the terms of the dividend, lining up the like terms vertically.

$$x + 2 \,\overline{\smash{\big)}\, \overset{\displaystyle x^2}{x^3 + 7\,x^2 + 13\,x + 6}} \qquad \text{Scrapwork: } x^2(x + 2) = x^3 + 2x^2$$
$$\; +x^3 + 2x^2$$

Step 5: Subtract the $x^3 + 2x^2$ from the like terms above them; but recall that to subtract we change the signs of the terms being subtracted and add the changed terms to the terms from which we are subtracting.

$$x + 2 \,\overline{\smash{\big)}\, \overset{\displaystyle x^2}{x^3 + 7x^2 + 13\,x + 6}}$$

Subtract:
$$x^3 + 2x^2$$
$$\overline{5x^2}$$

Scrapwork:
$$\begin{array}{cc} x^3 & 7x^2 \\ \underline{-x^3} & \underline{-2x^2} \\ 0 & 5x^2 \end{array}$$

Step 6: Bring down the other terms; divide, multiply and subtract as done in Steps 3, 4, and 5 above. Remember, always the first term of divisor divides the first term of dividend.

$$x^2 + 5x$$
$$x + 2 \mid \overline{x^3 + 7x^2 + 13x + 6}$$
$$+ x^3 + 2x^2$$

Scrapwork: $\dfrac{5x^2}{x} = 5x$

$$+ 5x^2 + 13x + 6$$

Subtract: $\qquad + 5x^2 + 10x$ $\qquad$ Scrapwork: $5x(x + 2) = 5x^2 + 10x$

$$+ 3x$$

Step 7: Bring down the 6, divide, multiply, and subtract.

$$x^2 + 5x + 3$$
$$x + 2 \mid \overline{x^3 + 7x^2 + 13x + 6}$$
$$+x^3 + 2x^2$$

$$+ 5x^2 + 13x + 6$$

Subtract: $\qquad + 5x^2 + 10x$

$$+ 3x + 6$$ $\qquad$ Scrapwork: $\dfrac{3x}{x} = 3$
$$+ 3x + 6$$
$$0 + 0$$

$\therefore \qquad \dfrac{x^3 + 7x^2 + 13x + 6}{x + 2} = x^2 + 5x + 3$

Normally, when we perform the division process, we show only Step 7 above.

Note: In the above division process, we divide, multiply, subtract, bring down the remaining terms of the dividend and repeat the process. Note also that always, the **first term** of divisor divides the **first term** of the dividend.

Case 2: The division process leaves a remainder.   38

Solution: The steps are exactly like those of Example 1; however we will show only the compact
          form in Step 7 of Example 1.

Divide $6x^4 + x^2 - x^3 + 30x + 7$ by $3 + 2x$

Step 1 : Arrange the divisor and dividend in descending powers of x; and set up the problem as in
        arithmetic long division.

$$
\begin{array}{r}
3x^3 - 5x^2 + 8x + 3 \text{ R -2} \\
2x + 3 \,\overline{\big)\, 6x^4 - x^3 + x^2 + 30x + 7}
\end{array}
$$

Step 2: (as above)

Scrapwork: $\dfrac{6x^4}{2x} = 3x^3$

$+6x^4 + 9x^3$

$\overline{}$

$-10x^3 + x^2 + 30x + 7$

$-10x^3 - 15x^2$

$\overline{}$

$16x^2 + 30x + 7$

$+16x^2 + 24x$

$\overline{}$

$+6x + 7$

$+6x + 9$

$\overline{}$

-2 <---------Remainder

There is a remainder of -2. The division process ends when the degree of the divisor is larger
than the degree of the dividend.

We may write the answer as $3x^3 - 5x^2 + 8x + 3$ R -2 or $3x^3 - 5x^2 + 8x + 3 + \dfrac{-2}{2x + 3}$

$$= 3x^3 - 5x^2 + 8x + 3 - \dfrac{2}{2x + 3}$$

To check the answer: Multiply $3x^3 - 5x^2 + 8x + 3$ by $2x + 3$ and add the remainder -2.
If you obtain the original dividend the answer is correct.

Lesson 10: Multiplication and Division of Polynomials

Case 3: The dividend has missing powers

Example   Divide $x^5 - 2x^3 - 10x^2 - 30$ by $x - 3$.

In setting up the problem vertically, we may either leave a column space for the missing powers or we may write zero-coefficient terms for the powers. We choose to write zero coefficient powers.

Then the dividend becomes $x^5 + 0x^4 - 2x^3 - 10x^2 + 0x - 30$

Solution   Arrange the divisor and dividend in descending powers of x and set up the problem as in arithmetic long division, and divide.

$$
\begin{array}{r}
x^4 + 3x^3 + 7x^2 + 11x + 33 \\
x - 3 \overline{) x^5 + 0x^4 - 2x^3 - 10x^2 + 0x - 30}
\end{array}
$$

$$+x^5 - 3x^4$$
$$\overline{\quad\quad + 3x^4 - 2x^3 - 10x^2 + 0x - 30}$$
$$+ 3x^4 - 9x^3$$
$$\overline{\quad\quad\quad + 7x^3 - 10x^2 + 0x - 30}$$
$$+ 7x^3 - 21x^2$$
$$\overline{\quad\quad\quad\quad + 11x^2 - 0x - 30}$$
$$+ 11x^2 - 33x$$
$$\overline{\quad\quad\quad\quad\quad + 33x - 30}$$
$$+ 33x - 99$$
$$\overline{\quad\quad\quad\quad\quad\quad 69}$$

Scrapwork

Divide: $\dfrac{x^5}{x} = x^4$

Subtract: $0x^4 - (-3x^4) = 3x^4$

Divide $\dfrac{3x^4}{x} = 3x^3$

Subtract: $-2x^3 - (-9x^3) = 7x^3$

Subtract: $-10x^3 - (-21x^2) = 11x^2$

Subtract: $-30 - (-99) = -30 + 99 = 69$
<--------Remainder

In this problem, the division process ends when the degree of the first term of the divisor is larger than the degree of the first term of the dividend (if the terms are in descending powers of x).

$$\therefore \frac{x^5 - 2x^3 - 10x^2 - 30}{x - 3} = x^4 + 3x^3 + 7x^2 + 11x + 33 \text{ R } 69$$

or

$$= x^4 + 3x^3 + 7x^2 + 11x + 33 + \frac{69}{x - 3}$$

Case 4: Division of polynomials involving two or more variables.

In this case, we may base the descending order arrangement on one of the variables.

Example Divide $-3y^3 + 2x^2y - 5xy^2$ by $2x + y$

We base the order of arrangement of the terms on x.

$$
\begin{array}{r}
xy - 3y^2 \\
2x + y \,\overline{\big)\, 2x^2y - 5xy^2 - 3y^3} \\
+\,2x^2y + xy^2 \\
\overline{} \\
-6xy^2 - 3y^3 \\
-6xy^2 - 3y^3 \\
\overline{} \\
0 + 0
\end{array}
$$

Note above that if we base the order of arrangement of the terms on y, we will obtain the same quotient, (except for the order of the terms) $-3y^2 + xy$.

Lesson 10 Exercises

A. **1**. Find the product of $-5ab^2$ and $(6b^4)$;   **2**. Simplify: $-4x(x-5)$;   **3**. $(x-3)(x+7)$.

4. Find the product: $(4x+2)(3x-6)$;   **5.** Simplify: $(x-y)(x+y)$

Answers: 1. $-30ab^6$; **2.** $-4x^2 + 20x$; **3.** $x^2 + 4x - 21$; **4.** $12x^2 - 18x - 12$; **5.** $x^2 - y^2$

B. Carry out the indicated operations:

1. $2x(bx)$ **2**. $-3x(-7x)$ **3**. $x(2x+6y)$

4. $2(3a-b)$ **5**. $(x+2)(x+3)$ **6**. $(x-2)(x+3)$

7. $(2a+3b)(a-b)$ **8**. $(a+b+c)(a-b+c)$ **9**. $(3x-2)(2x+5)$

10. $(2x-3)^2$ **11**. $(a+b)^2$ **12**. $2a^2(5a-5b-c)$

13. Simplify: $(x+4)(x-5) + (x-6)(x+3)$

Answers: **1**. $2bx^2$;   **2**. $21x^2$;   **3**. $2x^2 + 6xy$;   **4**. $6a - 2b$;   **5**. $x^2 + 5x + 6$;   **6**. $x^2 + x - 6$;
7. $2a^2 + ab - 3b^2$; **8**. $a^2 + 2ac - b^2 + c^2$; **9**. $6x^2 + 11x - 10$; **10**. $4x^2 - 12x + 9$;
11. $a^2 + 2ab + b^2$;   **12**. $10a^3 - 10a^2b - 2a^2c$;   **13**. $2x^2 - 4x - 38$.

C. Simplify the following **1**. $-(-4x + 6y + 6)$; **2**. $2 + 4(3x - 5y - 2)$; **3**. $-4(3a + c) - 3(5a - 4)$

4. $+(6x + 4y) + 4(7x + 8)$;   **5.** $4xy(5x^2y - 3y) - 7xy^2$

Answers: **1**. $4x - 6y - 6$; **2**. $12x - 20y - 6$; **3**. $-27a - 4c + 12$;
4. $34x + 4y + 32$;   **5**. $20x^3y^2 - 19xy^2$

D. Simplify and leave answers with positive exponents:

1. $\dfrac{10a^7}{2a^2}$ **2.** $\dfrac{12x^4y^6}{-2x^3y^2}$ **3.** $\dfrac{-18a^5b^7}{-3a^2b^4}$;   **4.** $\dfrac{16a^2bc}{14a^3bc^2}$

5. $\dfrac{-10x^5y^7}{20x^6y^6}$ **6.** $\dfrac{16c^3d}{-12cd^2}$ **7.** $\dfrac{.75a^4bc}{2.5abc}$

Answers: **1.** $5a^5$;   **2.** $- 6xy^4$;   **3.** $6a^3b^3$;   **4.** $\dfrac{8}{7ac}$;   **5.** $-\dfrac{y}{2x}$;   **6.** $-\dfrac{4c^2}{3d}$;   **7.** $.3a^3$

E. Simplify the following:

1. $\dfrac{-8x^5y^3}{2x^2y}$;   **2.** $\dfrac{15a^4b^2 - 20a^3b^2 - 5ab}{5ab}$;   **3.** $\dfrac{14x^3y^2 - 21x^2y}{7xy}$

Answers: **1.** $-4x^3y^2$;   **2.** $3a^3b - 4a^2b - 1$;   **3.** $2x^2y - 3x$

F. Divide:

1. $\dfrac{16x^2y - 8x}{2x}$ **2.** $\dfrac{16x^2y - 8x}{-2x}$ **3.** $\dfrac{14c^5d^2 - 21cd}{-7cd}$

4. $\dfrac{20x^4yz - 5xy}{5xy}$ **5.** $\dfrac{12a^2bc - 14ab^3c}{6}$

Answers: **1.** $8xy - 4$;   **2.** $-8xy + 4$;   **3.** $-2c^4d + 3$;   **4.** $4x^3z - 1$;   **5.** $2a^2bc - \dfrac{7ab^3c}{3}$

G. Carry out the indicated operation.

1. Divide $6x^2 + x^3 + 6 + 11x$ by $x + 2$ **2.** Divide $8x^4 + 9x^2 + 14x^3 + 19x + 12$ by $3 + 2x$

3. $(x^4 + 4x^2 + 5x + 2) \div (x + 3)$ **4.** $(4x^3y - 6x^2y^2 + 5y^4 + 6xy^3) \div (2x + y)$

Answers   **1.** $x^2 + 4x + 3$;   **2.** $4x^3 + x^2 + 3x + 5$ R - 3 ;
3. $x^3 - 3x^2 + 13x - 34$ R 104;   **4.** $2x^2y - 4xy^2 + 5y^3$

H. **1.** Divide $x^3 + 2x^2 - 5x - 6$ by $x - 2$ **2.** Divide $6x^2 - 8x + 5$ by $x + 3$

3. Divide $x^3 + 10x^2 + 33x + 36$ by $x + 4$ **4.** Divide $4x^4 - 2x^2 - 3$ by $x - 3$

5. Find the remainder when $x^3 + 6x^2 + 6x - 12$ is divided by $x + 2$

Answers: **1.** $x^2 + 4x + 3$;   **2.** $6x - 26$ R 83 ;   **3.** $x^2 + 6x + 9$;
4. $4x^3 + 12x^2 + 34x + 102$ R 303;   **5.** Remainder $= -8$

Lesson 10: Multiplication and Division of Polynomials

Test # 4 - Student's Self-Test (Always, **Test yourself before you are tested**)

Attempt all questions on clean sheets of paper, **Do not write in the book** Show all necessary work.

Add the like terms:

1. (a) $4x - 6y + 7x - 4y$

  (b) $2a - 4b - 2a - 5b$

Simplify:

2. (a) $-7x + 6y - 3x + 4$

  (b) $5a + 3b - 5a + 4b - 9$

3. (a) Subtract $-8x^3 + 2x - 5$ from $6x - 2$.

  (b) From $9x - 2$, subtract $-6x^3 + 4x - 4$

4. (a) Subtract 5 from 3.

  (b) Subtract 3 from 7.

5. (a) Subtract $5y - 4x + 2$ from $3x - 6y - 5$

  (b) Subtract $2a - 7b - 8$ from $a - 36$

6. (a) Simplify: $3a^2 - 4ab + b^2 - (5a^2 + 2ab - 3b^2)$

  (b) Subtract $a - 32$ from $2a - 7b - 8$

7. (a) Find the product of $-4ab^2$ and $(5b^4)$.

  (b) Find the product: $(3x + 2)(5x - 6)$.

8. Simplify :

  (a) $6x^2y^3 - (x^2y^3 + x^2y) + xy^2 - 5x^2y$;

  (b) $5a^4b^5 + 3ab - (5a^4b^5 + 3ab)$

9. Simplify:

  (a) $2(3a - b)$

  (b) $(x + 2)(x + 3)$

10. Simplify:

  (a) $(6x - 8y + 7) - (-3x + 5y - 1)$

  (b) $-4x(x - 5)$;

11. Simplify:

  (a) $(x - 4)(x + 6)$.

  (b) $(x - y)(x + y)$

12. Simplify: (a) $(a + b + c)(a - b + c)$

  (b) $(3x - 2)(2x + 5)$

13. Simplify: (a) $(4x - 3)^2$

  (b) $(a + b)^2$

14. Simplify:

  (a) $2a^2(5a - 5b - c)$

  (b) $(x + 2)(x - 5) + (x - 6)(x + 3)$

15. Carry out the indicated operations:

   (a) $(x - 2)(x + 5)$

   (b) $(4a + 3b)(a - b)$

16. Simplify: (a) $2x(bx)$

   (b) $-4x(-7x)$

17. Simplify the following
  (a) $- (-4x + 6y + 6)$

  (b) $2 + 5(3x - 5y - 2)$.

18. Simplify: (a) $-4(3a + c) - 3(5a - 4)$

   (b) $+(2x + 4y) + 3(7x + 8)$;

19. Simplify: (a) $3xy(5x^2y - 3y) - 7xy^2$

      (b) $7ab - 8ba + 2ab + 6b$

20 Simplify:
  (a) $6x - 7y + 3y - 4x$

  (b) $\frac{1}{2}a^2b + 8 + \frac{1}{4}a^2b$

21. Add the following:

  (a) $2x - 6y - 5$ and $2y - 4x + 3$

  (b) $3x^2y - 9xy^2$ and $5x^2y + 3xy^2$

22. Simplify: (a) $-3a^2b + 5ab^2 + 3a^2b + ab^2$

   (b) $- 4x^2y - 5xy + 2x^2y + y^2$

Simplify:

23. $\dfrac{.75a^4bc}{2.5abc}$

Simplify:

24.   (a) $\dfrac{10a^7}{2a^2}$; (b) $\dfrac{12x^4y^6}{-2x^3y^2}$

Simplify:

25. (a) $\dfrac{-18a^5b^7}{-3a^2b^4}$;   (b) $\dfrac{16a^2bc}{14a^3bc^2}$

Simplify:

Bonus: (a) $\dfrac{-10x^5y^7}{20x^6y^6}$; (b) $\dfrac{16c^3d}{-12cd^2}$

Answers: **1**. (a) $11x - 10y$; (b) $-9b$; **2**. (a) $-10x + 6y + 4$; (b) $7b - 9$;
3. (a) $8x^3 + 4x + 3$; (b) $6x^3 + 5x + 2$; **4**. (a) -2; (b) 4; ; **5**. (a) $7x - 11y - 7$; (b) $-a + 7b - 28$
; **6**. (a) $-2a^2 - 6ab + 4b^2$; (b) $a - 7b + 24$; **7**.(a) $-20ab^6$; (b) $15x^2 - 8x - 12$;
8 (a) $5x^2y^3 - 6x^2y + xy^2$; (b) 0; **9**. (a) $6a - 2b$; (b) $x^2 + 5x + 6$; **10**. (a) $9x - 13y + 8$;
(b) $-4x^2 + 20x$; **11**. (a) $x^2 + 2x - 24$; (b) $x^2 - y^2$; **12**. (a) $a^2 + 2ac - b^2 + c^2$; (b) $6x^2 + 11x - 10$;
13. (a) $16x^2 - 24x + 9$; (b) $a^2 + 2ab + b^2$; **14**. (a) $10a^3 - 10a^2b - 2a^2c$; (b) $2x^2 - 6x - 28$;
15. (a) $x^2 + 3x - 10$; (b) $4a^2 - ab - 3b^2$; **16**. (a) $2bx^2$; (b) $28x^2$; **17** (a) $4x - 6y - 6$;
(b) $15x - 25y - 8$; **18**. (a) $-27a - 4c + 12$; (b) $23x + 4y + 24$; **19**. (a) $15x^3y^2 - 16xy^2$;
(b) $ab + 6b$; **20**. (a) $2x - 4y$; (b) $\frac{3}{4}a^2b + 8$; **21**. (a) $-2x - 4y - 2$; (b) $8x^2y - 6xy^2$;
22. (a) $6ab^2$; (b) $-2x^2y - 5xy + y^2$; **23**. $.3a^3$ or $\frac{3}{10}a^3$; **24**. (a) $5a^5$; (b) $-6xy^4$;
25. (a) $6a^3b^3$; (b) $\frac{8}{7ac}$; **Bonus**: (a) $-\frac{y}{2x}$; (b) $-\frac{4c^2}{3d}$.

Test # 5 - Student's Self-Test (Always, Test yourself before you are tested)

Attempt all questions on clean sheets of paper, **Do not write in the book** Show all necessary work.

---

1. Simplify: $\dfrac{42x^3y^2 - 21x^2y}{7xy}$

2. Simplify: $\dfrac{30a^4b^2 - 20a^3b^2 - 5ab}{5ab}$

---

3. Divide:

 (a) $\dfrac{24x^2y - 8x}{2x}$; (b) $\dfrac{14x^2y - 8x}{-2x}$

Divide:

4. (a) $\dfrac{14c^5d^2 - 21cd}{-7cd}$; (b) $\dfrac{20x^4yz - 5xy}{5xy}$

---

5. Divide $6x^2 + x^3 + 6 + 11x$ by $x + 3$

6. Simplify: $(x^4 + 4x^2 + 5x + 2) \div (x + 3)$

---

7. Divide $6x^2 - 8x + 5$ by $x - 3$

8. Divide $x^3 + 10x^2 + 33x + 36$ by $x + 4$

---

9. Divide $4x^4 - 2x^2 - 3$ by $x - 2$

10. $(4x^3y - 6x^2y^2 + 5y^4 + 6xy^3) \div (2x + y)$

11.
Divide $8x^4 + 9x^2 + 14x^3 + 19x + 12$ by $3 + 2x$

12. Divide $x^3 + 2x^2 - 5x - 6$ by $x - 2$

Answers: **1**. $6x^2y - 3x$; **2.** $6a^3b - 4a^2b - 1$; **3.** (a) $12xy - 4$; (b) $-7xy + 4$; **4.** (a) $-2c^4d + 3$;
 (b) $4x^3z - 1$; **5.** $x^2 + 3x + 2$; **6.** $x^3 - 3x^2 + 13x - 34 + \dfrac{104}{x+3}$; **7.** $6x + 10 + \dfrac{35}{x-3}$; **8.** $x^2 + 6x + 9$;
9. $4x^3 + 8x^2 + 14x + 28 + \dfrac{53}{x-2}$; **10.** $2x^2y - 4xy^2 + 5y^3$; **11.** $4x^3 + x^2 + 3x + 5 - \dfrac{3}{2x+3}$;
12. $x^2 + 4x + 3$.

CHAPTER 5
First Degree Equations Containing One Variable
(Simple Linear Equations)

Lesson 11: **Axioms for Solving Equations**
Lesson 12: **Solving First Degree Equations**
Lesson 13: **Solving Decimal and Literal Equations**

Lesson 11

Axioms for Solving Equations

Axioms are general mathematical statements that we accept as true, without any proof, in order to deduce other less obvious statements. The following are very useful in constructing proofs and solving equations.

1. A quantity is equal to itself. (reflexive property of equality, also identity principle)
$a = a$

2. An equality may be reversed. (symmetric property of equality)
If $a = b$, then
$b = a$

3. Quantities equal to the same quantity are equal to each other.(transitive property of equality)
If $a = b$, and $b = c$, then
$a = c$.

4. A quantity may be substituted for its equal in any expression or equation. (substitution axiom)

5. A whole equals the sum of all its parts. (partition axiom)

6. If equal quantities are added to equal quantities, the sums are equal (addition axiom)

If $a = b$,		(If $a = b$, and
then $a + c = b + c$	**ALSO**	$c = d$, then
		$a + c = b + d$)

7. If equal quantities are subtracted from equal quantities, the differences are equal.(subtraction axiom)

If $a = b$,		(If $a = b$, and
then $a - c = b - c$	**ALSO**	$c = d$, then
		$a - c = b - d$)

8. If equal quantities are multiplied by equal quantities, the products are equal. (multiplication axiom)

If $a = b$,		(If $a = b$, and
then $ac = bc$	**ALSO**	$c = d$, then
		$ac = bd$.)

9. If equal quantities are divided by equal quantities (not zero), the quotients are equal. (division axiom) ^{4} 6

If $a = b$,   (If $a = b$, and

then $\dfrac{a}{c} = \dfrac{b}{c}$ **ALSO** $c = d$, then

$$\dfrac{a}{c} = \dfrac{b}{d})$$

10. Like powers of equals are equal . (powers axiom)

Example: If $a = 3$, then

$$a^2 = 3^2 \text{ or } a^2 = 9.$$

11. Like roots of equals are equal. (roots axiom)

Example: if $a^3 = 8$, then $\sqrt[3]{a^3} = \sqrt[3]{8}$

$$a = 2.$$

Lesson 12
Solving First Degree Equations

An equation is a statement of the equality between two expressions.

Example: $2x + 3 = x - 10$

The equality symbol "=' breaks up an equation into two sides or members , namely the left-hand side and the right-hand side of the equation. To solve an equation involving a single variable, say x, means we are to find values of x which satisfy the equation. **A value of x is said to satisfy an equation if this value when substituted in the equation makes both the left-hand side and the right-hand side of the equation equal to each other.**

A value of x which satisfies a given equation is said to be a solution or a root of the given equation. **To obtain a value for the variable, x, we will get x by itself on one side of the equation** . We agree that a value of x has been obtained if we have x by itself on one side of the equation and all the other quantities on the other side of the equation do **not** involve x. To get x by itself, we will use inverse operations. Addition and subtraction are inverse operations. Multiplication and division are inverse operations. Inverse operations "undo" each other. For example, to undo multiplication by a number, we will use division by the same the number. We will keep the above discussion in mind when we solve linear equations.

Example 1 Solve for x:

1. $x - 5 = 11$

To get x by itself on the left-hand side of the equation, we will remove the "-5" by adding its opposite to both sides of the equation.(This is the same as adding + 5 to both sides of the equation)

$$x - 5 = 11$$
$$\underline{+ 5 \quad + 5}$$
$$x = 16$$

The solution is 16.

Example 2 Solve for x: $\quad x + 8 = -11$

To get x by itself on the left-hand side of the equation, we remove the "+8" by adding -8 to (that is, subtracting 8 from) both sides of the equation.

$$x + 8 = -11$$
$$\underline{- 8 \quad - 8} \quad .$$
$$x = -19$$

(by stressing on **addition** of - 8, the right-hand side of the equation is done without confusion)

Note that to remove (undo) any number, use the inverse (opposite) operation.

Example 3 Solve for x: $\quad 5x = 20$

To obtain x by itself, we will undo (remove) the "5". We use division by 5 (since the "5" is multiplying the x)

$$\frac{5x}{5} = \frac{20}{5}$$
$$x = 4$$

The solution is 4.

Example 4 Solve for x:

$$- 6x = 24$$
$$\frac{-6x}{-6} = \frac{24}{-6}$$
$$x = -4$$

The solution is -4.

Example 5   Solve for x: $\dfrac{x}{6} = 4$

To remove the "6" which is a divisor, we will multiply
both sides of the equation by 6

$(6)\dfrac{x}{6} = (6)4$

$\cancel{(6)}\dfrac{x}{\cancel{6}} = (6)4$

$x = 24$. The solution is 24.

Example 6 Solve for x: $\dfrac{-x}{5} = 2$ (or $\dfrac{x}{5} = -2 \text{---}> x = 5(-2) = -10$)

$\cancel{(5)}\dfrac{(-x)}{\cancel{5}} = (5)2$

$-x = 10$ <- ------------------------

$(-1)(-1x) = (-1)10$ **Note:** $-x = -1x$

$x = -10$

OR   from this step , just change
the sign of the left-hand side and
change the sign of the right-hand side.

The solution is -10.

Example 7   Solve for x
$2x - 6 = 10$

We will remove the "- 6 " first, and then remove the "2" (i.e., we will always undo multiplication and division last)
$$2x - 6 = 10$$
$$\underline{+6 \quad +6}$$
$$2x = 16$$
$$\dfrac{2x}{2} = \dfrac{16}{2}$$
$$x = 8. \quad \text{The solution is 8.}$$

Example 8   Solve for x: $5x - 6 = x + 2$

In this case, we will collect all the x's on one side by removing either the x-term
on the right-hand side or the x-term on the left-hand side.
Let us remove the x-term on the right-hand side.

$$5x - 6 = x + 2$$
$$\underline{-x \qquad -x}$$
$$4x - 6 = +2$$
$$\underline{+6 \quad +6}$$
$$4x = 8$$
$$\dfrac{4x}{4} = \dfrac{8}{4}$$
$$x = 2. \quad \text{The solution is 2.}$$

Example 9   Solve for x
$7x + 2 - 3x = 2x + 18 + 3x$

In this case, first, simplify each side of the equation, and then proceed.

Step 1: $7x + 2 - 3x = 2x + 18 + 3x$
$4x + 2 = 5x + 18$

Step 2: $\underline{-5x \qquad -5x}$
$-x + 2 = 18$
$\underline{-2 \quad -2}$
$-x = 16$
$x = -16.$ The solution is -16.

Example 10 Solve for x

$$3(x + 2) = 5(x - 6)$$

First, remove the parentheses on both sides of the equation

$$3(x + 2) = 5(x - 6)$$
$$3x + 6 = 5x - 30$$
$$\underline{-5x \qquad -5x}$$
$$-2x + 6 = -30$$
$$\underline{\quad - 6 \quad - 6}$$
$$- 2x = -36$$
$$\frac{-2x}{-2} = \frac{-36}{-2}$$
$$x = 18$$

The solution is 18.

Checking Solutions of Equations (see also Lesson 35)

Example (a)   Determine if - 5 is a solution of the equation
$$4x + 18 - x = 8x + 32 + 2x.$$
- 5 will be a solution of the given equation if this value when substituted in the given equation makes the left-hand side of the equation equal to the right-hand of the equation.

Checking: Replace x by -5 in the equation.

$$\text{Then, } 4(-5) + 18 - (-5) \stackrel{?}{=} 8(-5) + 32 + 2(-5)$$
$$-20 + 18 + 5 \stackrel{?}{=} - 40 + 32 \ - 10$$
$$-2 + 5 \stackrel{?}{=} - 8 - 10$$
$$3 = -18 \quad \text{False}$$

The left-hand side of the equation and the right-hand side of the equation are **not** equal. Therefore, -5 is **not** a solution of the given equation.

(b) Determine if -2 is a solution of $4x + 18 - x = 8x + 32 + 2x.$

Checking: Replace x by -2 in the equation.

$$\text{Then, } 4(-2) + 18 - (-2) \stackrel{?}{=} 8(-2) + 32 + 2(-2)$$
$$-8 + 18 + 2 \stackrel{?}{=} -16 + 32 - 4$$
$$10 + 2 \stackrel{?}{=} 16 - 4$$
$$12 = 12 \quad \text{True}$$

Since the left-hand side of the equation equals the right-hand side of the equation, -2 is a solution of the given equation.

More Examples on Solutions of Linear Equations   50

In the following examples, attempt the problems before reading the solutions:

Example 1   Solve for x: $5x - 9 = 6$

Solution

$$5x - 9 = 6$$
$$\underline{+9 \quad +9}$$
$$5x + 0 = 15$$
$$\frac{5x}{5} = \frac{15}{5}$$
$$x = 3$$

Check: $5(3) - 9 \overset{?}{=} 6$

$$15 - 9 \overset{?}{=} 6$$
$$6 = 6 \quad \text{True}$$

Example 2   Solve for x: $6x - 3 = 2x + 8$

Solution :

$$6x - 3 = 2x + 8$$
$$\underline{-2x \qquad -2x}$$
$$4x - 3 = 0 + 8$$
$$\underline{+3 \qquad + 3}$$
$$4x = 11$$
$$\frac{4x}{4} = \frac{11}{4}$$
$$x = \frac{11}{4}$$

Check: $6(\frac{11}{4}) - 3 \overset{?}{=} 2(\frac{11}{4}) + 8$

$$33 - 6 \overset{?}{=} 11 + 16$$
$$27 = 27 \text{ True}$$

Example 3   Solve for x:

$$\frac{x}{6} = \frac{11}{12}$$

Method 1   To get x by itself on the left-hand side of the equation, we undo the "6" (which is a divisor) by multiplying each side of the equation by 6.

$$(6)\frac{x}{6} = (6)\frac{11}{12}_2$$

$$\cancel{(6)}\frac{x}{\cancel{6}}^1 = \cancel{(6)}\frac{11}{\cancel{12}}_2$$

$$x = \frac{11}{2}$$

Method 2   Multiply each term of the equation by the LCD of the fractions.   The LCD is 12.

$$(12)\frac{x}{6} = (12)\frac{11}{12}$$

$$2\,\cancel{(12)}\frac{x}{\cancel{6}}^1 = \cancel{(12)}^1\frac{11}{\cancel{12}}$$

$$2x = 11$$
$$\frac{2x}{2} = \frac{11}{2}$$
$$x = \frac{11}{2}$$

Method 3   **Cross-multiplication**

$$\frac{x}{6} = \frac{11}{12}$$

$$12x = 6(11)$$

$$1\frac{\cancel{12}x}{\cancel{12}}_1 = \frac{^1\cancel{6}(11)}{\cancel{12}}_2$$

$$x = \frac{11}{2}$$

Note: We can cross-multiply if there is only a single fraction on the left-hand side of the equation, and only a single fraction on the right-hand side of the equation. For instance, in the following example, we cannot cross-multiply immediately, but rather, we will have to combine the right-hand side into a single fraction before cross-multiplying.

Example 4   Solve for x: $\frac{x}{4} = \frac{x}{20} + \frac{2}{5}$

Method 1   The LCD method. Multiply each term of the equation by the LCD of the fractions. The LCD is 20.

$$(20)\frac{x}{4} = (20)\frac{x}{20} + (20)\frac{2}{5}$$

$$^5\cancel{(20)}\frac{x}{\cancel{4}}_1 = \cancel{(20)}\frac{x}{\cancel{20}}_1 + {}^4\cancel{(20)}\frac{2}{\cancel{5}}_1$$

$$5x = x + 8$$

$$\underline{-x \quad -x \qquad .}$$

$$4x = 8$$

$$\frac{4x}{4} = \frac{8}{4}$$

$$x = 2$$

The solution is 2.

Method 2   **By cross-multiplication**

   To apply cross-multiplication to the above problem, first, combine the two terms on the right-hand side into a single fraction. (We have a single fraction if we have a single denominator.)

continuing, $\frac{x}{4} = \frac{x}{20} + \frac{2}{5}$

$$\frac{x}{4} = \frac{x}{20} + \frac{2(4)}{5(4)}$$

$$\frac{x}{4} = \frac{x}{20} + \frac{8}{20}$$

$$\frac{x}{4} = \frac{(x + 8)}{20}$$

Now, we cross-multiply:

$$20x = 4(x + 8) \text{ <------ The parentheses are important}$$

$$20x = 4x + 32$$

$$\underline{- 4x \ - 4x \qquad .}$$

$$16x = 0 + 32$$

$$\frac{16x}{16} = \frac{32}{16}$$

$$x = 2$$

Method 3 You may undo one denominator at a time. 5 2

$$\frac{x}{4} = \frac{x}{20} + \frac{2}{5}$$

Step 1: Undo the "4" :
$$\frac{{}^1\cancel{(4)}x}{\cancel{4}_1} = \frac{{}^1\cancel{(4)}x}{\cancel{20}_5} + \frac{(4)2}{5}$$
(Multiply each term by 4)

$$x = \frac{x}{5} + \frac{8}{5}$$

Step 2: To undo the "5" multiply by 5:

$$(5)x = \frac{{}^1\cancel{(5)}x}{\cancel{5}_1} + \frac{\cancel{(5)}8}{\cancel{5}_1}$$

$$5x = x + 8$$
$$4x = 8$$
$$x = 2$$

Note: In method 3, multiplying the equation first by 4 and then by 5 is equivalent to multiplying the equation by 20 (as was done in Method 1).

Lesson 12 Exercises

A

Solve for x: **1.** $x - 7 = 12$; **2.** $x + 6 = 9$; **3.** $4x = 20$; **4.** $\frac{x}{6} = 8$; **5.** $4x + 3 = 35$

6. $7x - 6 = 2x + 24$; **7.** $5x - 7 + 3x + 9 = 6x + 5 + 8x$; **8.** $\frac{x}{6} + 4 = 9$;

9. $3(2x - 5) + 4 = -4(x + 7)$

Answers: **1.** $x = 19$; **2.** $x = 3$; **3.** $x = 5$; **4.** $x = 48$; **5.** $x = 8$; **6.** $x = 6$; **7.** $x = -\frac{1}{2}$

8. $x = 30$; **9.** $x = -\frac{17}{10}$

B Solve and check:

1. $x + 4 = 6$;   **2.** $x + 5 = -7$ **3.** $x - 3 = 9$

4. $x - 6 = -11$ **5.** $x - \frac{1}{3} = \frac{1}{4}$ **6.** $\frac{x}{2} = 8$

7. $\frac{x}{-2} = 7$ **8.** $\frac{-x}{4} = 9$ **9.** $3x = 18$

10. $-4x = 20$ **11.** $6x = 22$ **12.** $7x = -28$

13. $\frac{x}{4} - 6 = 14$ **14.** $\frac{-x}{3} + 5 = 7$ **15.** $2x + 4 = 9$

16. $-3x - 5 = 10$ **17.** $5x - 4 + x = 4x + 6 - 3x$

18. $2x - 5 + 3x + 2 = 7x - 5 - 3x$ **19.** $2(x - 4) + 3 = 15$;   **20.** $-3(x + 5) + 1 = 5(x + 4) - 6$

Answers: **1.** 2 ; **2.** -12 ; **3.** 12 ; **4.** -5 ; **5.** $\frac{7}{12}$; **6.** 16 ; **7.** -14 ; **8.** -36 ; **9.** 6 ; **10.** -5 ; **11.** $\frac{11}{3}$;

12. -4 ; **13.** 80 ; **14.** -6 ; **15.** $\frac{5}{2}$; **16.** -5 ; **17.** 2 ; **18.** -2 ; **19.** 10 ; **20.** $-\frac{7}{2}$

C

1. Is 2 a solution of $3x + 1 = 7$? **2.** Is -3 a solution of $4x - 1 = 15$?

3. Is 2 a solution of $5x + 2 = x - 4$?

Answers: **1.** Yes ; **2.** No ; **3.** No.

Solve for x: **1.** $\dfrac{x}{6} + \dfrac{x}{3} = \dfrac{3}{4}$ **2.** $\dfrac{x}{8} = \dfrac{x}{2} - 6$;   **3.** $\dfrac{x}{2} + 1 = \dfrac{x}{3}$;   **4.** $\dfrac{x-2}{4} + \dfrac{2}{3} = \dfrac{1}{2}$

Answers : **1.** $x = \dfrac{3}{2}$; **2.** $x = 16$; **3.** $x = -6$; **4.** $x = \dfrac{4}{3}$

E

Solve for x: **1.** $\dfrac{x+3}{4} + \dfrac{x-2}{8} = 2$ **2.** $\dfrac{x-2}{3} + 1 = \dfrac{x}{5}$ **3.** $\dfrac{x}{4} + \dfrac{1}{2} = \dfrac{x}{5} + 1$

4. $\dfrac{x}{3} = \dfrac{x}{6} + \dfrac{3}{5}$ **5.** $\dfrac{x}{2} + \dfrac{2}{3} = \dfrac{x}{4}$

Answers: **1.** 4 ; **2.** $-\dfrac{5}{2}$; **3.** 10 ; **4.** $\dfrac{18}{5}$; **5.** $-\dfrac{8}{3}$

Lesson 13
Solving Decimal and Literal Equations

Solving Decimal Equations

Discussion: There are a number of methods for solving decimal equations.

Method **1:** We can solve a decimal equation by strictly using decimals.

Method **2:** We can also change a decimal equation to an equivalent equation in which all the constants are integers, by multiplying every term of the equation by a power of 10. The power of 10 to use is based on the constant with the most number of decimal places or decimal digits.

Method **3:** We can make a partial conversion by multiplying the equation by a power of 10 so that only the numerical coefficients of the variables are necessarily changed to integers; some of the constant terms may consequently be changed partially or wholly, since we have to multiply every term of the equation by a power of 10. In multiplying by a power of 10, just move the decimal point accordingly.

Example 1 Solve for x:
$$.5x = 2$$

Solution: **Using Method 2** above, multiply both sides of the equation by 10.
$$5x = 20$$
$$x = 4$$

Example 2 Solve for x:
$$.02x - 6 = .1$$

Solution; Multiply every term by 100. (**Using Method 2**)
$$2x - 600 = 10$$
$$2x = 610$$
$$x = 305$$

Example 3 Solve for x:
$$.7x + .002 + .04 = .01x$$

Solution: **Method 3** , multiply every term 100.

$$70x + .2 + 4 = x$$
$$70x + 4.2 = x$$
$$69x = -4.2$$
$$x = \frac{-4.2}{69}$$
$$x = -.06 \text{ (to the nearest hundredth)}$$

Note that in Example 3, if we used **Method 2**. we would have
$$700x + 2 + 40 = 10x$$
$$700x + 42 = 10x$$
$$690x = -42$$
$$x = -\frac{42}{690} = x = -.06$$

Or if we used **Method 1** (strictly decimals) we would have
$$.7x - .01x = -.042$$
$$.69x = -.042 \text{----}> x = \frac{-.042}{.69} = -.06$$

Solving Literal Equations

Here, note that the letters represent numbers and therefore the principles and techniques we have used, so far, in solving linear equations are applicable.

Example 1   Solve for x: $4x + 3y = 7$

We want to get x by itself on one side of the equation. Let us use some common sense here. We will undo (remove) the $3y$ and the 4 on the left-hand side of the equation.

$$4x + 3y = 7$$
$$\underline{\quad -3y \qquad -3y}$$
$$4x + 0 = 7 - 3y$$

$$\frac{4}{4}x = \frac{7 - 3y}{4}$$

$$x = \frac{7 - 3y}{4} = \frac{-3y + 7}{4} = \frac{-3y}{4} + \frac{7}{4} = -\frac{3}{4}y + \frac{7}{4}$$

If the question is of the multiple-choice type, then, be on the look out for any of these equivalent forms of answers.

Example 2   Solve Example 1 for y.

i.e., Solve $4x + 3y = 7$ for y.

Note in this case that, we want y by itself one side of the equation.

$$4x + 3y = 7$$
$$\underline{-4x \qquad\qquad - 4x}$$
$$0 + 3y = 7 - 4x$$

$$\frac{3y}{3} = \frac{7 - 4x}{3}$$

$$y = \frac{-4x + 7}{3} = \frac{-4}{3}x + \frac{7}{3}$$

$$y = -\frac{4}{3}x + \frac{7}{3}$$

Example 3 Solve for A: $B = AC + CE$

We will obtain A by itself on one side of the equation.

$$B = AC + CE$$
$$B - CE = AC + CE - CE \qquad \text{(subtracting } CE \text{ from both sides)}$$
$$B - CE = AC$$
$$\frac{B - CE}{C} = \frac{AC}{C} \qquad \text{(dividing both sides by } C)$$
$$\frac{B - CE}{C} = A$$
$$\therefore A = \frac{B - CE}{C} \text{ or } \frac{B}{C} - \frac{CE}{C} = \frac{B}{C} - E$$

Lesson 13 Exercises

A   Solve for x:

1. $2x + 0.6 = 1$ **2.** $5x + 0.04 = 0.2$ **3.** $4x + 0.1 = 2.5$

Answers: **1.** 0.2 ; **2.** 0.032 ; **3.** 0.6

B

Solve for y: **1.** $3x + 8y = 32$;   **2.** $2y - 6x = 21$;   **3.** Solve for C: $F = \frac{9}{5}C + 32$

   Answers: **1.** $y = -\frac{3}{8}x + 4$ **2.** $y = 3x + \frac{21}{2}$;   **3.** $C = \frac{5}{9}F - \frac{160}{9}$ or $C = \frac{5}{9}(F - 32)$

C

1. Solve for x: $5x + 4y = 20$ **2.** Solve for y: $5x + 4y = 20$

3. Solve for x: $2x - 3y = 12$ **4.** Solve for b: $abc = x$

5. Solve for P $nRT = PV$ **6.** Solve for R: $nRT = PV$

  Answers: **1.** $x = -\frac{4}{5}y + 4$;   **2.** $y = -\frac{5}{4}x + 5$;   **3.** $x = \frac{3}{2}y + 6$;   **4.** $b = \frac{x}{ac}$;   **5.** $P = \frac{nRT}{V}$;   **6.** $R = \frac{PV}{nT}$

Test # 6 - Student's Self-Test (Always, Test yourself before you are tested)

Attempt all questions on clean sheets of paper, **Do not write in the book** Show all necessary work.

Problems 1- 23: Solve the equations unless otherwise instructed:

1. (a) $x - 5 = 12$

   (b) $x + 4 = -9$

2. (a) $4x + 5 = 35$

   (b) $2x + 3 = 9$

3. (a) $5x = 20$

   (b) $-2x = 20$

4. (a) $8x = -32$

   (b) $-8x = 32$

5. (a) $7x - 8 = 3x + 24$

6. (a) $\frac{x}{6} + 3 = 7$

   (b) $8x = 22$

7. (a) $\frac{x}{6} = 8$ (b) $3x = 18$

8. (a) $5x - 7 + 3x + 9 = 6x + 5 + 8x$;

   (b) $\frac{x}{4} - 6 = 14$

9. (a) $2(x - 4) + 3 = 15$

   (b) $3(x + 5) + 1 = 5(x + 4) - 6$

10. (a) $3(2x - 5) + 4 = -4(x + 7)$

    (b) $5x - 4 + x = 4x + 6 - 3x$

11. $2x - 5 + 3x + 2 = 7x - 5 - 3x$

12. $\frac{-x}{3} + 5 = 7$

13. Is 2 a solution of $5x + 2 = x - 4$?

14. Is -3 a solution of $4x - 1 = 15$?

15. Is 2 a solution of $3x + 1 = 7$?

16. (a) $x - \frac{1}{3} = \frac{1}{4}$

    (b) $\frac{x}{3} = \frac{x}{6} + \frac{3}{5}$

17. (a) $\frac{x}{-2} = 7$ (b) $\frac{-x}{4} = 9$

18. (a) $\frac{x}{6} + \frac{x}{3} = \frac{3}{4}$

    (b) $\frac{x}{8} = \frac{x}{2} - 6$

19. **(a)** $\frac{x}{2} + 1 = \frac{x}{3}$

(b) $\frac{x-2}{4} + \frac{2}{3} = \frac{1}{2}$

20. **(a)** $\frac{x+3}{4} + \frac{x-2}{8} = 2$

(b) $\frac{x}{4} + \frac{1}{2} = \frac{x}{5} + 1$

21. **(a)** $\frac{x-2}{3} + 1 = \frac{x}{5}$

(b) $\frac{x}{2} + \frac{2}{3} = \frac{x}{4}$

22. **(a)** $2x + .6 = 1$

(b) $5x + .04 = .2$

23. **(a)** Solve for x $\quad 4x + .1 = 2.$
(b) Solve for x: $\quad 2x - 3y = 12$

24. **(a)** Solve for C: $\quad F = \frac{9}{5}C + 32$

(b) Solve for y: $\quad 5x + 4y = 20$

25. **(a)** Solve for y: $\quad 3x + 8y = 32$

(b) Solve for x: $\quad 2y - 6x = 21$

Bonus: **(a)** Solve for P $\quad nRT = PV$

(b) Solve for R: $\quad nRT = PV$

(c) Solve for b: $\quad abc = x$

(d) Solve for x: $\quad 5x + 4y = 20$

Answers: **1.** (a) $x = 17$; (b) $x = -13$; **2.** (a) $x = \frac{15}{2}$; (b) $x = 3$; **3.** (a) $x = 4$; (b) $x = -10$;

4. (a) $x = -4$; (b) $x = -4$; **5.** $x = 8$; **6.** (a) $x = 24$; (b) $x = \frac{11}{4}$; **7.** (a) $x = 48$; (b) $x = 6$;

8.(a) $x = -\frac{1}{2}$; (b) $x = 80$; **9.**(a) $x = 10$; (b) $x = 1$; **10.** (a) $x = -\frac{17}{10}$; (b; $x = 2$; **11.** $x = -2$;

12. $x = -6$; **13.** No; **14.** No; **15.** Yes; **16.** (a) $x = \frac{7}{12}$; (b) $x = \frac{18}{5}$; **17.** (a) $x = -14$;

(b) $x = -36$; **18.** (a) $x = \frac{3}{2}$; (b) $x = 16$; **19.** (a) $x = -6$; (b) $x = \frac{4}{3}$; **20.** (a) $x = 4$; (b) $x = 10$;

21. (a) $x = -\frac{5}{2}$; (b) $x = -\frac{8}{3}$; **22.** (a) $x = .2$ or $\frac{1}{5}$; (b) $x = .032$ or $\frac{4}{125}$; **23.** (a) $x = .475$ or $\frac{19}{40}$;

(b) $x = \frac{3}{2}y + 6$; **24** (a) $C = \frac{5F-160}{9}$ or $C = \frac{5}{9}(F-32)$; (b) $y = -\frac{5}{4}x + 5$; **25.** (a) $y = -\frac{3}{8}x + 4$;

(b) $x = \frac{1}{3}y - \frac{7}{2}$; **Bonus:** (a) $p = \frac{nRT}{V}$; (b); $R = \frac{PV}{nT}$; (c) $b = \frac{x}{ac}$; (d) $x = -\frac{4}{5}y + 4$.

CHAPTER 6
WORD PROBLEMS

Lesson 14: **Words and Phrases for Translating Word Problems into Equations**
Lesson 15: **Solving Number Word Problems**
Lesson 16: **Perimeter Problems; Work-Time Problems**
Lesson 17: **Motion Problems** (Distance-Time-Speed Problems)
Lesson 18: **Consecutive Integer Problems; Age Problems; Percent Problems**
Lesson 19: **Mixture and Mixture-Like Problems**

To translate a word problem into an equation, we need an algebraic expression for the left-hand
 side of the equation and an algebraic expression for the right-hand side of the equation.
The equality symbol "=" is placed between these two expressions. **Note** that an expression could
consist of only a single term such as the integer 3.

Lesson 14
Words and Phrases for Translating Word Problems into Equations

Implication of Equality   "="

1. Equals
2. Is equal to
3. Result is
4. Equivalent to   } All these mean "="
5. One gets
6. Answer is
7. One obtains

Begin each of the following with " **the number which is**"
Example: **The number which is** the sum of 6 and 2 is: 6 + 2 = 8.

Implication of Addition:
1. the sum of 6 and 2 is **6+2=8**
2. 2 added to 6 is   **6+2**
3. 2 more than 6 is   **6+2**
4. 6 increased by 2 is **6+2**
5. 2 greater than 6 is **6+2**

Implication of Subtraction
1. the difference between 6 and 2 is **6- 2**=4; but
2. the difference between 2 and 6 is **2-6 = - 4**
3. the difference of 6 and 2 is **6-2**= 4 ; but   (Note that in subtraction, order is important)
5. 6 decreased by 2 is **6-2**
6. 6 less 2 is 6 - 2; **but also note that**
7. 2 less than 6 also is **6-2**
Note the difference between #1 and #2

Implication of Multiplication

1. the product of 6 and 2 is 6×2 (or 2×6)=12

2. 6 times 2 is 6×2

3. 6 times as much as 2 is 6×2

4. twice 6 is 2×6

5. double 6 is 2×6

Implication of Division

1. the quotient of 6 (divided) by 2 is $\frac{6}{2}$;**but**

2. the quotient of 2 (divided) by 6 is $\frac{2}{6}$ (Note that in division, order is important)

3. the ratio 6:2 is $\frac{6}{2}$

4. the ratio 2:6 is $\frac{2}{6}$

Note also that in the future, if you come across any words or phrases, you may add them to the above list.

Lesson 15
Solving Number Word Problems (Using a Single Variable)

1. Read the word problem twice, first quickly and then slowly and determine what is unknown (or are unknown).
2. Represent the unknown (or one of the unknowns) by a letter and express any other unknowns in terms of the same letter.
3. Read the whole problem again, and break it up into "parts" with the help of some of the phrases or words in the above table (or from any other source).
4. Put the parts together to form an equation.
5. Read the translated problem (equation) ,in your own words if possible, to see if it is equivalent to the original word problem.
6. Solve the equation for the unknown (variable) and check the solution in the original word problem.

Number Problems

Example 1   If five less than twice a number is 11, what is the number?

Solution   Let the number be x. Then, "twice the number" translates to $2x$.
"Five less than twice the number" translates to $2x - 5$.

"Five less than twice x is 11" translates to: $2x - 5 = 11$

$$\text{Solve for } x:$$
$$2x - 5 = 11$$
$$\underline{+ 5 \quad + 5}$$
$$2x = 16$$
$$\frac{2x}{2} = \frac{16}{2}$$
$$x = 8$$

The number is 8.

Example 2   If 6 is subtracted from five times a number, the result is 14
Find the number.

Solution   Let the number be x
Then, translating, $5x - 6 = 14$
Solve for x: $5x - 6 = 14$
$$\underline{+6 \quad +6}$$
$$5x = 20$$
$$x = 4$$

The number is 4.

Example 3 If the product of four and a number is subtracted from 14, the result is equal the number diminished by 1. Find the number.

Solution   Let the number be x
Then, $14 - 4x = x - 1$

Solve for x: $14 - 5x = -1$
$$-5x = -1 - 14$$
$$-5x = - 15$$
$$x = 3$$

The number is 3.

Example 4 62

The sum of two numbers is 72. One of the numbers is 12 more than five times the other number. Find these numbers.

Solution Let "the other number" be x.

Then "one of the numbers" $= 5x + 12$

"The sum of the two numbers is 72" translates to: $x + (5x + 12) = 72$

Solve for x:

$$x + (5x + 12) = 72$$
$$x + 5x + 12 = 72$$
$$6x + 12 = 72$$
$$6x = 60$$
$$x = 10$$

Therefore, "the other number" is 10 and

"one of the numbers" $= 5(10) + 12$
$$= 50 + 12 = 62$$

The numbers are 62 and 10.

Note: The use of the phrases "the other number" and "one of the numbers" was to help direct translation. This technique is useful in problems in which it is not easy or we do not want to define the smaller or the larger number at the beginning of solution of the problem. If it is easy to define the smaller or the larger number at the beginning of the problem, do so. In the above problem, we could have done this: Let the smaller number be x. Then the larger number $= (5x + 12)$. The equation needed : $x + (5x + 12) = 72$, which is the same equation as before.

Example 5 A number is three more than another number. Twice their sum is 38. Find these numbers.

Let the smaller number $= x$
Then the larger number $= x + 3$
The sum of the numbers $= x + x + 3$
Twice their sum $= 2(x + x + 3) = 2(2x + 3)$

The equation needed: $2(2x + 3) = 38$
Solve for x:
$$4x + 6 = 38$$
$$4x = 32$$
$$x = 8$$

The smaller number is 8 and the larger number is 11.

Lesson 15 Exercises

A. If twice the difference of a number and six is increased by four, the result is 10. Find the number.

Answer: 9

B

1. If six more than twice a number is 48, find this number.

2. If five times a number is subtracted from 8, the result is 68. Find this number.

3. A number is 6 less than twice another number. If the sum of these numbers is 42, find them.

4 In a chemistry class, there are 32 students. The number of females is 4 more than three times the number of males. How many males and how many females are in this class?

Answers: **1**. 21 ; **2**. -12 ; **3**. 26 and 16; **4.** There are 7 males and 25 females in this class.

Lesson 16

Perimeter Problems; Work-Time Problems

Perimeter Problems

The **perimeter** of a two-dimensional figure is the total distance around the figure.
(That is, the sum of the lengths of all the sides enclosing the figure)

We will approach perimeter problems in much the same way as we approached the "number" problems above. However, here ,we need to know some geometric facts which will not usually be stated in the problem. For example, for a rectangle, we need to know how the length and the width are related to the perimeter.

Perimeter of a Rectangle

Let the length of a rectangle = L; and let its width (or altitude) = W
Then the perimeter, P, of a rectangle is given by the formula, $P = 2(L + W)$ or $2L + 2W$

Example 6 We will reword Example 5 as a perimeter problem.

The length of a rectangle is three units more than its width. If the perimeter of this rectangle is 38 units, find the dimensions (length and width) of this rectangle.

Solution
Let the width of this rectangle = x
Then the length = $x + 3$
The perimeter formula is $P = 2(L + W)$ or $2L + 2W$

The equation needed: $2(x + x + 3) = 38$ or $2x + 2(x + 3) = 38$
Solve for x: $4x + 6 = 38$
                                        $4x = 32$
                                          $x = 8$
The width is 8 units and the length is 11 units.

Work and Time Problems

Example 7

Working alone, Maria can paint an apartment in 4 hours, while James can paint the same apartment in 6 hours, working alone. How long it will take Maria and James to paint the same apartment if they both paint together?

Method 1: Using Algebra

Let the painting work $= w$.
Let the rate (per hr.) of doing the work be r.   (i.e., speed of working or how fast the work is done)
Let the time taken to do the work $= t$.
These three quantities are related as follows:

1. $r = \dfrac{w}{t}$;   **2.** $t = \dfrac{w}{r}$;   **3.** $w = r \times t$

The rate (per hr.) at which Maria can do the work $= \dfrac{w}{4}$.

The rate (per hr.) at which James can do the work $= \dfrac{w}{6}$.

Working together, the rate (per hr.) at which Maria and James do the work $= \dfrac{w}{4} + \dfrac{w}{6} = \dfrac{5w}{12}$.

Now, r (for Maria and James) $= \dfrac{5w}{12}$; $t = \dfrac{w}{r}$ or $w \div r$ and

$\therefore t$ (for Maria and James) $= w \div \dfrac{5w}{12} = \dfrac{w}{1} \times \dfrac{12}{5w} = \dfrac{12}{5} = 2\dfrac{2}{5}$ hrs.

Method 2: Using Algebra

Let the time taken when both Maria and James work together $= t$.
Let the painting work $= w$.

Part of the work done by Maria $= r \times t = \dfrac{w}{4} \times t$

Part of the work done by James $= r \times t = \dfrac{w}{6} \times t$

Total work done by Maria and James in t hours when working together $= \left(\dfrac{w}{4} \times t \right) + \left(\dfrac{w}{6} \times t \right) = W$ .....(1)

Dividing through equation (1) by w, we obtain $\boxed{\dfrac{t}{4} + \dfrac{t}{6} = 1}$ ....(2)

Multiplying equation (2) by 12 (the LCM of the denominators), we obtain

$$3t + 2t = 12$$
$$5t = 12$$
$$t = \dfrac{12}{5} = 2\dfrac{2}{5} \text{ hours.}$$

Method 3: Compact formula (Derived from the general form equation (2) of Method 2)

Let the time Maria takes to do the work alone $= t_1$

Let the time James takes to do the work alone $= t_2$

Let the time both Maria and James take working together $= t$

Then $\boxed{t = \dfrac{t_1 t_2}{t_1 + t_2}}$ (i.e., t = **Product** of the individual times **divided** by the **sum** of the individual times.)

Substituting $t_1 = 4$, $t_2 = 6$ in the above compact formula,

$$t = \frac{4 \times 6}{4 + 6}$$

$$t = \frac{24}{10} = \frac{12}{5} = 2\frac{2}{5} \text{ hours.}$$

Note above that in Example 7, three quantities are involved (t, t_1, and t_2). Therefore, given any two of these quantities, we can find the third quantity by substituting the given quantities in the compact formula (or in the general form of equation (2) of Method 2, given below *) and solving for the unknown quantity.

* The general form of $\frac{t}{4} + \frac{t}{6} = 1$ is $\boxed{\dfrac{t}{t_1} + \dfrac{t}{t_2} = 1}$ OR $\boxed{\dfrac{1}{t} = \dfrac{1}{t_1} + \dfrac{1}{t_2}}$

where t is the "working-together" time; and t_1, and t_2 are the individual times.

Lesson 16 Exercises

A

1. The width of a rectangle is four units less than its length. If the perimeter of this rectangle is 20 units, find the dimensions of this rectangle.

2. Maria fenced a rectangular farm with 600 ft of wire. If this farm is twice as long as it is wide, what are the dimensions of this farm?

Answers: **1.** Length = 7 units, width = 3 units; **2.** Length = 200 ft, width = 100 ft

B

The length of a rectangle is 3 units more than the width. If the perimeter of this rectangle is 66, find the length and width of this rectangle.

Answer: $w = 15$; $L = 18$

C

Working alone, John can sort a quantity of mail in 5 hours, while Betty working alone can sort the same quantity of mail in 2 hours. How long will it take John and Betty to sort the same quantity of mail if they work together?

Answer : $1\frac{3}{7}$ hours.

Lesson 17
Motion Problems

(Distance-Time-Speed-Problems)

Let the distance traveled $= d$. Let the time taken to travel $d = t$. Let the constant speed at which d was traveled $= s$. These three quantities are related as follows:

$$d = st \quad \text{(distance = speed × time)} \; ; \quad s = \frac{d}{t} \; \text{(i.e., speed} = \frac{\text{distance}}{\text{time}}; \quad t = \frac{d}{s} \; \text{(i.e., time} = \frac{\text{distance}}{\text{speed}})$$

In the problems covered below, we may base our **main equation** on the quantity (time, speed, or distance) which is the same (sometimes with some minor adjustments) for say, both car A and car B. The main equation may also be based on the following: sum of distances; equality of distances; sum or difference of elapsed times or equality of elapsed times or equality of speeds. We will consider five main cases.

Case 1 Two Cars Start at the Same Time and Travel Towards Each Other
Example 1a

Two cars A and B, initially d miles apart, travel towards each other in opposite directions along a straight road. They both start at the same time. Car A's speed is a mph and Car B's speed is b mph Obtain an expression for where they meet. (Note: mph means miles per hour.)

Solution

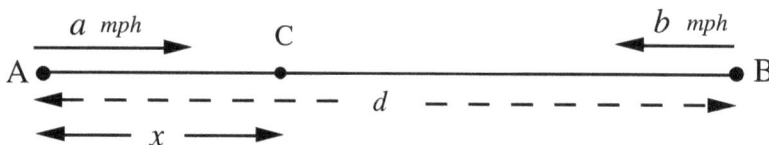

Step 1: In this problem, the time they take to travel to meet is the same for A and B since they both start at the same time. Therefore, our **main equation** is $t_A = t_B$. (t_A is A's time and t_B is B's time)

Step 2: Let Car A and Car B meet at point C which is x miles from Car A's starting point (Point A). Then the distance Car A covered is x miles and the distance Car B covered $= (d - x)$ miles. Now, we will obtain expressions for t_A and t_B in terms of a, b, d and x.

Step 3: $t_A = \dfrac{x}{a}$　　$\left(\dfrac{\text{distance A traveled}}{\text{A's speed}} \right)$

$t_B = \dfrac{d - x}{b}$　　$\left(\dfrac{\text{distance B traveled}}{\text{B's speed}} \right)$

Step 4: Since $t_A = t_B$ (since they both started traveling towards each other at the same time)

$$\frac{x}{a} = \frac{d - x}{b}$$

Step 5: Solve for x: $bx = ad - ax$

$ax + bx = ad$

$x(a + b) = ad$

$x = \dfrac{ad}{a + b}$

Step 6: The two cars meet at a distance $\dfrac{ad}{a + b}$ miles from A.

(Similarly, if they meet y miles from B, then $y = \dfrac{bd}{a + b}$ miles)

Lesson 17: Motion Problems (Distance-Time-Speed Problems)

Example 1b (Numerical Example)

Two cars A and B, initially 60 miles apart, travel towards each other in opposite directions along a straight road. They both start at the same time. Car A's speed is 25 mph and Car B's speed is 15 mph Obtain an expression for where they meet. (Note: mph means miles per hour.)

Solution

Step 1: In this problem, the time they take to travel to meet is the same for A and B since they both start at the same time. Therefore, our **main equation** is $t_A = t_B$. (t_A is A's time and t_B is B's time)

Step 2: Let Car A and Car B meet at point C which is x miles from Car A's starting point (Point A). Then the distance Car A covered is x miles and the distance Car B covered = (60 - x) miles. Now, we will obtain expressions for t_A and t_B.

Step 3: t_A, A's travel time = $\dfrac{x}{25}$ $\left(\dfrac{\text{distance A traveled}}{\text{A's speed}} \right)$

t_B, B's travel time = $\dfrac{60 - x}{15}$ $\left(\dfrac{\text{distance B traveled}}{\text{B's speed}} \right)$

Step 4: Since A's travel time = B's travel time (They started traveling towards each other at the same time)

$$\frac{x}{25} = \frac{60 - x}{15}$$

Step 5: Solve for x: $15x = 1500 - 25x$

$$x = \frac{1500}{40} = 37.5$$

Step 6: The two cars meet at a distance 37.5 miles from A.

(Similarly, if they meet y miles from B, then $y = \dfrac{15(60)}{25 + 15} = 22.5$ miles)

Lesson 17: Motion Problems (Distance-Time-Speed Problems)

Example 2a   Two Cars Start at Different Times and Travel Towards Each Other

Two cars A and B, initially d miles apart, travel towards each other (in opposite directions) along a straight road. Car A starts first and travels at a mph; and e hours later, Car B starts traveling at b mph. Obtain an expression for where they meet.

Solution

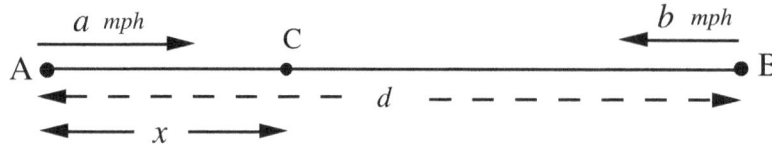

Step 1. In this problem, Car A's travel time before meeting Car B is given by:

$t_A = t_B + e$ (where t_A is Car A's time and t_B is Car B's time), since Car A started e hours before Car B.

Step 2: Let Car A and Car B meet at point C which is x miles from Car A's starting point (Point A). Then the distance Car A covered is x miles and the distance Car B covered $= (d - x)$ miles. Now, we obtain expressions for t_A and t_B in terms of a, b, d, x, and e.

Step 3: $t_A = \dfrac{x}{a}$ $\left(\dfrac{\text{distance A traveled}}{\text{A's speed}}\right)$

$t_B = \dfrac{d - x}{b}$ $\left(\dfrac{\text{distance B traveled}}{\text{B's speed}}\right)$

Step 4: Since $t_A = t_B + e$

$\dfrac{x}{a} = \dfrac{d - x}{b} + e$ (Substituting for t_A, and t_B, from Step 3)

Step 5: Solve for x:

$\dfrac{x}{a} = \dfrac{d - x + be}{b}$

$bx = a(d - x + be)$ (by cross-multiplication)

$bx = ad - ax + abe$ <----Now, solve this equation for x.

$ax + bx = ad + abe$

$x(a + b) = ad + abe$

$x = \dfrac{ad + abe}{a + b}$

Step 6 . The two cars meet $\dfrac{ad + abe}{a + b}$ miles from Car A's starting point.

(Similarly, if they meet y miles from Car B's starting point then $y = ?$)

Example 2b   (Numerical Example)

Two cars A and B, initially 60 miles apart, travel towards each other (in opposite directions) along
a straight road. Car A starts first and travels at 30 mph; and one hour later, Car B starts traveling
at 20 mph. Obtain an expression for where they meet.

Solution

Step 1. Let Car B's travel time $= t$.
  Then Car A's travel time $= t + 1$ (Since Car A started one hour before Car B.

Step 2: Let Car A and Car B meet at point C which is x miles from Car A's starting point (Point A).
  Then the distance Car A covered is x miles and the distance Car B covered $= (60 - x)$ miles.

Step 3:   Car A's travel time $= \frac{x}{30}$ $\left(\frac{\text{distance A traveled}}{\text{A' s speed}}\right)$

  Car B's travel time $= \frac{60 - x}{20}$ $\left(\frac{\text{distance B traveled}}{\text{B' s speed}}\right)$

Step 4: Since Car A's travel time = Car B's travel time + 1

$$\frac{x}{30} = \frac{60 - x}{20} + 1$$

Step 5: Solve for x:

$$\frac{x}{30} = \frac{60 - x + 20}{20}$$

$$20x = 30(60) - 30x + 30(20)$$

$$20x = 1800 - 30x + 600$$

$$50x = 2400$$

$$x = \frac{2400}{50}$$

$$x = 48$$

Step 6 . The two cars meet 48 miles from Car A's starting point.

Case 3: **Two cars start at different times from the same point and travel in the same direction** 7 0

Example 3a

Two cars A and B embark on a journey in the same direction and from the same point. Car A leaves first and travels at a mph; and e hours later, Car B follows at b mph ($b > a$) and travels to catch-up with Car A. After how many hours, and where, did Car B catch-up with Car A.?

Solution

Step 1: Car B will catch-up with Car A when both Car A and Car B have traveled the same distance from the starting point. Therefore, the quantity which is the same for Car A and Car B is the distance traveled.

Therefore, $d_A = d_B$ (equality of distances)

(d_A and d_B are the distances traveled by Car A and Car B respectively.)

Step 2: Let the travel time for Car A $= t$ hours.

Then the travel time for Car B $= t - e$ (since Car A left e hours earlier)

$d_A = at$ ($d = st$)

$d_B = b(t - e)$

Step 3: Since $d_A = d_B$

$at = b(t - e)$

Step 4: Solve for t:

$at = bt - be$

$at - bt = -be$

$t(a - b) = -be$

$$t = -\frac{be}{a - b} = \frac{be}{b - a}$$

Answer: Car B caught-up with Car A $\dfrac{be}{b - a}$ hours from the time Car A started the journey.

Car B caught-up with Car A $a \cdot \dfrac{be}{b - a} = \dfrac{abe}{b - a}$ miles from the starting point. ($d = st$)

Note that the time Car B took to catch-up with Car A $= \dfrac{be}{b - a} - e = \dfrac{be - be + ae}{b - a} = \dfrac{ae}{b - a}$ hours, which we could also obtain directly as follows:

If Car B's travel time is t, then $bt = ae + at$ (distance Car B traveled = distance Car A traveled), which on solving yields $t = \dfrac{ae}{b - a}$, as before.

Example 3b (Numerical Example)

Two cars A and B embark on a journey in the same direction and from the same point Car A
leaves first and travels at 20 mph; and 2 hours later, Car B follows at 50 mph and travels to
 catch-up with Car A. After how many hours, and where, did Car B catch-up with Car A.?

Solution

Step 1: Car B will catch-up with Car A when both Car A and Car B have traveled the same distance
   from the starting point. Therefore, the quantity which is the same for Car A and Car B is the
   distance traveled.

Step 2: Let the travel time for Car A = t hours.
   Then the travel time for Car B = $t - 2$ (Car B's time is 2 hours less than Car A's time)
   Car A's travel distance = $20t$ ($d = st$)
   Car B's travel distance = $50(t - 2)$

Step 3: Since Car A's travel distance = Car B's travel distance
$$20t = 50(t - 2)$$
Step 4: Solve for t:
$$20t = 50t - 100$$
$$-30t = -100$$
$$t = \frac{-100}{-30}$$
$$t = 3\tfrac{1}{3}$$

Distance from starting point at which Car B caught-up with car A $= 20 \times \dfrac{10}{3} = 66\tfrac{2}{3}$ ($S = 20; t = 3\tfrac{1}{3}; d = st$)

Answer: Car B caught-up with Car A $3\tfrac{1}{3}$ hours from the time Car A started the journey.

Car B caught-up with Car A at a distance of $66\tfrac{2}{3}$ miles from the starting point.

Note: If we were asked to find Car B travel time, t, then we could obtain t directly from
$$50t = 20(2) + 20t \quad \text{(distance Car B traveled = distance Car A traveled)}$$
$$30t = 40$$
$$t = 1\tfrac{1}{3} \text{ hours}$$

Lesson 17: Motion Problems (Distance-Time-Speed Problems)

Example 4a Car A and Car B leave point C at the same time and travel in opposite directions .
Car A's speed is a mph and Car B's speed is b mph. After how many hours are they d miles apart.?

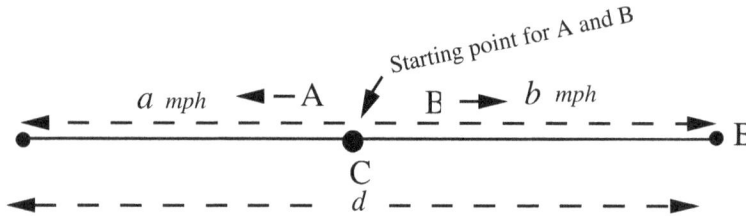

Solution

Let the time taken for the cars to be d miles apart be t hours.
Distance Car A traveled at a mph in t hours is at, and the distance Car B traveled at b mph in t hours is bt miles (Note: distance = speed × time)
The equation needed is based on the total distance traveled by Car A and Car B.
Therefore, $at + bt = d$ <---------Solve for t.

$$t (a + b) = d$$

$$t = \frac{d}{a + b} \text{ hours}$$

Answer: Car A and Car B are d miles apart after $\dfrac{d}{a + b}$ hours.

Example 4b (Numerical Example)

Car A and Car B leave point C at the same time and travel in opposite directions .
Car A's speed is 40 mph and Car B's speed is 30 mph. After how many hours are they 210 miles apart.?

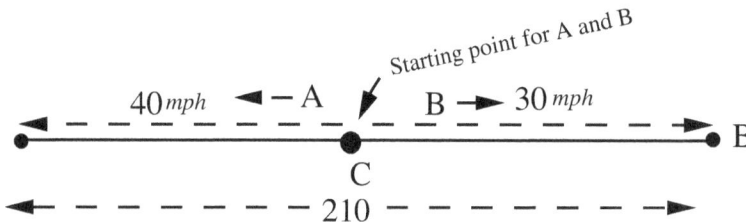

Solution

Let the time taken for the cars to be 210 miles apart be t hours.
Distance Car A traveled at 40 mph in t hours is $40t$; and the distance Car B traveled at 30 mph in t hours is $30t$ miles (Note: distance = speed × time)
The equation needed is based on the total distance traveled by Car A and Car B.
Therefore, $40t + 30t = 210$ <---------Solve for t.

$$70t = 210$$

$$t = \frac{210}{70}$$

$$t = 3$$

Answer: Car A and Car B are 210 miles apart after **3** hours.

Lesson 17: Motion Problems (Distance-Time-Speed Problems)

Example 5a Car A and Car B leave point C and travel in opposite directions.
Car A departs 2 hours before Car B. Car A's speed is a mph and Car B's speed is
b mph. After how many hours from the time Car A departed are they d miles apart.?

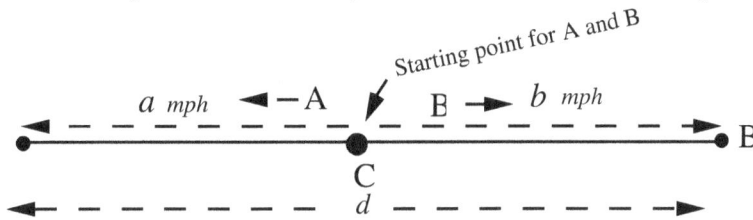

Solution Let the travel time for Car B = t hours.
Then the travel time for Car A = $(t + 2)$ hours (Since Car A started 2 hours before Car B)
Distance Car A traveled = $a(t + 2) = at + 2a$ ($2a$ is the distance A traveled before B started)
Distance Car B traveled = bt.
The equation needed is based on the total distance traveled.
$(at + 2a) + bt = d$ <---------Solve for t.
$at + 2a + bt = d$
$at + bt = d - 2a$
$t(a + b) = d - 2a$
$t = \dfrac{d - 2a}{a + b}$

Answer: The two cars are d miles apart after $\left(\dfrac{d - 2a}{a + b} + 2\right)$ hours from the time Car A started.

Lesson 17: Motion Problems (Distance-Time-Speed Problems)

Example 5b **Two cars start at different times and travel from the same point in opposite**
directions (Numerical Example)

Car A and Car B depart point C and travel in opposite directions.
Car A departs 2 hours before Car B. Car A's speed is 50 mph and Car B's speed is
40 mph. After how many hours from the time Car A departed are they 280 miles apart?

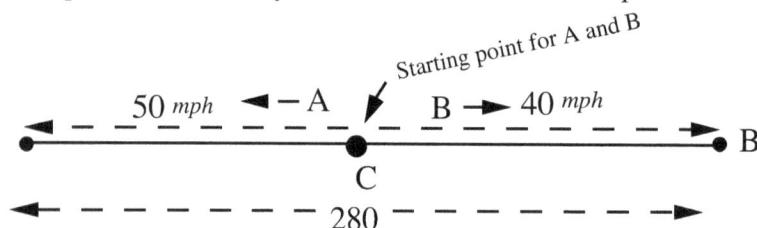

Solution Let the travel time for Car B $= t$ hours.

Then the travel time for Car A $= (t + 2)$ hours   (Since Car A started 2 hours before Car B)

Distance Car A traveled $= 50(t + 2) =$

Distance Car B traveled $= 40t$

The equation needed is based on the total distance traveled.

Car A's distance + Car B's distance = 280

$$50(t + 2) + 40t = 280 \text{<--------Solve for } t.$$

$$50t + 100 + 40t = 280$$

$$90t + 100 = 280$$

$$90t = 180$$

$$t = \frac{180}{90}$$

$$t = 2$$

The two cars are 280 miles apart after $(t + 2) = 2 + 2$ or 4 hours from the time Car A started.

Answer: The two cars are 280 miles apart after **4** hours from the time Car A started.

Example 6a
Maria rode her bicycle at b mph from point A to point B, and returned by car at c mph.
If her total traveling time was t hours, find the distance between point A and point B

We will base our (main) equation on: **car's time + bicycle's time = total time.** Also, $t = \frac{d}{s}$.

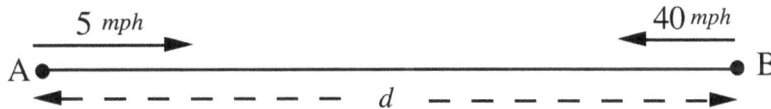

$$\overset{\text{b \ mph}}{\xrightarrow{\hspace{2cm}}} \qquad\qquad \overset{\text{c \ mph}}{\xleftarrow{\hspace{2cm}}}$$

A •━━━━━━━━━━━━━━━━━━━━━━━━• B

←─ ─ ─ ─ ─ ─ d ─ ─ ─ ─ ─ ─→

Solution: Step 1: $\frac{d}{b} + \frac{d}{c} = t$ (where d is the distance between point A and point B; b is the bicycle's speed;

c is the car's speed and t is the total time for the bicycle and the car.)

Step 2: Solve for d:

$$\frac{cd + bd}{bc} = t$$

$$cd + bd = bct$$

$$d(c + b) = bct$$

$$d = \frac{bct}{b + c} .$$

The distance between point A and point B is $\frac{bct}{b + c}$ miles.

Example 6b (Numerical Example)
Maria rode her bicycle at 5 mph from point A to point B, and returned by car at 40 mph.
If her total traveling time was 3 hours, find the distance between point A and point B

(We will base our (main) equation on: **car's time + bicycle's time = total time.** Also, $t = \frac{d}{s}$)

$$\overset{\text{5 \ mph}}{\xrightarrow{\hspace{2cm}}} \qquad\qquad \overset{\text{40 \ mph}}{\xleftarrow{\hspace{2cm}}}$$

A •━━━━━━━━━━━━━━━━━━━━━━━━• B

←─ ─ ─ ─ ─ ─ d ─ ─ ─ ─ ─ ─→

Solution: Step 1: car's time + bicycle's time = total time.

$$\frac{d}{5} + \frac{d}{40} = 3 \qquad\qquad \text{(where } d \text{ is the distance between point A and point B)}$$

Step 2: Solve for d:

$$8d + d = 120$$

$$9d = 120$$

$$d = \frac{120}{9}$$

$$d = 13\tfrac{1}{3}$$

The distance between point A and point B is $13\tfrac{1}{3}$ miles.

Distance-Time-Speed Problems Involving Travel by Water or by Air   76

In problems of travel by water (such as **boat rowing)** or by air (such as **aircraft fligh**t or
bird flight) in which we have to consider the effect of the direction of the flow of water or air,
there is an additional variable involved, namely, the water (river) current or the wind current.
We can solve these problems using a single variable. However, because of the additional
variable, we will prefer to solve these problems using two variables in **Chapter 19.**

Lesson 17 Exercises

A
1. Two cars A and B, initially 100 miles apart, travel towards each other in opposite directions
along a straight road. They both start at the same time. Car A's speed is 20 mph and Car B's
speed is 30 mph Where do they meet?

Answer: They meet 40 miles from point A;

B

Two cars A and B, initially 100 miles apart, travel towards each other (in opposite directions)
along a straight road. Car A starts first and travels at 40 mph; and one hour later, Car B starts
traveling at 30 mph. Where do they meet?

Answers: They meet $74\frac{2}{7}$ miles from point A.

C

Two cars A and B embark on a journey in the same direction and from the same point Car A
leaves first and travels at 30 mph; and 3 hours later, Car B follows at 50 mph and travels to
catch-up with Car A. After how many hours, and where, did Car B catch-up with Car A.?

Ans: Car B caught-up with Car A $4\frac{1}{2}$ hours after Car B started, and at a point 225 miles from the starting poin**t.**

D
1. Car A and Car B leave point C at the same time and travel in opposite directions .Car A's speed
is 60 mph and Car B's speed is 50 mph. After how many hours are they 440 miles apart.?

Answer: After 4 hours;

E

Car A and Car B depart point C and travel in opposite directions. Car A departs 1 hour before
Car B. Car A's speed is 50 mph and Car B's speed is 30 mph. After how many hours from the
time Car A departed are they 290 miles apart?

Answer: After 4 hours.

F

Maria rode her bicycle at 10 mph from point A to point B, and returned by car at 60 mph. If her total traveling time was 4 hours, find the distance between point A and point B.

Answer: The distance between point A and point B is $34\frac{2}{7}$ miles.

Lesson 18
Consecutive Integer Problems; Age Problems; Percent Problems

Consecutive Integer Problems

Consecutive integers are integers written in increasing or decreasing order such that each integer differs from the integer immediately before it or immediately after it by 1. Example: 5, 6, and 7 are consecutive ntegers.

Example   The sum of three consecutive integers is 21. Find the numbers.

Solution   Let the first consecutive integer be x.
Then the second consecutive integer is $(x + 1)$
and the third consecutive integer is $(x + 1 + 1)$.
"Their sum is 21" translates to:
$$x + (x + 1) + (x + 1 + 1) = 21$$
$$x + x + 1 + x + 2 = 21$$
Solve for x. $$3x + 3 = 21$$
$$\underline{\quad -3 \quad -3}$$
$$3x = 18$$
$$\frac{3x}{3} = \frac{18}{3}$$
$$x = 6$$

∴   The first consecutive integer is 6.
The second consecutive integer is $6 + 1 = 7$.
The third consecutive integer $6 + 1 + 1 = 8$.

$$\text{Check: } 6 + 7 + 8 \stackrel{?}{=} 21$$
$$21 = 21 \quad \text{True}$$

∴   The three consecutive integers are 6, 7 and 8 (**note** that the integers are consecutive and their sum is 21)
Note also that in the above problem, if we know a given integer, to obtain next consecutive integer we just add 1.

Consecutive Even Integer Problems

Consecutive **even** integers are even integers written in increasing or decreasing order such that each integer differs from the integer immediately before it or immediately after it by 2. Example: 6, 8, 10, and 12 are consecutive even integers.

Here, if we know the first consecutive even integer, to obtain
the next consecutive integer we add 2.

Example 2 The sum of three consecutive even integers is 36. What are these even integers?
Solution:

Step 1:   Let the first consecutive even integer be x.
Then the second consecutive even integer is $(x + 2)$
and the third consecutive even integer is $(x + 2 + 2)$
"Their sum is 36" translates to:
$$x + (x + 2) + (x + 2 + 2) = 36$$
$$x + x + 2 + x + 4 = 36$$
$$3x + 6 = 36$$
$$\underline{\quad -6 \quad -6}$$
$$3x + 0 = 30$$
$$\frac{3x}{3} = \frac{30}{3}$$
$$x = 10$$

Step 2:
The first consecutive even integer is 10
The second consecutive even integer is 12
The third consecutive even integer 14
Check: $10 + 12 + 14 = 36$
$$36 = 36 \quad \text{True}$$
The three consecutive even integers are
10 12 and 14

Consecutive Odd Integer Problems

Consecutive **odd** integers are odd integers written in increasing or decreasing order such that each integer differs from the integer immediately before it or immediately after it by 2.
Example: $5, 7, 9$ and 11 are consecutive odd integers.

Here **also**, as in the case of even integer problems, if we know the first odd integer, to obtain the next odd integer, we add 2. For example, if the first odd integer is $5,$ then the next odd integer is $5 + 2 = 7,$ and the next odd integer is $7 + 2 = 9.$

Example 3 The sum of three consecutive odd integers is 33. Find the integers.
Solution Let the first consecutive odd integer be x.
Then the second consecutive odd integer is $(x + 2)$
and the third consecutive odd integer is $(x + 2 + 2)$

"Their sum is 33" translates to:
$$x + (x + 2) + (x + 2 + 2) = 33$$
$$x + x + 2 + x + 4 = 33$$
$$3x + 6 = 33$$
$$\underline{\quad -6 \quad -3\quad}$$
$$3x = 27$$
$$\frac{3x}{3} = \frac{27}{3}$$
$$x = 9$$

$\therefore$ The first consecutive odd integer is 9
The second consecutive odd integer is $9 + 2 = 11$
The third consecutive odd integer $11 + 2 = 13$

Check: $9 + 11 + 13 \overset{?}{=} 33$
$33 = 33$ True

$\therefore$ The three consecutive odd integers are 9 ,11 and 13.

A note about consecutive even and odd integer problems. Sometimes, because of the use of the word "odd" in consecutive odd integer problems, some students think that if given an odd integer, to obtain the next odd integer, they would add odd integers $1, 3, 5$ etc. **This notion is incorrect.** For example, if the first odd integer is 5, the next odd integer 7 is obtained by adding 2; and similarly, the next odd integer, 9, is obtained by adding 2 to 7.
Also $5 + 1 = 6, 5 + 3 = 8$ are all even.
Therefore, note that whether an integer is odd or even, to obtain the next consecutive odd or even integer we **add 2.** The difference in these types of problems will be that if the problem states that the integers are even, then the answers we obtain must be even integers, otherwise, there is no solution. On the other hand, if the problem states that the integers are odd, then the solutions we obtain should be odd integers, otherwise there is no solution.

Age Problems

Example 1   A father is six times as old as his daughter. Last year he was seven times as old as his daughter. How old are they?

Solution: Let the daughter's present age be x
Then the father's present age is $6x$.
The father's age last year $= (6x - 1)$
The daughter's age last year $ = (x - 1)$
Last year, their ages were related as follows :

Father's age = seven times daughter's age.

$(6x - 1) = 7(x - 1)$
$6x - 1 = 7x - 7$
$\underline{\quad +1 \qquad + 1\quad}$
$6x = 7x - 6$
$\underline{-7x \quad -7x\quad}$,
$-x = - 6$
$x = 6$

The daughter is 6 years old.
The father is $6(6) = $ 36 years old.

Extra: How old was the father when the daughter was born?

Percent Problems

We use both arithmetic and algebraic methods.

Finding the Original Number (Finding the base)

Example   If 20% of a number is 64, what is the number?

Solution

Method 1   Let the number be x

Then, "20% of the number is 64" translates to $\dfrac{20x}{100} = 64$

i.e. $\dfrac{20x}{100} = 64$

Solve for x: $\dfrac{\overset{1}{\cancel{(100)}}}{\underset{1}{\cancel{(20)}}} \dfrac{\overset{1}{\cancel{20x}}}{\cancel{100}} = 64\dfrac{\overset{5}{\cancel{(100)}}}{\underset{1}{(\cancel{20})}}$

$x = 320$

The number is 320 .
Note also that since 20% of the number is 64, the number must be greater than 64.

Method 2: **Using Proportions**

20% is to 64 as 100% is to x, where 100% of the
original number (the base) is x.
Translating the proportion,

$$\frac{20\%}{64} = \frac{100\%}{x} \qquad (\text{ or } \frac{.20}{64} = \frac{1}{x})$$

Solve for x: 20% x= 100% (64)   (or .20x = 64)

$$x = \frac{100\%}{20\%}(64) \qquad (\text{or } x = \frac{64}{.20})$$
$$x = 320$$

Note that 20% as a fraction is $\frac{20}{100} = \frac{1}{5}$

Therefore, in the above problem, instead of asking "if 20% of a number is 64, what is the number?",
we could have asked "if $\frac{1}{5}$ of a number is 64, what is the number? "

Example   If $\frac{1}{5}$ of a number is 64, what is the number? We can solve this by any of the
methods discussed above. Let us use the algebraic method.
   Solving:
        Let the number be x
        Then $\frac{x}{5} = 64$ (Note that $\frac{1}{5}x = \frac{x}{5}$)

        $(5)\frac{x}{5} = 64(5)$
           $x = 320$
        The number is 320.

Example   Mary spends 20% of her weekly income on food. If she spends $64 on
food every week, what is her weekly income?

Solution : We could reword this problem in the familiar form as " If 20% of a number is 64,
   what is the number?"; and then use exactly the same method as in the example, above.

Answer: $320 (numerically, the same answer as for the last example).

Finding the Rate %

Example 1 What rate % of 24 is 15?

Solution

Let $x\%$ be the rate %. We will write an equation in terms of x, and solve for x.
From the wording, "$x\%$ of 24 is 15" translates to:

$$\frac{x}{100} \times \frac{24}{1} = 15$$

$$\frac{24x}{100} = \frac{15}{1}$$

$24x = 100 \times 15$ (by cross-multiplication or by multiplying both sides of the equation by 100).

$$x = \frac{100 \times 15}{24}$$ (by dividing both sides of the equation by 24)

$$x = 62\frac{1}{2}$$

$\therefore$ the required rate $= 62\frac{1}{2}\%$.

Example 2 What rate % of 18 is 44?

Solution

Let the required rate be $x\%$.
Then, "$x\%$ of 18 is 44" translates to:

$$\frac{x}{100} \times \frac{18}{1} = 44$$

$$\frac{18x}{100} = \frac{44}{1}$$

$18x = 100 \times 44$ (by cross-multiplication or by multiplying both sides of the equation by 100)

$$x = \frac{100 \times 44}{18}$$ (by dividing both sides of the equation by 18)

$$x = 244\frac{4}{9}$$

$\therefore$ the required rate $= 244\frac{4}{9}\%$.

Example 3 A family's annual income last year was $20,000. This year, the income is $33,000. What is the percent increase in the annual income?

Step 1: The increase in income = $33,000 - $20,000
$$= \$13,000$$

Step 2: The percent increase in income $= \frac{13000}{20000} \times 100\%$ (Finding the rate percent)
$$= 65\%$$

Note: In Step 2 , the question could have been posed as: What percent of 20,000 is 13,000 ?

More Applications of Base Finding and Percentage Finding

In these applications, the questions are **not** worded in forms such as "find 20% of a number; "if 30% of a number is 45, what is the number?". A good approach is to reword the problem in any of these familiar forms and then proceed accordingly.

Note 1. The fraction involved in the problem may very likely be the **rate** (but it may **not** be the rate).

Note for example that $\frac{3}{5} = 60\%$

Example 1   3 out of 5 students at a certain college study Biology.

(a) If 600 students at this college study Biology, how many students are there at this college?

(b) If one were to collect a sample of 200 students at this college, how many of these students would study Biology?

Solution:

Note that 3 out of 5 means $\frac{3}{5}$

Part (a): We can reword this part of the question as: if $\frac{3}{5}$ of a number is 600, what is the number?

Solution   Let the number be x

Then $\frac{3x}{5} = 600$

$$3x = 3000$$
$$x = 1000$$

There are 1000 students at this college.
We solve part(b) using arithmetic, since after rewording the calculation becomes straightforward.

Part (b) We can reword this part of the question as: Find $\frac{3}{5}$ of 200.

$$\frac{3}{5} \text{ of } 200 = \frac{3}{5} \times \frac{200}{1} = 120$$

That is, of a sample of 200 students, 120 students would study Biology.

Note: In finding the base and the rate% , the use of algebra facilitates the solution process.

In finding the percentage as in part (b) of the last problem, perform a straightforward arithmetic calculation.

Lesson 18 Exercises

A

1. The sum of three consecutive integers is 48.   Find these integers.

2. The sum of two consecutive integers is 35 less than four times the larger. Find these integers.

Answers: **1**.15,16 and 17 : **2**.   16 and 17

B   The sum of three consecutive even integers is 60 . Find these integers.

Answer: 18, 20 and 22.

C

Find three consecutive even integers such that the sum of the second and third integers is 18 less than three times the first integer.

Ans: 24, 26 and 28.

D   Find three consecutive odd integers whose sum is 63.

Answer: 19,21 and 23

E

Find two consecutive integers such that the difference of twice the larger integer and the smaller integer is 51

Answers : 49 and 50

F

1. James is 3 years older than Mary. Five years ago, James was twice as old as Mary . Find their ages.

2. Four years from now, Paul's age will be $3x + 3$. Express his age five years ago in terms of x.

3. A father is 20 years older than his daughter. In 3 years, the sum of their ages will be 40 years. Find their present ages.

1. Answers: Mary is 8 years old;   James is 11 years old.;   **2** Five years ago, Paul's age was $3x - 6$ years.

3. The daughter's age is 7 years and the father's age is 27 years.

G

1. If 20% of a number is 72, what is the number?

2. 53 is 25% of what number?

3. 16% of what number is 82?

Answers **1.**360 ; **2.** 212; **3.** 512.5 or $512\frac{1}{2}$

H

   The monthly rent for Maria's apartment is $800. If Maria spends 25% of her monthly income on this rent, what is her monthly income?

Answer: $3,200

I

1. What rate percent of 32 is 12?

2. What rate% of 20 is 80?

3. On the last CUNY test, there were 40 questions. James answered 32 questions correctly. What was his grade in percent?

4. Repeat Step 2 of Example 3, above using algebra.

Answers   **1.** 37.5% or $37\frac{1}{2}$% **2.** 400% ; **3.** 80%

Lesson 19

Mixture and Mixture-Like Problems, Problems on Averages

We will call a problem a mixture problem if the solution method conforms to the methods to be presented below. There are problems which may not appear to be mixture problems, but which on close examination of the solution method will reveal that they are mixture problems

Example 1 A publishing company has 120 workers of whom 40% are women. How many new women workers must be hired such that 50% of the workers are women? Assume that no workers are fired.

Solution
Method 1   Using Arithmetic

Step 1: Find the present number of women workers.
The present number of women workers is 40% of 120.

$$\text{i.e.} \frac{40}{100} \times \frac{120}{1} = 48$$

Therefore, there are at present 48 women and $(120 - 48) = 72$ men

Step 2: Now, we want 50% or ($\frac{1}{2}$) of the workers to be women.

This is equivalent to saying that we want the number of women to be equal to the number of men.
Since there are at present 72 men and 48 women, we need $72 - 48 = 24$ women.
Therefore 24 new women must be hired.

Note: If 24 new women are hired, the number of men would be 72; the number of women would be 72, and the total number of workers (both men and women) would be 144.

Method 2: Using Algebra
The present number of women = 40% of 120 = 48
Let the number of the new women workers hired be x.
Then after this hiring, the total number of women would be $48 + x$, and
the total number of all the workers would be $120 + x$
From the problem, "the number of women should be 50% of the number of workers" translates to:

$$48 + x = \frac{50}{100}(120 + x)$$

Solve for x: $48 + x = \frac{1}{2}(120 + x)$

$2(48 + x) = 120 + x$ (multiplying both sides of the equation by 2)
$96 + 2x = 120 + x$
$\underline{-96 \qquad\quad -96}$
$2x = 24 + x$
$\underline{\quad -x \qquad\quad - x}$
$x = 24$

Again, the number of new women to be hired is 24.

Example 2   A chemist has 120 liters of a 40% solution of alcohol. How many liters of pure alcohol should be added to obtain a 50% solution of alcohol?
(Assume the solution is composed of only pure alcohol and water)
Solution: Use exactly the same procedure as in Example 1, above).
Answer: 24 liters (Numerically, same answer as in Example 1)

A rather more involved problem is the next example.

Example 3   A hospital has 120 doctors of whom 40% are surgeons. Assuming that no doctors are fired, how many new surgeons must be hired such that 75% of the doctors are surgeons?

We will solve this problem both algebraically and arithmetically.

Algebraic Method

The present number of surgeons = 40% of 120 = $\frac{40}{100} \times 120 = 48$

Thus, there are at present 48 surgeons. Let the number of new surgeons be x. Then after hiring x new surgeons the total number of surgeons would be $(48 + x)$ and the total number of all the doctors would be $120 + x$.
From the problem, the number of surgeons is to be equal to 75% of all the doctors.

$$48 + x = \frac{75}{100} \ (120 + x) \qquad\qquad \textbf{Note: } 75\% = \frac{75}{100} = \frac{3}{4}$$

Now solve this equation for x.

$$48 + x = \frac{75}{100}(120 + x) \qquad \text{or } 48 + x = \frac{3}{4}(120 + x)$$
$$4800 + 100x = 9000 + 75x \qquad \text{or } 192 + 4x = 360 + 3x$$
$$25x = 4200 \qquad\qquad\qquad x = 168$$
$$x = 168$$

∴ 168 new surgeons must be hired.
After this new hiring ,the total number of surgeons would be 48 + 168 = 216, and the total number of all the doctors would be 168 + 120 = 288.

Arithmetic Method

Step 1: The present number of surgeons = 40% of 120 = $\frac{40}{100} \times 120 = 48$

Step 2: The present number of non-surgeons = 120 - 48 = 72.
Since, we want 75% of the doctors to be surgeons,
we want 25% of the doctors to be non-surgeons.

Step 3:   If 25% of the number of doctors is (and is to be) 72,
then , the total number (100%) of all the doctors $= \frac{72}{1} \times \frac{100}{25} = 288$

(Step 3 could have been posed as: if 25% of a number is 72, what is the number?; the calculation is based on the quantity that remains unchanged in the process).

Since, the total number of all the doctors is to be 288,
the total number of the surgeons is to be 288 -72 = 216.
At present, there are 48 surgeons; and
therefore, the number of new surgeons to be hired = 216 - 48 = 168
Again, we obtain the same answer as by the algebraic method.

Generalization of Mixture Problems

Example: Mixing solutions of alcohol

Consider mixing two kinds of alcohol solutions. Let q_1 = volume of alcohol #1 and let its concentration = r_1%. Similarly, let q_2 = volume of alcohol #2, and let its concentration = r_2%. Let the volume of the mixture (made up of alcohol #1 and alcohol #2) = q, and let its concentration = r%

Then we may write a general equation as:

$$q_1\, r_1\% \; + q_2\, r_2\% \; = qr\%$$

Note #1: A 20% alcohol solution, for example, means 100 units of the alcohol solution contains 20 units of pure alcohol and 80 units of water, assuming the solution contains alcohol and water only.

Note #2: 100% alcohol (or pure alcohol) means $r\% = 100\% = \dfrac{100}{100} = 1$

Note: Adding pure water means $r\%$ for alcohol = 0, and therefore, $q_{alcohol}\, r_{alcohol}\% = q_{alcohol} \times 0 = 0$

General Formula for mixture Problems

A modified form of the above equation, (given below) may be applied to other mixture problems.

$$q_1\, r_1 \; + q_2\, r_2 = qr$$

where q_1, q_2, q (for quantity) may be mass, weight, volume or a number ; r_1, r_2, r may be a fraction, a rate %, a price or cost per unit, or a value per item, etc. Note that sometimes the product qr on the right-hand side of the equation may be given and you may not have to perform the operation $q \times r$.

The author found this formula to be surprisingly applicable to most (if not all) of the popular mixture problems. Moreover, it is easy to memorize and recall. The applications include the hiring of men and women to obtain a certain percent of workers who are men or women; mixing of alcohol solutions; mixing of ingredients of different values (cost, price, etc.) to obtain a desired mixture of a specified value (cost, price, etc.); investing money at different rates in different accounts; and coin problems. (See also Chapter 12)

Problems on Averages

Example 1   During the semester, James scored 85, 70, 75 on the first three tests. He wants his class average for the term to be 80. What is the score he needs on the next and last test?

Solution We will cover three methods.

Method 1: Using algebra.

Let x be the next and last score.

Then, the total score for the four tests will be $85 + 70 + 75 + x$

$$\text{Average score} = \frac{\text{Total score}}{\text{Number of tests}} = \frac{85 + 70 + 75 + x}{4}$$

Now we want this average score to be 80

$$\frac{85 + 70 + 75 + x}{4} = 80 \qquad \text{Scrapwork: } 85+70+75 =230$$

$$\frac{230 + x}{4} = 80$$

We solve for x.
$$\frac{4(230 + x)}{4} = 80\,(4)$$

$$230 + x = 320$$
$$\underline{-230 \qquad\quad -230}$$
$$x = 90$$

∴ James needs a score of 90 on the next and last test.

Method 2: Arithmetic approach.

This method can follow from Method 1 (the algebraic approach).

Step 1:   For an average of 80 on 4 tests, total score must be $80 \times 4 = 320$

Step 2:   Total score on the first three tests is $75 + 70 + 75 = 230$

Step 3:   Score needed on the 4th test to make total score 320 is
$320 - 230 = 90$

∴   James needs a score of 90 on the next test.

Method 3: Layperson's approach
Step 1:

1st **test**	2nd **est**	3rd **test**	4th **test**
85	**70**	**75**	**?**
Here, there are an extra 5 points to spare	Here, we need 10 points	Here, we need 5 points	Here, we need 80 points

Step 2: If we add the 5 extra points from the first test to the 3rd test, we would have

1st **test**	2nd **test**	3rd **test**	4th **test**
80	**70**	**80**	**?**
This is O.K	Here, we need 10 more points	This is O.K	Here, we need 80 points

Since there is only one more test to be taken, James must make up the 10 points needed for the 2nd test, and obtain 80 points needed for the 4th test , all on the last test.
Hence, James needs a score of $10 + 80 = 90$ on the last test.

Lesson 19 Exercises

A. A chemist has 120 liters of a 40% solution of alcohol. How much pure alcohol must be added in order to obtain a 75% solution of alcohol?
(Assume the solution is composed of only pure alcohol and water)

Answer: 168 liters

B. 1..An engineering company has 260 workers of whom 30% are women. How many new women must be hired such that 50% of the workers are women? Assume that no workers are fired.

2.. A school system has 400 teachers of whom 70% are males. Assuming that no teachers are fired, approximately, how many new female teachers must be hired such that 60% of the teachers are males?

3. A company has 240 workers of whom 100 are males. Assuming that no workers are fired, how many new male workers must be hired such that $\frac{3}{4}$ of the workers are males?

4. Betty deposited a total of $20,000 in two bank accounts. The interest rate for one of the accounts was 6.5%, and the rate for the other account was 4%. If the total interest Betty was paid, at the end of last year, was $1100, how much money was invested at 6.5% and at 4% ?

Ans: **1**. 104 women; **2**. 67 females; **3**. 320 males; **4**. $12,000 at 6.5% and $8,000 at 4%.

C. Betty has $850, $700, $750 in three different bank accounts. This morning, she has decided to open a fourth bank account, today. What is the total amount of money Betty needs to deposit so that at the end of the day, she would have exactly $800 in each of the four bank accounts?

Answer: $900

D. During the semester, Maria scored 92 and 84 on the first two math tests. If there is one more test to take, what is the score she needs on the third test so that her class average is 90? Ans: 94

Test # 7-Student's Self-Test (**Always, Test yourself before you are tested**)

Attempt all questions on clean sheets of paper, **Do not write in the book** Show all necessary work.

1. If three tines the difference of a number and two is increased by eight, the result is 17. Find the number.

2. If five more than twice a number is 48, find this number.

3. If four times a number is subtracted from 20, the result is 28. Find this number.

4 A number is 4 less than twice another number. If the sum of these numbers is 32, find them.

5. The length of a rectangle is 4 units more than the width. If the perimeter of this rectangle is 64, find the length and width of this rectangle.

6. Maria fenced a rectangular farm with 1200 ft of wire. If this farm is twice as long as it is wide, what are the dimensions of this farm?

7. The width of a rectangle is three units less than its length. If the perimeter of this rectangle is 30 units, find the dimensions of this rectangle.

8. During the semester, Maria scored 82, 78 and 84 on the first three math tests. If there is one more test to take, what is the score she needs on the fourth test so that her class average is 85?

9. Working alone, John can sort a quantity of mail in 2 hours, while Betty working alone can sort the same quantity of mail in 3 hours. How long will it take John and Betty to sort the same quantity of mail if they work together?

10. The sum of three consecutive integers is 99. Find these integers.

11. The sum of two consecutive integers is 49 less than five times the larger. Find these integers.

12. The sum of three consecutive even integers is 84 . Find these integers.

13. Find three consecutive even integers such that the sum of the first and third integers is 12 less than the second integer.

14. Find three consecutive odd integers whose sum is 135.

15. Find two consecutive integers such that the difference of three times the larger integer and the smaller integer is 39.

16. James is 4 years older than Mary. Three years ago, James was twice as old as Mary . Find their present ages.

Lesson 19: Mixture and Mixture-Like Problems

17. If 30% of a number is 45, what is the number?

18. 60 is 30% of what number?

19. 18% of what number is 126?

20. What rate percent of 16 is 12?

21 The monthly rent for Maria's apartment is $600. If Maria spends 30% of her monthly income on this rent, what is her monthly income?

22. What rate% of 80 is 240?

23. On the last CUNY test, there were 40 questions. James answered 4 questions incorrectly. What was his grade in percent?

24. Betty has $820, $780, $840 in three different bank accounts. This morning, she has decided to open a fourth bank account, today. What is the total amount of money Betty needs to deposit so that at the end of the day, she would have exactly $850 in each of the four bank accounts?

25. The difference between two positive numbers is 40. The larger of the two numbers is 6 times the smaller. Find these two numbers.

Bonus:
In a chemistry class, there are 35 students. The number of females is 5 less than three times the number of males. How many males and how many females are in this class?

Answers: **1**. The number is 5 ;   **2**. The number is $\frac{43}{2}$; **3**. The number is -2 ;
4. The numbers are 12 and 20 ; **5**. The width is 14 units and the length is 18 units;
6. The length is 400 ft and the width is 200 ft ;   **7**. The width is 6 units and the length is 9 units ;
8. Maria needs a score of 96;   **9**. 1$\frac{1}{5}$ hours; **10**. The numbers are 32, 33, and 34 ;
11. The numbers are 15 and 16;   **12**. The numbers are 26, 28 and 30;
13. The numbers are -14, -12 and -10 ;   **14**. The numbers are 43, 45 and 47 ;
15. The numbers are 18 and 19;   **16**. James is 11 years old and Mary is 7 years old;
17. The number is 150;   **18**; The number is 200;   **19**. The number is 700; **20**. 75% ;
21. Maria's monthly income is $2,000 ;   **22**. 300%;
23. James' grade was 90%;   **24**. Betty needs to deposit $960;
25. The smaller number is 8, and the larger number is 48;
  Bonus: There are 25 females and 10 males

CHAPTER 7
Ratio

Lesson 20: **Definition; Reduction of Ratios to Lowest Terms**
Lesson 21: **Using Ratios to Compare Quantities**
Lesson 22: **Using Ratios to Divide a Quantity into Parts**

Lesson 20

Definition and Reduction of Ratios to Lowest Terms

Whenever you go shopping to buy apples, you always want to know the selling price, and the first question you usually ask the seller is "How much are these apples?" A typical answer may be "4 apples for one dollar". As soon as you have agreed to this price, you and the seller have established a ratio between the number of apples the seller is willing to give you and the number of dollars you are willing to pay. Simply, the ratio you have agreed to is 4 apples to 1 dollar, and we can symbolize this as 4:1. The seller might also have replied "1 dollar for 4 apples", but in this case, the ratio of dollars to apples is 1 to 4 or 1:4. Note the order in which the terms 1 and 4 of the ratio follow "dollar and apples".

If your grandmother left $1000, in her will, to be divided between you and your brother so that you receive $600 and your brother receives $400, then the ratio of your share to your brother's share is $600 to $400 and in simplest terms, this ratio is 3 **to** 2, which means that whenever you receive $3, your brother receives $2.

If the simple interest rate for your savings account at a bank is 5% (5 percent or 5 per hundred), then after one year, the ratio of the interest to the deposit you had in your account is 5 to 100 (That is, your interest is $5 for every $100 in your account.)

We can observe from the above examples that, in everyday life, we are always dealing with ratios, consciously or unconsciously. Uses of ratios include comparison of quantities, division of a quantity into parts, and formation of proportions.

Definition
The **ratio** of two quantities a and b or of a to b (written $a:b$) is a comparison of the two quantities and written as a fraction is $\frac{a}{b}$ (i.e., is the first quantity mentioned divided by the second quantity mentioned), where a and b may or may not be of the same units. The quantities a and b are called the terms of the ratio. However, the ratio of b to a is $\frac{b}{a}$.

Examples

1. The ratio, 3:4 is $\frac{3}{4}$ but the ratio 4:3 is $\frac{4}{3}$. **Note** that order is important.

2. The ratio, 4 inches to 12 inches is the fraction $\frac{4 \text{ inches}}{12 \text{ inches}} = \frac{1}{3}$
   (Terms have the same units and cancel out)

3. The ratio, 4 inches to 1 foot is the fraction $\frac{4 \text{ inches}}{1 \text{ foot}} = \frac{4 \text{ inches}}{12 \text{ inches}} = \frac{1}{3}$ (1 ft = 12 in.)
   (One unit can be converted to the other)

4. The ratio, 4 apples to 1 dollar is the fraction $\frac{4 \text{ apples}}{1 \text{ dollar}}$
   (Terms do not have the same units and one unit cannot be converted to the other)

5. The ratio, 50 miles to 2 hours is the fraction $\frac{50 \text{ miles}}{2 \text{ hours}} = \frac{25 \text{ miles}}{1 \text{ hour}}$ (25 mph) <-different units

6. The ratio of the number of boys to the number of girls in a class is the fraction $\dfrac{\text{number of boys}}{\text{number of girls}}$.

Note that since ratios can be written as fractions, ratios can be reduced to lowest terms in the same way that we reduce fractions to lowest terms.

Since ratios are used for comparing quantities, a ratio may have three or more terms.

Example The number of pencils, the number of pens, and the number of notebooks are in the ratio of 2:3:4. (i.e., 2 pencils to 3 pens to 4 notebooks)

Rate

If a and b are **not** of the same units **and** one unit **cannot** be converted to the other unit, some authors call the ratio a **rate.**

1. The rate, 50 miles to 1 hour is the fraction $\dfrac{50 \text{ miles}}{1 \text{ hour}}$ (50 mph)

Even though **rate** is a common terminology for quantities with different units, in the following example, note that it is actually a ratio, since the terms have the same units or no units.

2. An interest rate of 5% for a savings account (at a bank in the US.) means that the depositor receives \$5 for every \$100 in the account. Note also that 5% is also the fraction $\dfrac{5}{100}$ or

the ratio, 5:100, which is a ratio. Perhaps we should call this interest rate interest ratio.

Example: Reduce the ratio 12:8 to lowest terms

Solution: $12:8 = \dfrac{12}{8} = \dfrac{3}{2}$

Answer: In lowest terms 12:8 = 3:2.

Lesson 20 Exercises

Reducing Ratios to Lowest Terms

Reduce to lowest terms (Two terms)
(a). The ratio 2:4
(b). The ratio 6:3
(c). The ratio 8:12
(d) The ratio 12:8
(e). The ratio 5 inches to 10 inches
(f) The ratio 6 apples to 3 dollars

Answers: (a) 1:2; (b) 2:1; (c) 2:3; (d) 3:2; (e) 1:2; (f) 2 apples:1 dollar;

Reduce to lowest terms (Three Terms)
(a) .The ratio 2:4:6   (b) The ratio 6:3:18
(c) The ratio 8:12:6
(d) The ratio 12:8:6
(e) . The ratio 5 inches to 10 inches to 30 inches
(f) The ratio 6 apples to 10 apples to 24 apples
Answers: (a) 1:2:3; (b) 2:1:6; (c) 4:6:3; (d) 6:4:3; (e) 1:2:6; (f) 3:5:12;

Lesson 21

Using Ratios to Compare Quantities

Example 4

At a certain college, there are 2,000 male students and 6,000 female students.

(a) What is the ratio of the number of male students to the number of female students at this college?

(b) What is the ratio of number of female students to the number of male students at this college?

(c) What is the ratio of the number of female students to the total number of students at this college?

Solution

(a) The ratio of males to females is $2000{:}6000 = \frac{2000}{6000} = \frac{1}{3}$ or $1:3$

(b) The ratio of females to males is $6000{:}2000 = \frac{6000}{2000} = \frac{3}{1}$ or $3{:}1$

(c) Total number of students $= 2000 + 6000 = 8000$
The ratio of the number of females to the total number of students is $6000{:}8000$
$= \frac{6000}{8000} = \frac{3}{4}$ or $3:4$

Example 5

Now let us verify the results of the above problem by doing it backwards.

There are altogether 8,0000 students at a certain college. The ratio of the number of males to the number of females at this college is 1:3.

(a) Find the number of males.   **(b)** Find the number of females at this college.

Solution

(a) Fraction of students who are males $\frac{1}{1+3} = \frac{1}{4}$ (Fraction corresponds to first term, 1, of ratio)

The number of males $= \frac{1}{4} \times \frac{8000}{1} = 2{,}000$

(b) Fraction of students who are females $\frac{3}{1+3} = \frac{3}{4}$ (Fraction corresponds to second term, 3)

The number of females $= \frac{3}{4} \times \frac{8000}{1} = 6{,}000$

(c) Let us recalculate the number of females given that the ratio of number of females to the total number of students is $3{:}4$.

Solution: Fraction of students who are females $= \frac{3}{4}$.

The number of females $= \frac{3}{4} \times \frac{8000}{1} = 6{,}000$ <-----This is the same solution as is (b)

Note the difference between how the denominator, 4, of the fraction in (c) was obtained by using only the second term of the ratio, and how the denominator of the fraction in (b) was obtained by adding 1 and 3, the terms of ratio. Note also that in (b) and (a), we added the terms of the ratio to obtain the denominators of the fractions, since the terms, 1 and 3 correspond to the individual genders involved in the problem; while in (c), the second term, 4, of the ratio corresponds to the total number of students. Note particularly that we did **not** add the 3 and the 4 to obtain the denominator of the fraction in (c), but we rather used only the 4 as the denominator. See also p.112. The problem in (c) could have been posed as " 3 out of 4 students at a certain college are females. If there are 8,000 students at this college, how many of the students are females. The solution method is the same as in (b). We could also algebraically, use proportion to obtain the same solution: $\frac{3}{4} = \frac{x}{8000}$, where x is the number

of females. $4x = 3 \times 8000$ and $x = \frac{3}{4} \times \frac{8000}{1} = 6{,}000$.

As we shall soon observe, Example 5 actually belongs to the next lesson.

Lesson 21 Exercises

Comparing two quantities

1. At a certain college, there are 18,000 male students and 32,000 female students.

(a) What is the ratio of the number of male students to the number of female students at this college?

(b) What is the ratio of number of female students to the number of male students at this college?

c) What is the ratio of the number of female students to the total number of students at this college?   **Answer** (a) 9:16; (b) 16:9 ; (c) 16:25

2. In a math class, class, there are 15 females and 8 males.

(a) What is the ratio of the number of males to the number of females in this class?

(b) What is the ratio of the number of females to the number of males?

(c) What is the ratio of the number of males to the total number of students?

Answer: (a) 8:15; (b) 15:8; (c) 8:23

3. Rameltha and Albert are to divide $45,000

If Rameltha receives $18,000, what is the ratio of Albert's share to Rameltha's share?

Answer: 3:2

4. A piece of black tape is 4 inches long, and a similar piece of red tape is 2 feet long..

What is the ratio of the length of the black tape to the length of the red tape?

Answer: 1:6

5. Out of 20 games that a team played, the team won 12 games and lost the rest.

(a) What is the ratio of the number of games won to the number of games played?

(b) What is the ratio of the number of games won to the number of games lost?

Answer: (a) 3:5; (b) 3:2

Comparing three quantities

6. In a bag, there are 16 red marbles, 20 black marbles and 12 green marbles.

(a) What is the ratio of red marbles to black marbles to green marbles?

(b) What is the ratio of black marbles to red marbles to green marbles?

(c) What is the ratio of black marbles to the sum of the red and green marbles?

(d) What is the ratio of green marbles to the total number of marbles in the bag?

Answer: (a) red:black:green = 4:5:3 ; (b) black:red:,green = 5:4:3; (c) 5:7; (d) 1:4

7. Rameltha, Albert and James are to divide $45,000. If Rameltha receives $18,000, and Albert receives $17,000, what is the ratio of Rameltha's share to James' share to Albert's share?   Answer: 18:10:17

8. A piece of black tape is 4 inches long, and similar pieces of red and green tapes are 2 feet , and 3 feet long respectively.

(a) What is the ratio of the length of the black tape to green tape to red tape?

(b) What is the ratio of the length of green tape to red tape to black tape?

Answer: (a) 1:9:6   (b) 9:6:1

9. Out of 20 games that a team played, the team won 12 games, lost 4 and drew the rest

((a) What is the ratio of games won to games lost to games drawn?

(b) What is the ratio of games lost to games won to games played?

(c) What is the ratio of games drawn to games won?

Answer: (a) 3:1:1 ; (b) 1:3:5 ; (c) 1:3

Lesson 22

Using Ratios to Divide a Quantity into Parts

Example 1 Divide 28 in the ratio 3: 4

Method 1:   Using Arithmetic

Step 1:   Fraction of the number corresponding to the first term: $\frac{3}{3+4} = \frac{3}{7}$

Fraction of the number corresponding to the second term: $\frac{4}{3+4} = \frac{4}{7}$

Step 2: The number corresponding to the first term (of the ratio) $= \frac{3}{7} \times \frac{28}{1} = 12$

The number corresponding to the second term (of the ratio) $= \frac{4}{7} \times \frac{28}{1} = 16$

The numbers are 12 and 16.

Method 2: Using algebra

Let the number corresponding to the first term (of the ratio) $= 3x$
Let the number corresponding to the second term (of the ratio) $= 4x$
"The sum of the two numbers is 28" translates to

$3x + 4x = 28$
$7x = 28$
$x = 4$

One of the numbers $= 3x = 3(4) = 12$, and
the other number $= 4x = 4(4) = 16$

The numbers are 12 and 16.

Method 3: Using Algebra

See Method 3 of Example 2, below.

Example 2 A bag contains red and white marbles in the ratio 3:4. There are altogether 28 marbles in this bag. How many of these marbles are red?

Solution Method 1:   Using arithmetic

The number of red marbles $= \frac{3}{7} \times \frac{28}{1} = 12$

Method 2: Using algebra

Let the number of red marbles $= 3x$
Let the number of white marbles $= 4x$

$3x + 4x = 28$
$7x = 28$ The number of red marbles $= 3x = 3(4) = 12$
$x = 4$

Method 3: Using Algebra

Let the number of red marbles $= x$
Then the number of white marbles $= (28 - x)$
The proportion "3:4 as $x : (28 - x)$ translates to: (ratios of red marbles to white marbles)

$\frac{3}{4} = \frac{x}{(28 - x)}$

Solve for x:

$3(28 - x) = 4x$
$84 - 3x = 4x$
$84 = 7x$
$12 = x$

The number of red marbles $= 12$

Lesson 22 Exercises

Dividing a quantity into two parts

1. The ratio of male students to female students at a certain college is 2:5. if there are 2,800 students at this college, how many of the students are females? Answers: **1.** 2,000;

2. If the ratio (by weight) of hydrogen to oxygen in water is 1:8, find the weight of hydrogen in 27 grams of water. Answer: 3 grams

3. The sum of two numbers is 40. The numbers are in the ratio 2:3. Find these numbers. Answers: **1.** 16 and 24

4. James and Mary are to divide $72 in the ratio 5:7. How much does each receive? Answer: James: $30, Mary: $42

5. A grandmother left $45,000 in her will to be divided between two granddaughters, Maria and Diana in the ratio 2 to 3. How much does each receive? **Answer** Maria: $18,000; Diana: $27,000

6. John invested $2,000 in a company while James invested $3,000 in the same company. If after 3 years, a profit of $2,000 is to be shared between them in the ratio of their investments. How much does each receive? **Answer** John: $800; James: $1200

7. Mary wants to prepare a solution consisting of alcohol and water in the ratio 1:2 by volume. If the total volume of the solution is 12 liters, find the volumes of alcohol and water to be used **Answer** alcohol: 4 liters; water: 8 liters

9. A farmer wants to prepare 24 liters of gasohol which consists of alcohol and gasoline in the ratio 1:9. How many liters of alcohol and gasoline are to be mixed? **Answer** alcohol: 2.4 liters; gasoline: 21.6 liters

9. A bag contains p fruits, made up of oranges and apples only in the ratio q:r. Obtain an expression for the number of oranges in terms of p, q, and r. Answer: $\frac{pq}{q+r}$

10. In a math class, the ratio of males to females is 1:3.: If there are 28 students in this class, how many males are in this class? Answer: 7 males

Dividing a quantity into three parts

11. A grandmother left $45,000 in her will to be divided between three granddaughters, Ana, Elizabeth, and Diana in the ratio 1 to 2 to 3. How much does each receive? **Answer:** Ana: $7,500, Elizabeth: $15,000, Diana: $22,500

12. John invested $3,000 in a company while James and Alexander invested $4,000 and $5,000 respectively in the same company. If after 4 years, a profit of $4,800 is to be shared between them in the ratio of their investments. How much does each receive? **Answer** : John: $1,200, James: 1,600, Alexander: $2,000

13. Mary wants to prepare a solution consisting of alcohol, water and orange juice in the ratio 1:2:3 by volume. If the total volume of the solution is 36 liters, find the volumes of alcohol, water and juice to be used. **Answer** Alcohol: 6 liters, water: 12 liters, orange juice: 18 liters

14. A farmer wants to prepare 44 liters of fuel blend which consists of fuel A, fuel B and fuel C in the ratio 1:1:9. How many liters of each type of fuel are to be mixed? **Answer** : fuel A: 4 liters, : fuel B: 4 liters, fuel C: 36 liters

15. A bag contains p fruits, made up of oranges, grapefruits, and apples in the ratio q:r:s. Obtain an expression for the number of oranges in terms of p, q, r, and s. Answer: $\frac{pq}{q+r+s}$

16. On a math test, the number of A's, B's, and C's are in the ratio of 1:4:2 If there are 28 students in this class, determine the number of (a) A's, (b) B's, and (c) C's. ? **Answer**: A's: 4, B's: 16, C''s: 8:

CHAPTER 8
PROPORTION (VARIATION)

Lesson 23: Introduction to Direct and Indirect Proportion
Lesson 24: Methods for Solving Direct Proportion Problems
Lesson 25: Methods for Solving Inverse Proportion Problems
Lesson 26: Compound Proportion (Compound Variation) Problems

Lesson 23
Introduction to Direct and Inverse Proportion
(Definitions, Terminology, Types of Proportion and Principles)

Basically, there are two main **types** of **proportion** (or **variation**):
1.   **Direct proportion** or simply, **proportion** (or direct variation).
2.   **Inverse proportion** (or indirect variation).
   There are also problems which involve both direct and inverse proportion simultaneously.

Direct proportion
Discussion: In the introduction of the lesson on ratios, we discussed an example involving buying apples. There, the seller quoted the price of apples as "4 apples for 1 dollar", and we established the ratio of the number of apples to the number of dollars as 4 to 1. Now, if the buyer wants to buy exactly 4 apples, then the buyer pays exactly 1 dollar. However, if the number of apples is different from 4, then the number of dollars to pay would be different from 1, and the seller has to determine how much the buyer pays for the new number of apples. For example, if the buyer wants 8 apples (instead of 4 apples), then the seller will determine how many dollars the buyer pays (in this case, $2), or if the buyer wants 2 dollars worth of apples, the seller has to determine how many apples to give the buyer for $2 (in this case, 8 apples). We note above that if the buyer wants to buy exactly 4 apples, then only two quantities are involved in the problem, namely 4 apples and 1 dollar.

However, if the buyer wants 8 apples, then four quantities would be involved in the problem, namely 4 apples, 1 dollar, 8 apples, and the price for 8 apples. We will determine the fourth quantity, namely, the price for 8 apples using these three known quantities and a relationship called a **direct proportion**. In this relationship, as the number of apples increases, the number of dollars also increases; or as the number of apples decreases, the number of dollars also decreases. When the number of apples is zero, the number of dollars is also zero. Uniform or constant changes in the number of apples result in uniform changes in the number of dollars. For example, as the number of apples is doubled, the number of dollars is also doubled; but if he number of apples is halved, the number of dollars is also halved.

   Another application of proportion is as follows: Suppose the roof of your house has a triangular shape, and you want to build a much bigger or smaller house such that the corresponding roof has the same shape as your present house, then **direct proportion** can be used to determine the lengths of the sides of the new roof.

Direct proportion occurs very often in the arithmetic of everyday life. It may not be the number of apples and number of dollars, but it may be the number of miles traveled and number of gallons of gasoline used, or the number of miles traveled and number of hours, or the number of grams of hydrogen and the number of grams of oxygen. Direct proportion is usually termed **proportion** in most textbooks, and it is this type that you meet very often.

We should note above that problems stated in terms of proportion can also be stated in terms of variation and conversely. In the examples covered in this chapter, we cover both cases.

Simple Direct Proportion (Direct Variation)

Definition: A **proportion** is a statement that two ratios are equal.
The direct proportion "a is to b as c is to d" (or $a:b::c:d$) is the equality

$$\frac{a}{b} = \frac{c}{d} \quad \textbf{(quotient = quotient)}$$

In the above proportion, a and d are called the extremes and b and c are called the means of the proportion. We may also define a proportion as a mathematical statement indicating the equality between two **equivalent fractions.** Example: $\frac{2}{3} = \frac{4}{6}$ is a proportion

Proportion Principles

1. We can invert both ratios of a proportion: From (1) above, we obtain after the inversion,

$$\frac{b}{a} = \frac{d}{c}$$

2. We can interchange the extremes of a proportion: From (1) above we would obtain:

$$\frac{d}{b} = \frac{c}{a}$$

3. We can also interchange the means to obtain:

$$\frac{a}{c} = \frac{b}{d}$$

Also $ad = bc$ (That is the product of the extremes = the product of the means) <-- also called cross-multiplication

The proportion principles can be used to obtain desired ratios. For example, if we have $\frac{3}{2}$ as the ratio of one side of a proportion, and we desire the ratio $\frac{2}{3}$ then proportion principle #1 above (inversion of both sides of a proportion) can be used to obtain this ratio, $\frac{2}{3}$, from $\frac{3}{2}$.

More Principles (From $\frac{a}{b} = \frac{c}{d}$, we obtain the following:

4. $\frac{a+b}{b} = \frac{c+d}{d}$ (by addition). Note: $\frac{a+b}{b} = \frac{a}{b} + \frac{b}{b} = \frac{a}{b} + 1$ **and** $\frac{c+d}{d} = \frac{c}{d} + \frac{d}{d} = \frac{c}{d} + 1$

5. $\frac{a-b}{b} = \frac{c-d}{d}$ (by subtraction) | **8.** If $\frac{a}{b} = \frac{c}{d} = \frac{e}{f}$, then $\frac{a+c+e}{b+d+f} = \frac{a}{b} = \frac{c}{d} = \frac{e}{f}$

6. $\frac{a-b}{a+b} = \frac{c-d}{c+d}$ (by subtraction and addition) | e.g.,$(\frac{2}{10} = \frac{3}{15} = \frac{4}{20} = \frac{2+3+4}{10+15+20} = \frac{9}{45} = \frac{1}{5})$

7. $\frac{a+b}{a-b} = \frac{c+d}{c-d}$ (by addition and subtraction) | To show that **8** is true, see Appendix, p.306

Problems Involving Proportion

Example 1 Find x from the proportion x is to 4 as 3 is to 2.

Solution
$$\frac{x}{4} = \frac{3}{2}$$
$$2x = 12 \quad \text{(cross multiplication)}$$
$$x = 6$$

(You could also solve the above equation by multiplying both sides of the equation by 4)

Example 2 Find x from the proportion: 6 is to x as 4 is to 12.

Solution
$$\frac{6}{x} = \frac{4}{12}$$
$$4x = 72; \text{ and } x = 18$$

Simple Inverse Proportion (Inverse Variation)

Discussion: Another type of a relationship called **inverse proportion** (or indirect proportion) also occurs in everyday life. For example, suppose that when you drive your car from City A to City B at a speed of 25 mph, it takes you 4 hours. Now, if you always drive at a speed of 25 mph from City A to City B, and it takes you 4 hours, then only two quantities are involved, namely 25 mph and 4 hours. However, if you want to travel the same distance at a speed of say, 50 mph, then you would like to know the time needed to drive from City A to City B at 50 mph, and in this case, four quantities would be involved in the problem, namely, 25 mph, 4 hours, 50 mph, and the time, t, for 50 mph. Since for the same distance, as the speed is increased, the time taken decreases, or if the speed is decreased, the time taken increases, speed is said to be inversely proportional to the time.

Recall the main difference between direct proportion and inverse proportion: for direct proportion, both quantities increase or decrease at the same time; whereas for inverse proportion, as one quantity increases, the other quantity decreases and vice versa.
For example, as one quantity is doubled, the other quantity is halved; but if one quantity is halved, the other quantity is doubled.

Example 1 The time taken by a number of people to do a piece of work is **inversely proportional** to the number of people (assuming that each person works at the same rate as everyone else). Thus, as the number of people increases, the time taken to do the work decreases but as the number of people decreases, the time taken increases (i.e., more people, less time; and less people, more time)

Example 2 At constant temperature, the volume of a given mass of a gas is inversely proportional to the pressure on the gas. (This relationship is known as Boyle's law.)

Definition

If a is inversely proportional to b. (same as a is proportional to the **reciprocal** of b)

Then a_1 corresponds to b_1, and a_2 corresponds to b_2. and

$$a_1 : \frac{1}{b_1} = a_2 : \frac{1}{b_2}$$
$$a_1 \div \frac{1}{b_1} = a_2 \div \frac{1}{b_2}$$
$$a_1 b_1 = a_2 b_2 \quad (\textbf{product = product})$$

Or if we use the familiar terms a, b, c, d as the terms of the inverse proportion

Then $a : \frac{1}{b} = c : \frac{1}{d}$
$$a \div \frac{1}{b} = c \div \frac{1}{d}$$
$$ab = cd \quad (\textbf{product = product}) \; \texttt{<------} \; \text{inverse proportion}$$

We will use the product = product form since this form is easy to remember

Compare:

Direct proportion $\frac{a}{b} = \frac{c}{d}$ <-------- (**quotient = quotient**)

Indirect proportion $ab = cd$ <------(**product = product**)

Determining the type of Proportion (Variation) from the Wording

Explicit Specification of direct proportion (variation)

There are a number of ways of **explicitly** specifying direct proportion (variation)
If y is a function of x, then the following are equivalent to one another:

(a) y is directly proportional to x, or simply
(b) y is proportional to x. (The word "directly" is omitted.)
(c) y varies directly as x, or simply,
(d) y varies as x, (The word "directly" is omitted.)

The methods we will cover for solving direct proportion problems can be used to solve any of the problems worded as in (a), (b), (c) and (d) above.

Explicit Specification of inverse proportion (variation)

If y is a function of x, then the following are equivalent to one another:

(a) y is inversely or indirectly proportional to x.
(b) y varies inversely as x,

Any of the methods (we will cover) for solving indirect proportion problems can be used to solve any of the problems stated as in (a) and (b) above.

Implicit Specification of proportion (variation)

In most problems, the terms " directly proportional to" , "inversely proportional to", " varies directly as" or "varies inversely as" and similar terms are not used in wording the problem; but rather, we have to deduce from the problem whether it is a direct or an indirect proportion from experience. For example, if 15 oranges cost \$3, what is the cost of 10 oranges?. From life experience, we know that at the same price per an orange, as the **number** of oranges **increases**, the **cost** of oranges **increases** accordingly. Therefore, the relationship in this problem is that of direct **proportion** (variation).

Lesson 23 Exercises A

A. Solve for x: 1. $\frac{x}{8} = \frac{3}{2}$; 2. $\frac{5}{8} = \frac{20}{x}$; 3. $\frac{9}{x} = \frac{3}{5}$; 4. $\frac{6}{8} = \frac{x}{2}$; 5. $\frac{x}{9} = \frac{3}{2}$;

6. $\frac{10}{x} = \frac{3}{5}$; 7. $\frac{x}{8} = \frac{4\frac{1}{2}}{2}$; 8. $\frac{x}{8\frac{2}{3}} = \frac{3\frac{2}{3}}{2\frac{3}{4}}$; 9. $\frac{x}{1.8} = \frac{3}{2.1}$; 10. $\frac{x}{8} = \frac{12\%}{96}$

Answers: 1. $x = 12$; 2. $x = 32$; 3. $x = 15$; 4. $x = \frac{3}{2}$; 5. $13\frac{1}{2}$; 6. $16\frac{2}{3}$; 7. $x = 18$; 8. $11\frac{5}{9}$;

9. $2\frac{4}{7}$; 10. 1% or $\frac{1}{100}$

Lesson 23 Exercises B

Determine which of the following are direct proportion or inverse proportion:

1. If 10 gallons of gasoline cost 15 dollars, find the cost of 24 gallons of gasoline.

2. If on 2 gallons of gasoline, a car can travel 24 miles, how many miles can a car travel on 20 gallons of gasoline?

3. If 8 oranges sell for 2 dollars, find the selling price for 24 oranges.

4. If 2 people can paint a house in 6 days, how many people would paint the house in 3 days?

5. At a speed of 60 miles per hour, a car takes 2 hours to travel from City A to City B. At a speed of 40 miles per hour, how long will it take the car to ravel from City A to City B?

6. If 6 mail men can sort a given quantity of mail in 4 hours, how many mail men would be needed to sort the same quantity of mail in 2 hours?

Answers , 1, 2, 3 are direct proportion; 4, 5 and 6 are inverse proportion

Lesson 24
Methods for Solving Direct Proportion Problems
(Direct Variation Problems)

There are a number of methods or approaches for solving **direct proportion** problems. In simple proportion problems, four quantities are involved, and we are (usually) given three quantities and we are required to find the fourth quantity.

We cover **five** methods with examples, and then discuss the relative merits of the methods.

Method 1: Quotient = Quotient or Ratio = Ratio method (Algebraic)

In this method, we word or reword the problem as discussed below.

Consider a direct proportion problem involving four quantities such that a_1 corresponds to b_1, and a_2 corresponds to b_2. We word the proportion as follows:

The proportion a_1 is to b_1 as a_2 is to b_2 translates into the equation

$$\frac{a_1}{b_1} = \frac{a_2}{b_2} \quad \textbf{(quotient = quotient or } \text{ratio = ratio)} \textbf{ or}$$

Using the familiar notation "a is to b as c is to d", we obtain

$$\frac{a}{b} = \frac{c}{d} \quad \textbf{(quotient = quotient)}$$

Example 1

If 2 dollars can buy 8 apples, how many dollars are needed to buy 24 apples?

Method 1: Quotient = Quotient Method (Algebraic)

We are required to find the number of dollars needed to buy 24 apples.

Step 1: Let the number of dollars needed to buy 24 apples $= x$.

(we represent the unknown by a variable)
2 dollars corresponds to 8 apples
x dollars corresponds to 24 apples.

(Note the order in which the units in the second correspondence is mentioned.
That is, dollars before apples, and not apples before dollars)

Step 2: 2 dollars are to 8 apples as x dollars are to 24 apples

Step 3: Translating into an equation, $\dfrac{2 \text{ dollars}}{8 \text{ apples}} = \dfrac{x \text{ dollars}}{24 \text{ apples}}$

Step 4: Cross-multiplying, $8x = 2(24)$ ($8x(\text{dollars})(\text{appless}) = 2(24)(\text{dollars})(\text{apples})$)

Step 5 Solve for x: $\qquad x = \dfrac{2(24)}{8}$ (Note above that the units cancel out)

$\qquad\qquad\qquad = 6$

Conclusion: 6 dollars are needed to buy 24 apples.

Note in the above example that the order in which we mention the units, dollars and apples is important.
Note also, however, that in the above example, we could word the proportion as follows:

The proportion, 8 apples are to 2 dollars as 24 apples are to x dollars" translates to $\dfrac{8}{2} = \dfrac{24}{x}$

$$8x = 2(24)$$
$$x = \frac{2(24)}{8}$$
$$= 6 \qquad \text{(Again, we obtain the same solution as before.)}$$

Method 2   Units Label Method or Dimensions Method   (Arithmetic method)   104
 This method is highly recommended in applications such as converting from one unit to
another unit in measurements as well as calculations in chemistry. In Chemistry, some
authors call this method " The factor label method". In this method, we use the units of
the given quantities to guide us in obtaining an expression which on simplification yields
 the desired result.

We repeat the question: If 2 dollars can buy 8 apples, how many dollars are needed to buy
                    24 apples?

Solution: We can consider this problem as converting 24 apples to dollars. The justification
           of this method is in parentheses.

Step 1: $\dfrac{24 \text{ apples}}{1} \times \dfrac{?}{?}$

Step 2: Multiply $\dfrac{24 \text{ apples}}{1} \times \dfrac{?}{?}$ by a fraction formed by using the quantities 2 dollars and 8

 apples as the terms of the fraction such that the denominator has the same units as the 24 apples
(the quantity in the numerator), and under such conditions, we can divide out (cancel) the common
units in the numerator and the denominator, leaving us the units, dollars.

        Then, $\dfrac{24 \text{ apples}}{1} \times \dfrac{2 \text{ dollars}}{8 \text{ apples}}$

   Step 3: $\dfrac{24 \text{ } \cancel{\text{apples}}}{1} \times \dfrac{2 \text{ dollars}}{8 \text{ } \cancel{\text{apples}}}$ (= Number of apples $\times$ cost per each apple)

        = 6 dollars

Conclusion: 6 dollars are needed to buy 24 apples.

Let us do the above problem backwards, with the justification of this method in parentheses
If 2 dollars can buy 8 apples, how many apples can one buy for 6 dollars?
Solution   Here, we are converting 6 dollars to apples.

   $\dfrac{6 \text{ dollars}}{1} \times \dfrac{?}{?}$

= $\dfrac{6 \text{ dollars}}{1} \times \dfrac{8 \text{ apples}}{2 \text{ dollars}}$ $\left(= \text{total cost} \div \text{cost per 1 apple} = \text{total cost} \times \dfrac{1}{\text{cost per 1 apple}}\right)$

= 24 apples, which is what was given in the original problem.

Method 3 Unitary Method (Arithmetic)
We repeat the question: If 2 dollars can buy 8 apples, how many dollars are needed to buy
24 apples?

Solution: Cost of 8 apples = 2 dollars

     Cost of 1 apple = $\frac{2 \text{ dollars}}{8}$ (Divide to obtain the cost for 1 apple))

     Cost of 24 apples = $\frac{2}{8} \times \frac{24}{1}$ dollars (Multiply to obtain the cost for other than 1)
               = 6 dollars

Conclusion: 6 dollars are needed to buy 24 apples.

Method 4 "More or Less Method" (Arithmetic method. This method the shorter form of Method 3.
We repeat the question: If 2 dollars can buy 8 apples, how many dollars are needed to buy 24 apple?

If cost of 8 apples = 2 dollars

Then cost of 24 apples = $\dfrac{2 \text{ dollars}}{1} \times \dfrac{24 \cancel{\text{ apples}}}{8 \cancel{\text{ apples}}}$

= 6 dollars

Conclusion: 6 dollars are needed to buy 24 apples.

In method 4, we used a very useful principle which states that " If more, the smaller divides, and if less, the larger divides". That is, since we expect 24 apples to cost more than 2 dollars, in forming the fraction involving 8 apples and 24 apples, the smaller of 8 apples and 24 apples is the divisor (smaller divides), and hence we used the fraction,

$\dfrac{24 \text{ apples}}{8 \text{ apples}}$ <----- " smaller divides" (We multiplied by an **improper fraction**)

However, if we had expected the number (answer) to be less than 2 dollars, we would have used the fraction

$\dfrac{8 \text{ apples}}{24 \text{ apples}}$ <----- "larger divides" (i.e., we would have multiplied by a **proper fraction**)

The More or Less Method above is very powerful once you have mastered it, because it can be applied to inverse proportion problems as well as to problems involving both inverse and direct proportion and any number of quantities (variables), especially in science.

Method 5 Variation Method (Algebraic)
We reword the problem: The number of apples a person buys varies directly as the number of dollars paid. If when the number of apples is 8, the number of dollars paid is 2, find the number of dollars paid when the number of apples is 24.

Step 1: Let the number of apples = A
Let the number of dollars paid = D
Then $A = kD$, where k is called the proportionality constant.
Step 2: Substituting $A = 8, D = 2$ in $A = kD$,
$8 = 2k$
$4 = k$
Substituting $k = 4$ in $A = kD$, we obtain the formula,
$A = 4D$.
Step 3: When $A = 24$
$24 = 4D$ (substituting 24 for A in the formula, $A = 4D$)
$6 = D$ (solving for D)
$D = 6$
Therefore, 6 dollars would be paid for 24 apples.

Example 2 On 3 gallons of gasoline, one can drive 60 miles. How far can one drive on 8 gallons of gasoline?

Solution
Method 1 Quotient = Quotient Method (Algebraic)
Let the number of miles for 8 gallons of gasoline = x
Proportion statement: 3 gallons are to 60 miles as 8 gallons are to x miles.

Translating, $\dfrac{3 \text{ gallons}}{60 \text{ miles}} = \dfrac{8 \text{ gallons}}{x \text{ miles}}$

$\dfrac{3}{60} = \dfrac{8}{x}$

$3x = 60(8)$

$$x = \frac{60(8)}{3}$$

$$x = 160$$

Therefore, on 8 gallons of gasoline, one can drive 160 miles.

Note above that the ratio, gallons to miles = the ratio, gallons to miles; or
the ratio miles to gallons = the ratio miles to gallons
Incorrect: the ratio miles to gallons = the ratio gallons to miles. (violation of the order of the units)

All other proportion problems involving four quantities can be reworded and translated into equations by imitating the examples covered.. Only, the units involved might be different. We may have apples and pounds as the units involved, or hours and dollars, or years and people.

Method 2 Units Label Method. **We repeat the problem:** On 3 gallons of gasoline, one can drive 60 miles. How far can one drive on 8 gallons of gasoline?
Solution: We can consider this problem as converting 8 gallons to dollars.

Step 1: $\frac{8 \text{ gallons}}{1} \times \frac{?}{?}$

Step 2: Multiply $\frac{8 \text{ gallons}}{1} \times \frac{?}{?}$ by a fraction formed by using the quantities 3 gallons and 60 miles

$\frac{8 \text{ gallons}}{1} \times \frac{60 \text{ miles}}{3 \text{ gallons}}$

Step 3: $\frac{8 \text{ \cancel{gallons}}}{1} \times \frac{60 \text{ miles}}{3 \text{ \cancel{gallons}}}$

$= 160 \text{ miles}$

Therefore, on 8 gallons of gasoline, one can drive 160 miles.

Method 3 Unitary Method (Arithmetic)
We repeat the problem: On 3 gallons of gasoline, one can drive 60 miles. How far can one drive on 8 gallons of gasoline?
Solution
Number of miles driven on 3 gallons of gasoline = 60 miles

Number of miles driven on 1 gallon of gasoline = $\frac{60}{3}$ miles

(Divide to obtain the miles for 1 gallon)

Number of miles driven on 8 gallons of gasoline = $\frac{60}{3} \times 8$ miles (Multiply for other than 1 gallon)

$= 160 \text{ miles}$

Therefore, on 8 gallons of gasoline, one can drive 160 miles.

Method 4 More or Less Method **We repeat the problem:**
On 3 gallons of gasoline, one can drive 60 miles. How far can one drive on 8 gallons of gasoline?

Solution   Number of miles driven on 3 gallons of gasoline = 60 miles

Number of miles driven on 8 gallons of gasoline = $\frac{60}{1} \times \frac{8}{3}$ miles

$= 160 \text{ miles}$

Therefore, on 8 gallons of gasoline, one can drive 160 miles.

Method 5 Variation Method (Algebraic)   We reword the problem:
The number of miles one drives is directly proportional to the number
of gallons of gasoline used. If when the number of gallons of gasoline
is 3, the number of miles is 60, find the number of miles when the number of gallons
of gasoline is 8.

Solution
Step 1: Let the number of gallons of gasoline $= n$
Let the number of miles driven $= d$
Then since n is directly proportional to $d,$
$d = kn$, where k is the proportionality constant
Step 2: Substitute for $d = 60$, $n = 3$, in $d = kn$ and solve for k.
Then $60 = k(3)$ and from which $k = 20$
Step 3: Substitute for $k = 20$ in $d = kn$ to obtain a formula relating d and n.
Then $d =20n$
Step 4: Substitute for $n = 8$ in $d =20n$ and determine d
$d = (20)(8)$
$= 160$
Therefore, on 8 gallons of gasoline, one can drive 160 miles.

Example 3   The ratio of an astronaut's weight on the earth to weight on the moon is 6:1. If
an astronaut weighs 180 lb (on earth), what is the astronaut's weight on the
moon?

Solution
Method 1 Quotient = Quotient Method (Algebraic)
Let the weight on the moon be x lb, (i.e., x lb corresponds to 180 lb on the earth).

Step 1:   The ratio of the weight on earth to weight on the moon is 6:1 or $\frac{6}{1}$.
Step 2:   Also, the ratio of the weight on earth to the weight on the moon is 180 lb to x lb
or $\frac{180}{x}$.
Step 3 :   Equate the two ratios and solve for x.
$$\frac{6}{1} = \frac{180}{x}$$
$$6x = 180(1)$$
$$x = \frac{180(1)}{6}$$
$$x = 30$$
Therefore, the astronaut 's weight on the moon is 30 lb.

Method 2 Reword the original problem as a proportion:
6 is to 1 as 180 is to x.
Solution
Translating, $\frac{6}{1} = \frac{180}{x}$
Solving, $x = 30$, and we obtain the same solution as by Methods 1 and 2.

Method 3 Unitary Method (Arithmetic)

Whenever the astronaut weighs 6 lb on the earth, the astronaut weighs 1 lb on the moon.

Whenever the astronaut weighs 1 lb on the earth, the astronaut weighs $\frac{1}{6}$ lb on the moon; and

Whenever the astronaut weighs 180 lb on the earth, the astronaut weighs $\frac{1}{6}$ × 180 lb = 30 lb on the moon.

Method 4: More or Less Method We reword the problem
The weight of an astronaut on earth is directly proportional to the weight on the moon. When the weight on earth is 6 lb, the weight on the moon is 1 lb, find the weight on the moon when the weight on earth is 180 lb..

Solution Weight on the moon when the weight on earth is 6 lb = 1 lb

Weight on moon when weight on earth is 180 lb = 1 ℓb × $\frac{180\ \ell b}{60\ \ell b}$ = 30 lb

Therefore, the astronaut 's weight on the moon is 30 lb.

Method 5   Variation Method (Algebraic)
We reword the problem   The weight of an astronaut on earth is directly proportional to the weight on the moon. When the weight on earth is 6 lb, the weight on the moon is 1 lb, find the weight on the moon when the weight on earth is 180 lb..

Solution
Step 1: Let the weight on earth = h lb
Let the weight on moon = n lb
Then since h is directly proportional to n,
$h = kn$
Step 2: Determine the value of k by substituting $h = 6, n = 1$ in $h = kn$
Then $6 = k(1)$ and from which $k = 6$
Step 3: Substitute $k = 6$ in $h = kn$ to obtain a formula relating h and n.
Then we obtain $h = 6n$
Step 4: Substitute $h = 180$ in the formula, $h = 6n$ and solve for n.
Then $180 = 6n$
$30 = n$
Therefore, the astronaut 's weight on the moon is 30 lb.

More Examples on the Variation Method
(Note that these examples can also be solved using Methods 1, 2, 3 and 4)

Example A: y varies directly as x. When $y = 48, x = 3$. Find x when $y = 96$
Solution
Step 1: $y = kx$ (1)
Step 2: Substitute $y = 40, x = 2$ in equation (1) and solve for k.
$48 = k(3)$
$48 = 3k$
$16 = k$
Substitute $k = 16$ in equation (1) above to obtain
the formula $y = 16x$ (2)
Step 3: Now, to find x when $y = 96$, replace y by 96 in (2)
Then $96 = 16x$
$6 = x$
Therefore when $y = 96, x = 6$.

Example B The distance y an object falls from rest is directly proportional to the square of the time x. When $y = 128, x = 4$. (*a*) Find x when $y = 200'$ (*b*) Find y when $x = 6$.

Solution

Step 1: $y = kx^2$ ....(1)

Step 2: Substitute $y = 128, x = 4$ in equation (1) and solve for k.

   Then $128 = k(4)^2$

   $128 = 16k$

   $k = 8$

   Substitute $k = 8$ in equation (1) above to obtain

   the formula $y = 8x^2$ .... (2)

Step 3: (a) To find x when $y = 200$. Replace y by 200 in equation (2) and solve for x.

   $200 = 8x^2$

   $\dfrac{200}{8} = x^2$

   $25 = x^2$

   $5 = x$

   $x = 5$ (Since the time taken must be positive in this problem, we reject -5 as a solution)

   (b) To find y when $x = 6$, replace x by 6 in equation (2) and evaluate.

   $y = 8(6)^2$

   $y = 8(36)$

   $y = 288$.

Therefore when $x = 6, y = 288$.

EXTRA

Example 4 Assume volume V is directly proportion to temperature T.

   Then V_1 corresponds to T_1 as V_2 corresponds to T_2

   Since this is direct proportion, we apply **quotient = quotient**

   Then $\dfrac{V_1}{T_1} = \dfrac{V_2}{T_2}$ (quotient = quotient)

   The above equation is a familiar formula for gases in physics.

Relative Merits of the Methods for Solving Proportion Problems

Method 1: Quotient = quotient Method

The quotient = quotient method such as $\frac{a}{b} = \frac{c}{d}$ is very popular and is efficient for a one-step problem involving only four quantities, but is not efficient in a problem such as in Method 2 below, since the method has to be repeated in steps with the introduction of a new variable in each step.

Method 2: Units Label Method (Arithmetic)

The Units Label Method, an arithmetic method, can readily do what Method 1 can do, if the quantities involved have units, and is more efficient than Method 1 in doing repeated calculations. This method is particularly very efficient in calculations involving measurements and in chemistry as in the following example.

Example How many yards are there in 96 inches?

Solution We reword this problem as: Convert 96 in. to yd.

Plan: This plan is based on what is available in the conversion chart or table.
(Using a conversion chart is analogous to using a map to go from one location to another)
The sequence is inches--> feet---> yards.
We use these relationships: 12 in. = 1 ft; 3 ft = 1 yd (We could use 36 in = 1 yd)
  (Note: inch = in.)

Step 1: $\frac{96 \text{ in.}}{1} \times \frac{?}{?}$

Step 2: Multiply the $\frac{96 \text{ in.}}{1}$ by a fraction formed by using the quantities 12 in. and 1 ft as

the terms of the fraction such that the denominator has the same units as the 96 in.; followed by similar multiplication by a fraction using 3 ft and 1 yd as the terms of the fraction. Then, we obtain

$$\frac{96 \text{ in.}}{1} \times \frac{1 \text{ ft}}{12 \text{ in.}} \times \frac{1 \text{ yd}}{3 \text{ ft}} \qquad \left(\frac{96 \text{ in.}}{1} \times \frac{1 \text{ ft}}{12 \text{ in.}} = \frac{96 \text{ in.}}{1} \times \frac{1 \text{ ft}}{12 \text{ in.}} \times \frac{1 \text{ yd}}{3 \text{ ft}} = \frac{96 \times 1 \times 1 \text{ yd}}{1 \times 12 \times 3} \right)$$

$$= 2\tfrac{2}{3} \text{ yd}$$

$\therefore$ 96 in. = $2\tfrac{2}{3}$ yd

Justification of the Units Label Method in the above problem:

From 12 in. = 1 ft, If we divide both sides of this equality by 12 in., we obtain

$$\frac{12 \text{ in.}}{12 \text{ in.}} = \frac{1 \text{ ft}}{12 \text{ in.}}$$

$$1 = \tfrac{1 \text{ ft.}}{12 \text{ in.}}$$

Similarly, from 3 ft = 1 yd, If we divide both sides by 3 ft., we obtain

$$\frac{3 \text{ ft}}{3 \text{ ft}} = \frac{1 \text{ yd}}{3 \text{ ft}}$$

$$1 = \frac{1 \text{ yd}}{3 \text{ ft}}$$

In step 2, multiplying by $\frac{1 \text{ ft.}}{12 \text{ in.}}$ or by $\frac{1\ yd}{3\ ft}$ is equivalent to multiplying by 1, and therefore, the given value remains equivalent.

Note: In the above example, if we were to use the quotient = quotient method, we would have to apply this method twice, with the introduction of a variable in each step. Note above also that it is easy to generalize the justification of the "Units Label Method" when we have equality such as 12 in. = 1 ft, than when we have equivalence such as 8 apples are equivalent to 2 dollars.

Method 3: Unitary Method (Arithmetic)
This is an arithmetic method and does not require any knowledge of algebra and can be taught as the first arithmetic method for solving proportion problems.

Method 4: More or Less Method (Arithmetic)
This method is a shorter form of Method 3, as one step in Method 3 is skipped. When mastered well, the "More or Less Method" can be extended to solve inverse proportion as well as compound proportion problems, and very efficient in stoichiometric calculations in chemistry as well as conversions in measurements. **Master this method.**

Method 4: Variation Method (Algebraic)
This has applications at all levels of mathematics. Some textbooks ignore this method at the elementary level and cover it at the intermediate level.

EXTRA: Two Cases of the Statement of Ratio and the Corresponding Proportion Problems

In solving problems involving ratios and proportion, we must distinguish between the cases in which the terms of a given ratio refer to **disjoint sets** and the case in which the ratios refer to sets which are **not** disjoint.

Case 1

The terms of the ratios refer to disjoint sets (example: the set of females and the set of males in a school)
Example: The ratio of the number of females to the number of males is 3 to 2.
Proportion: The ratio 3 **females** to 2 **males** = the ratio 6 **females** to 4 **males**
or ratio 2 **males** to 3 females = ratio 4 **males** to 6 females

(Two sets are disjoint if the sets do not have any members common to both sets)
Problems such as " If 8 apples cost 2 dollars, what is the cost of 24 apples" are of case 1

Case 2

The terms do **not r**efer to disjoint sets (example: the set of females and the set of the total number of students at a school)

Example

The ratio of the number of **females** to the **total number** of students is 3 to 5.
Proportion: The ratio 3 **females** to 5 **students** = the ratio 6 **females** to 10 **students.**
Case 2 is sometimes worded as **a number** out of a **total number**
For example, 3 out of 5 students are females or out of every 5 students, 3 are females, which
is the same as $\frac{3}{5}$ of the students are females.

Problems such as " if 3 out of 5 students in a school are females, and there are 600 females in this school, how many students are in this school?" can also be set up using
$$\frac{\text{what number out of}}{\text{what number}} = \frac{\text{what number out of}}{\text{what number}}$$ as well as other methods.

Example for Case 2
Example 1   3 out of 5 students at a certain college are females.

     (a) If 600 students at this college are females, how many students are there at this college?

     (b) If one were to collect a sample of 200 students at this college, how many of these students would be females?

Solution (a):
Method 1 Using ratio = ratio method
     Let the number of students at this college = x
Then the proportion 3 females are to 5 students as 600 females are to x students
translates to: $\frac{3 \text{ females}}{5 \text{ students}} = \frac{600 \text{ females}}{x \text{ students}}$

$$\frac{3}{5} = \frac{600}{x}$$
$$3x = 5(600)$$
$$x = \frac{5(600)}{3} ; \ x = 1000$$

Therefore, there are 1000 students at this college.

Method 2: Using $\dfrac{\text{what number out of}}{\text{what number}} = \dfrac{\text{what number out of}}{\text{what number}}$

Let the number of students at this college $= x$

$$\dfrac{3 \text{ out of}}{5} = \dfrac{600 \text{ out of}}{x}$$

$$\dfrac{3}{5} = \dfrac{600}{x}$$

$$3x = 5(600)$$

$$x = \dfrac{5(600)}{3}$$

$$x = 1000$$

Therefore, there are 1000 students at this college

Method 3 Since 3 out of 5 means $\dfrac{3}{5}$, we can reword this part of the question as: if $\dfrac{3}{5}$ of a number is 600, what is the number?

Let the number be x

Then $\dfrac{3x}{5} = 600$

$$3x = 3000$$
$$x = 1000$$

There are 1000 students at this college.

Method 4 Divide 600 by $\dfrac{3}{5}$ (If $\dfrac{3}{5}$ of a number is 600, what is the number?)

$$600 \div \dfrac{3}{5} = \dfrac{600}{1} \times \dfrac{5}{3}$$

$$= 1000$$

There are 1000 students at this college.

Method 5: The Units Label Method

Note: 3 out 5 students are females means that for every 5 students, there are 3 females.

Step 1: $\dfrac{600 \text{ females}}{1} \times \dfrac{?}{?}$

Step 2: $= \dfrac{600 \text{ females}}{1} \times \dfrac{5 \text{ students}}{3 \text{ females}}$

$$= 1000 \text{ students}$$

Part (b)

Method 1: We can reword this part of the question as: Find $\dfrac{3}{5}$ of 200.

$$\dfrac{3}{5} \text{ of } 200 = \dfrac{3}{5} \times \dfrac{200}{1} = 120$$

That is, of a sample of 200 students, 120 students would be females.

Method 2:: $\dfrac{\text{what number out of}}{\text{what number}} = \dfrac{\text{what number out of}}{\text{what number}}$

Let the number of females $= x$

$$\frac{3 \text{ out of}}{5} = \frac{x \text{ out of}}{200}$$

$$\frac{3}{5} = \frac{x}{200}$$

$$5x = 3(200)$$

$$x = \frac{3(200)}{5}$$

$$x = 120$$

Therefore, there are 120 females.

Method 3: The Units Label Method

Step 1: $\dfrac{200 \text{ students}}{1} \times \dfrac{?}{?}$

Step 2: $\dfrac{200 \text{ students}}{1} \times \dfrac{3 \text{ females}}{5 \text{ students}}$

$= 120$ females

Therefore, there are 120 females.

Method 4

Let the number of females. $= x$

Then the proportion 3 females are to 5 students as x females are to 200 students translates to:

$$\frac{3 \text{ females}}{5 \text{ students}} = \frac{x \text{ females}}{200 \text{ students}}$$

$$\frac{3}{5} = \frac{x}{200}$$

$$5x = 3(200)$$

$$x = \frac{3(200)}{5} = 120$$

Therefore, there are 120 females..

Extra Method

We repeat the question in **Example 1**   3 out of 5 students at a certain college are females.

      (a) If 600 students at this college are females,
how many students are there at this college?

      (b) If one were to collect a sample of 200 students at this
college, how many of these students would be females?

This method follows the algebraic method for dividing a quantity into parts, given the ratio.
Since the ratio of females to the total number of students is 3:5,
let the number of females $= 3x$, and
let the total number of students $= 5x$.
(a) Then $3x = 600$ (there are 600 females); and solving, $x = 200$
Therefore, total number of students $= 5x = 5(200) = $**1000**
Therefore, there are 1000 students at this college.

(b) $5x = 200$ (there are 200 students) , and
    $x = 40$
Therefore, the number of females $= 3x = 3(40) = 120$.

Lesson 24 Exercises A

Problems on direct proportion in which the terms refer to distinct sets

1. If 1 gallon of gasoline costs 2 dollars, how many gallons of gasoline can one buy for 15 dollars?   Answer: $7\frac{1}{2}$ gallons

2. If 10 gallons of gasoline cost 15 dollars, find the cost of 24 gallons of gasoline.
                                            Answer: 36 dollars

3. If 10 gallons of gasoline cost 15 dollars, how many gallons of gasoline can one buy for 45 dollars?   Answer: 30 gallons

4. If x gallons of gasoline cost y dollars, what is the cost of p gallons of gasoline
    (Express your answer in terms of x y and p)   Answer: $\frac{py}{x}$ dollars

5. If on a gallon of gasoline, a car can travel 24 miles, how many miles can the car travel on 10 gallons of gasoline?   Answer: 240 miles

6. If on 2 gallons of gasoline, a car can travel 24 miles, how many miles can a car travel on 20 gallons of gasoline?   Answer: 240 miles

7. If on 3 gallons of gasoline, a bus can travel 36 miles, how many gallons of gasoline are needed to travel 60 miles?   Answer: 5 gallons

8. If on p gallons of gasoline, a bus can travel r miles, how many gallons of gasoline are needed to travel m miles? (Express your answer in terms of the variables.)   Answer: $\frac{mp}{r}$ gallons

9. If 4 oranges cost 1 dollar, how many oranges can one buy for 20 dollars?
                                            Answer: 80 oranges

10. If 8 oranges sell for 2 dollars, find the selling price for 24 oranges.   Answer: 6 dollars

11. If 12 oranges cost 3 dollars, what is the cost of 60 oranges?   Answer: 15 dollars

12. If r oranges cost p dollars, what is the cost of m oranges?   Answer: $\frac{mp}{r}$ dollars
    (Express your answer in terms of the variables.)

13. If a household uses one pound of sugar in 3 weeks, how many pounds of sugar would be used in 6 weeks?   Answer: 2 pounds

14. If a household uses 2 pounds of sugar in 3 weeks, how many pounds of sugar would be used in 51 weeks?   Answer: 34 pounds

15. If a household uses 2 pounds of sugar in 3 weeks, 12 pounds of sugar would be used in how many weeks?   Answer: 18 weeks

16. If a household uses x pounds of sugar in r weeks, how many pounds of sugar would be used in y weeks? (Express your answer in terms of the variables.)   Answer $\frac{xy}{r}$ pounds

17. If one teacher is needed for a class of 22 students, how many teachers would be needed to teach 88 students, assuming that each class has 22 students?   Answer: 4 teachers

18. If 6 teachers are needed to teach 102 students, how many students would be taught by 3 teachers?   Answer: 51 students

19. If t teachers are needed to teach s students, how many students would be taught by b teachers? (Express your answer in terms of the variables.)   Answer: $\frac{bs}{t}$ students

20. If 4 grams of hydrogen are required to produce 36 grams of water, how many grams of hydrogen would be required to produce 108 grams of water?   Answer:: 12 grams

21, If from 4 grams of hydrogen, 36 grams of water are produced, how many grams of hydrogen are needed to produce 216 grams of water?   Answer: 24 grams

22. If from t grams of hydrogen, s grams of water are produced, how many grams of hydrogen are needed to produce b grams of water (Express your answer in terms of the variables   Answer: $\frac{bt}{s}$ grams

23. If p gallons of water are needed to produce x pounds of a concrete mixture, how many gallons of water are needed to produce y pounds of this mixture?   Answer: $\frac{py}{x}$ gallons

24. 1f each room in a building has 3 windows, how many windows are there in 6 rooms,?
                          Answer: 18 weeks

25. If 3 rooms in a building have 12 windows, how many windows are in 36 rooms, assuming that each room has the same number of windows?
Answer: 144 windows

26. If p rooms in a building have q windows, how many windows are in r rooms, assuming that each room has the same number of windows? (Express answer in terms of the variables.)
                          Answer: $\frac{qr}{p}$

27. A buyer purchased 20 items for 60 dollars. How many items can be purchased for 30 dollars, assuming that the cost is the same for each of the items.   Answer: 10 items

28. A buyer purchased r items for s dollars. How many items can be purchased for t dollars, assuming that the cost is the same for each of the items. (Answer in terms of the variables.)
                          Answer: $\frac{rt}{s}$

29. In 2 hours, a car can travel 120 miles. Find the distance the car can travel in 3 hours, assuming the same speed   Answer:: 180 miles

30. In x hours, a car can travel y miles. Find the distance the car can travel in k hours, assuming the same speed.   Answer: $\frac{ky}{x}$

31. In 2 hours, a car can travel 120 miles, At the same speed, how long will it take the car to travel 180 miles?.   Answer: 3 hours

32. In x hours, a car can travel y miles. At the same speed, how long will take the car to travel t miles? (Express your answer in terms of the variables)   Answer: $\frac{xt}{y}$

33. If 40 liters of air are needed to ventilate a patient for 2 minutes, how many liters of air are needed to ventilate this patient for 24 minutes? Answer: 480 liters

34. If a liters of air are needed to ventilate a patient for b minutes, how many liters of air are needed to ventilate this patient for c minutes? Answer $\frac{ac}{b}$

35. If $4\frac{1}{2}$ gallons of gasoline cost 6 dollars, how many gallons of gasoline can one buy for 16 dollars?   Answer: 12 gallons

36. If there are 12 inches in 1 foot, how many inches are there in 4 feet?   Answer: 48 inches

37. If there are 12 inches in I foot, how many inches are there in x feet?
(answer in terms of x) Answer: $12x$ inches

38. If a person's heart beats 4 times every 3 seconds, how many times will it beat in 21 seconds.   Answer: 28 times

39. To prepare an optimum mixture of gasohol which consists of alcohol and gasoline, Alexander used 20 liters of alcohol and 180 liters of gasoline. How many liters of alcohol are needed if 270 liters of gasoline are used and the composition ratio of the mixture remains unchanged?   Answer: 30 liters

40. Charles' Law states that at constant pressure, the volume of a gas is directly proportional the temperature of the gas. If when the volume of a given mass of a gas is 4.2 L, the temperature is $309°K$, find the volume when the temperature is $297°K$ Answer : 4.0 L

41. Gay-Lussac's Law states that at constant volume, the pressure on a gas is directly proportional to the temperature of the gas. If when the pressure on a given mass of a gas is 720 mm Hg, the temperature is $303°K$, find the pressure when the temperature is $309°K$?   Answer: 734 mm Hg

4 2 The ratio of a man's weight on the earth to his weight on the moon is 6:1. If the man weighs (on earth) 240 lb, what is his weight on the moon?   Answer: 40 l

Lesson 24 Exercises B
Problems on direct proportion in which some terms refer to non distinct sets

1. If 2 out of 5 students at a college are males and there are 2,800 students at this college, how many of the students are males?   Answers: 1,120

2. If 2 out of 5 students at a college are males and there are 2,800 students at this college, how many of the students are females?   Answers: 1,680

3. If 2 out of 5 students at a college are males and there are 1,120 males at this college, how many students are at this college?   Answers: 2,800

4.. If 40% students at a college are males and there are 1,120 males at this college, how many students are at this college?   Answers: 2,800

5. If 40% of the students at a college are males and there are 2,800 students at this college, how many of the students are males?   Answers: 1,120

6. Is $\frac{2}{5}$ of a number is 16, what is the number?   Answer: 40

7. James and Mary are to divide some amount of money and James is to receive $\frac{5}{12}$ of this amount. If James receives 30 dollars, what is the amount of money to be divided?
Answer: 72 dollars

8. If 2 out 3 dollars in a bank account are to be given to Diana, If Diana was given $30,000. how much money is in this bank account?   Answer: $45,000

9. If 1 out of 3 liters of an alcohol-water solution is pure alcohol, and this solution contains 12 liters of pure alcohol, what is the total volume of this alcohol-water solution?
Answer 36 liters

10. If q out of p fruits are oranges, and there are r oranges, how many fruits are there. Obtain an expression for the number of fruits in terms of p, q, and r.   Answer: $\frac{pr}{q}$

11. In a math class, 2 out of 4 students passed the final exam. If 32 students passed the exam, how many students are in this class?   Answer: 64

Lesson 24 Exercises C

1. y varies as x. When $y = 12$, $x = 4$. Find y when $x = 10$.

2. The distance S an object falls from rest is directly proportional to the square of the time t. If when $S = 64$, $t = 2$, find S when $t = 4$.

3. W is proportional to V. When $W = 120$, $V = 3$, find W when $V = 8$.

4.. S is directly proportional to t.. When $S = 64$, $t = 2$. Find t when $S = 288$.

5. Express by means of an equation: $(a - b)$ is proportional to $(c - d)$.

6. The distance S an object falls from rest varies directly as the square of the time t. If when $S = 32$, $t = 1$, find S when $t = 5$.

Answers: 1. 30;   **2.** 256;   **3.** 320;   **4.** 9;   **5.** $a - b = k(c - d)$, where k is a constant; **6.** 800.

Lesson 24 Exercises D

1. At constant pressure, the volume V of a perfect gas varies as the absolute temperature T.
If when the temperature is 2730 Absolute, the volume of one gram of a gas is 1.7 cm^3, find the volume when the absolute temperature is 4830°.

2. According to Hooke's law, the force required to stretch a spring is directly proportional to the elongation of the spring. If a 20 lb force stretches a spring 6 in., what force will be required to stretch it 8 in.?

3. The cost C of gasoline is proportional to the number of gallons N of gasoline. If 10 gallons of gasoline cost $12, what is the cost of 25 gallons of gasoline? How many gallons of gasoline can $18 purchase?

Answers: 1. 3.0 cm^3; **2.** 27 lb; **3.** $30; 15 gallons

Lesson 25

Methods for Solving Inverse Proportion Problems

It is very interesting but not surprising that the operations in the steps for solving inverse proportion problems are the opposite operations (inverse operations) in the steps for solving direct proportion problems. By operations, we mean multiplication and division are opposite (inverse) operations. Thus if we know the operations in the steps for solving a direct proportion problem, we can deduce the operations in the steps for solving an inverse proportion problem (by using the opposite operations). For every method for solving direct proportion problems, there is a corresponding method for solving inverse proportion problems, with the steps in the inverse method obtained from the direct proportion method by using opposite operations.

Method 1: **Product = Product Method** (Algebraic)

Assume a is inversely proportional to b.

(same as a is proportional to the **reciprocal** of b)

Then a_1 corresponds to b_1, and a_2 corresponds to b_2. and

$$a_1 : \frac{1}{b_1} \text{ as } a_2 : \frac{1}{b_2}$$

$$a_1 \div \frac{1}{b_1} \text{ as } a_2 \div \frac{1}{b_2}$$

$$a_1 b_1 = a_2 b_2 \qquad (\text{product = product})$$

Or if we use the familiar terms a, b, c, d as the terms of the inverse proportion

Then $a : \frac{1}{b}$ as $c : \frac{1}{d}$

$$a \div \frac{1}{b} \text{ as } c \div \frac{1}{d}$$

$$ab = cd \qquad (\textbf{product = product}) \ \text{<------ inverse proportion}$$

We will use the product = product form since this form is easy to remember.

Compare: **Direct proportion** $\quad \frac{a}{b} = \frac{c}{d}$ <------(**quotient = quotient**)

Inverse proportion $ab = cd$ <------(**product = product**)

Example 1 2 people can paint a house in 6 days. How many people would be required to paint the house in 3 days? Derive a formula relating the number of people and the number of days; and use this formula to find the number of people required to paint the house in 3 days.

Method 1 Product = Product Method

It is known that as the number of people increases, the number of days required decreases, and vice versa.

Let the number of people $= P$,
Let the number of days D.
Then P is indirectly proportional to D,
P_1 corresponds indirectly to D_1 as P_2 corresponds indirectly to D_2.
Since this is indirect proportion, $P_1 D_1 = P_2 D_2$.
From the problem, $P_1 = 2$, $D_1 = 6$, $P_2 = ?$. $D_2 = 3$.
Substituting these values in the equation

$$2(6) = P_2(3)$$

$$\frac{2(6)}{3} = P_2$$

$$4 = P_2$$

Therefore, 4 people are required to paint the house in 3 days.

Method 2: Units Label Method for Inverse Proportion
By reversing the steps in the corresponding direct proportion method (Method 2), we obtain the approach covered below.

We repeat the question: 2 people can paint a house in 6 days. How many people would be required to paint the house in 3 days?

Step 1: Find the product, 2 people $\bullet$ (6 days)

Step 2: Divide : $\dfrac{2 \text{ people} \bullet (6 \text{ days})}{3 \text{ days}}$

$$\dfrac{2 \text{ people} \bullet (\cancel{6 \text{ days}}^{2})}{\cancel{3 \text{ days}}}$$

$= 4$

Therefore, 4 people are required to paint the house in 3 days.

Method 3: Unitary Method (Arithmetic)

We repeat the question: 2 people can paint a house in 6 days. How many people would be required to paint the house in 3 days?

Solution

The number of people required for 6 days = 2

The number of people required for 1 day = 2×6 (Multiply. More people are needed)

The number of people required for 3 days = $\dfrac{2 \times 6}{3}$ (We divide. Less people are needed)

$= 4$

Therefore, 4 people are required to paint the house in 3 days.

Method 4: More or Less Method

We repeat the question: 2 people can paint a house in 6 days. How many people would be required to paint the house in 3 days?

Solution

Step 1: The number of people needed to paint the house in 6 days = 2

Step 2: The number of people needed to paint the house in 3 days = $2 \times \dfrac{6}{3}$

$= 4$ (more people needed)

Therefore, 4 people are required to paint the house in 3 days.

(Note that in Step 2, in forming a fraction using 3 and 6, we used an improper faction, $\dfrac{6}{3}$ (fraction greater than or equal to 1), since we expect the number of people to be greater than 2. For the painting to be done faster, more people are needed. If we had expected the number of people to be less than 2, we would have used the proper fraction $\dfrac{3}{6}$). Master this arithmetic method.

Method 5: Variation Method (Algebraic) **We reword the question:**
The number of people needed to paint a house varies inversely as the number of days taken to paint the house. If when the number of people is 2, the number of days is 6, find the number of people when the number of days is 3.

Step 1: Let the number of people = n
Let the number of days = d

Then since n varies inversely as d, (or n is proportional to the **reciprocal** of d

$$n = \frac{k}{d}$$

Step 2: Determine the value of k by substituting $n = 2, d = 6$ in $n = \dfrac{k}{d}$

Then $2 = \dfrac{k}{6}$ and from which $k = 12$.

Step 3: Substitute $k = 12$ in $n = \dfrac{k}{d}$ to obtain a formula relating n and d.

Then we obtain $n = \dfrac{12}{d}$

Step 4: Substitute for $d = 3$ in the formula . $n = \dfrac{12}{d}$ and solve for n

Then $n = \dfrac{12}{3} = 4$

Thus when $d = 3$, $n = 4$

Therefore, 4 people are required to paint the house in 3 days.

One more example on Method 5

Example 2 The number of people required to sort a quantity of letters varies inversely as the time aken to sort the letters. If 8 people can sort a quantity of letters in 3 hours, how many people would be required to sort the same quantity of letters in 2 hours?

Let the number of people required to sort the letters in x hours be y

Step 1: Then $y = \dfrac{k}{x}$ (1)

Step 2: When $x = 3$, $y = 8$. Substitute these values in equation (1) and solve for k.

$8 = \dfrac{k}{3}$.

Solving, $k = 24$.

Substitute for $k = 24$ in equation (1) above to obtain

the formula $y = \dfrac{24}{x}$ ....(2)

Step 3. when $x = 2$, $y = \dfrac{24}{2}$ (Replacing x by 2 in equation (2))

$y = 12$

Therefore, to sort the letters in 2 hours would require 12 people.

Example 4 Assume pressure P is indirectly proportional to volume V

Then P_1 corresponds to V_1 as P_2 corresponds to V_2

Since P is inversely proportional to V,

$P_1 : \frac{1}{V_1}$ as $P_2 : \frac{1}{V_2}$ and from which

Then $P_1 V_1 = P_2 V_2$ (product = product)

The above equation is a familiar formula for gases in physics.

Lesson 25 Exercises A (Inverse Proportion)

1. If 2 people can paint a house in 6 days, how many people would paint the house in 3 days? Answer: 4 people

2. At a speed of 60 miles per hour, a car takes 2 hours to travel from City A to City B. At a speed of 40 miles per hour, how long will it take the car to ravel from City A to City B? Answer: 3 hours

3. If 6 mail men can sort a given quantity of mail in 4 hours, how many mail men would be needed to sort the same quantity of mail in 2 hours? Answer: 12 men

4. According to Boyle's law, at constant temperature, the volume of a given mass of a gas is inversely proportional to pressure on the gas. When the volume of a gas is 50 cu. in, the pressure is 10 lb per sq. in. What is the pressure (a) when the volume is 80 cu. in.? ; (b) when the volume is 15 cu. in.?; (c) What is the volume when the pressure is 40 lb per sq. in.? Answer: (a) $6\frac{1}{4}$ lb per sq. in.; (b) $33\frac{1}{3}$ lb per sq. in; (c) $12\frac{1}{2}$ u. in;

5. The speed of rotation of meshed gears is inversely proportional to the number of teeth. A gear with 12 teeth rotates at 840 rpm. What is the rotation speed of a 16-tooth gear? Answer: 630 rpm

6. The electric current in a circuit is inversely proportional to the resistance of the circuit. If the current is 22 amps when the resistance is 5 ohms, find the current when the resistance is 10 ohms. Answer: 11 amps

Lesson 25 Exercises B

1. At constant pressure, the volume V of a perfect gas varies as the absolute temperature T. If when the temperature is 2730° Absolute, the volume of one gram of a gas is 1.7 cm^3, find the volume when the Absolute temperature is 4830°.

2. According to Hooke's law, the force required to stretch a spring is directly proportional to the elongation of the spring. If a 20 lb force stretches a spring 6 in., what force will be required to stretch it 8 in.?

3. The cost C of gasoline is proportional to the number of gallons N of gasoline. If 10 gallons of gasoline cost $12, (a) what is the cost of 25 gallons of gasoline? and (b) how many gallons of gasoline can $18 buy?

Answers: **1**. 3.0 cm^3; **2**. $26\frac{2}{3}$ lb; **3**. (a) $30, (b) 15 gallons

Lesson 25: Methods for Solving Inverse Proportion Problems

Lesson 25 Exercises C

1. y varies inversely as x. If when $y = 6, x = 2$, determine y when $x = 8$.

2. F is indirectly proportional to the square of D. When $F = 30, D = 3$. Find F when $D = 2$.

3. W is indirectly proportional to V. When $W = 15, V = 45$ find W when $V = 60$.

Answers: **1.** $\frac{3}{2}$; **2.** $67\frac{1}{2}$; **3.** $11\frac{1}{4}$;

Lesson 25 Exercises D

1 According to Boyle's law, at constant temperature, the volume of a given mass of a gas varies inversely as the pressure on the gas. When the volume of a gas is 50 in.^{3}, the pressure is 10 lb per $in.^2$. What is the pressure (a) when the volume is 80 in.^{3}? ; (b) when the volume is 15 in.^{3} ?; (c) Find the volume when the pressure is 40 lb per $in.^2$.?

2. If 8 people can complete a piece of work in 40 days, how many people will complete the same piece of work in 10 days? Assume that all the people work at the same rate as one another.

3. At constant temperature, the volume of a given mass of a gas is inversely proportional to the pressure on the gas. When the volume is 150 cu. in,. the pressure is 30 lb per $in.^2$.
  Find the pressure (*a*) when the volume is 60 in.^{3} ?; (*b*) when the volume is 300 in.^{3}.

Ans: 1. (a) $6\frac{1}{4}$ lb; (b) $33\frac{1}{3}$ lb per in.^{2}; (c) $12\frac{1}{2}$ in.^{3} **2.** 32 people; **3.** (*a*) 75 lb per in.^{2}.; (*b*) 15 lb per in.^{2}.

Lesson 26

Compound Proportion (Variation) Problems

Deriving formulas for any proportion problem, with any number of variables and involving both direct and inverse relationships

Case 1: Joint Proportion (Joint Variation)

If a quantity varies **directly** as **two** or more other quantities (i.e. as the product of two or more other quantities), we call such a proportion **joint proportion or joint variation**.

Example 1: If z varies directly as x and y, then

$$z = kxy \ \text{ or } \ \frac{z}{xy} = k \text{ (joint variation)}$$

or using subscripted variables, $\dfrac{z_1}{x_1 y_1} = \dfrac{z_2}{x_2 y_2}$ <-- Faster form since you do not have to find k first.

Case 2: Joint Inverse Proportion (Joint Inverse Variation)

The following type of proportion may not be found in current textbooks. However, the author believes that for completeness, this proportion should be included. See the two application examples below.

If a quantity varies **indirectly** as **two** or more other quantities (i.e. as the inverse of the product of two or more other quantities), we will call such a proportion inverse **joint proportion or** inverse **joint variation**.

Example   If z varies **indirectly** as x and y, then

$$z = \frac{k}{xy} \ \text{ or } \ xyz = k \text{ (inverse joint variation)}$$

or using subscripted variables, $z_1 x_1 y_1 = z_2 x_2 y_2$ <-- Faster form

Examples

1. A weight watcher exercising 3 hours per day, 5 times a week lost 10 pounds in 4 months. How many months will it take the weight watcher to lose 10 pounds exercising 2 hours per day, 6 times a week?   **Answer:** 5 months

2. A car traveling at 48 mph and for 6 hours per day covered a certain distance in 10 weeks. How many weeks will it take the car traveling at 60 mph and for 4 hours per day to cover the same distance?   **Answer:** 12 weeks

Case 3 Combined Proportion (Combined Variation)

If a quantity varies **directly** as one quantity (or as two or more quantities) and **inversely** as another quantity (or other quantities), we call such a variation **combined variation**.

Example 2: If z varies directly as x and inversely as y, then

$$z = k\frac{x}{y} \text{ or } \frac{zy}{x} = k \text{ (combined variation)}$$

or using subscripted variables, $\dfrac{z_1 y_1}{x_1} = \dfrac{z_2 y_2}{x_2}$ <-- Faster form since you do not have to find k first.

Example 3: Assume that pressure P is directly proportional to the temperature T, and inversely proportional to the volume V.

Step 1: Then P_1 corresponds to T_1 directly, and to V_1 inversely as P_2 corresponds to T_2 directly, and to V_2 inversely.

Step 2: Write down P_1 (the first quantity.)

Step 3: For each of the other quantifies, if it is directly proportional to P_1, then it divides P_1 (i.e., write it as a divisor), but if it is inversely proportional to P_1, it multiplies P_1 (that is write it as a factor in the numerator). Similarly, repeat the process beginning with P_2.

$$\text{Then } \frac{P_1 V_1}{T_1} = \frac{P_2 V_2}{T_2}, \text{ a familiar gas law in physics.}$$

Example 4: Assume that A is directly proportional to B and directly proportional to C.

Step 1: Then A_1 corresponds to B_1 directly, and to C_1 directly as A_2 corresponds to B_2 directly, and to C_2 directly.

Step 2: Write down A_1 (the first quantity).

Step 3: For each of the other quantities if it is directly proportional to A_1, then it divides A_1 (That is write it as a divisor) but if it is inversely proportional to A_1, it multiplies A_1 (i.e., write it as a factor in the numerator). Similarly, on the right side of the equality symbol, repeat the process beginning with A_2.

$$\text{Then } \frac{A_1}{B_1 C_1} = \frac{A_2}{B_2 C_2}$$

Example 5 Assume that the temperature T is directly proportional to the pressure, P, and directly proportional to the volume, V.

Step 1: Then T_1 corresponds to P_1 directly, and to V_1 directly as T_2 corresponds to P_2 directly, and to V_2 directly.

Step 2: Write down T_1 (the first quantity).

Step 3: For each of the other quantifies if it is directly proportional to T_1, then it divides T_1 (i.e., write it as a divisor) but if it is inversely proportional to T_1, it multiplies T_1 (i.e., is write it in the numerator as a factor) Similarly, repeat the process beginning with T_2.

$$\text{Then } \frac{T_1}{P_1 V_1} = \frac{T_2}{P_2 V_2}$$

By inverting both sides of the above equation, we obtain

$$\frac{P_1 V_1}{T_1} = \frac{P_2 V_2}{T_2}, \text{ the familiar gas law.}$$

Question: Can we obtain the same equation in above example, given the following information?:
Assume volume V is directly proportional to temperature T. and inversely proportional to the pressure P. Then V_1 corresponds to T_1 directly, to P_1 inversely as V_2 corresponds to T_2 directly, and to P_2 inversely.

Translating into an equation $\frac{V_1 P_1}{T_1} = \frac{V_2 P_2}{T_2}$ or equivalently, $\frac{P_1 V_1}{T_1} = \frac{P_2 V_2}{T_2}$. Answer: Yes.

Example 6: Assume that A is inversely proportional to B and inversely proportional to C.
Then A_1 corresponds to B_1 inversely, and to C_1 inversely as A_2 corresponds to B_2 inversely, and to C_2 inversely.
Then, following the agreement made in the previous examples, we obtain
$A_1 B_1 C_1 = A_2 B_2 C_2$ (**product = product**)

Solving Compound Proportion Problems Using Arithmetic   126

In the previous examples, we covered how to solve compound proportion problems using the variation method. However, we can also solve compound proportion problems using **arithmetic.** We can apply **Method 4** of Lesson 18 (p.66) and Lesson 19 (p.81), (The "More or Less Method")

Example 7. If 8 men can make 20 tables in 6 days, how many men would be required to make 60 tables in 9 days?

Solution:

The number of men required to make 20 tables in 6 days = 8 men

The number of men required to make 60 tables in 9 days = 8 men $\times \dfrac{60 \text{ tables}}{20 \text{ tables}} \times \dfrac{6 \text{ days}}{9 \text{ days}}$

(See p.66 and p.81 for setting up the fractions) $= 8 \text{ men} \times \dfrac{60}{20} \times \dfrac{6}{9} = 16 \text{ men}$

Lesson 26 Exercises A
(Compound Proportion)

1. If 5 people working 6 hours per day can complete a job in 8 days, how many people working 2 hours per day can complete the same job in 10 days?   **Answer:** 12 people

2   If 12 people working 4 hours per day, 3 days per week can complete a job in 4 years, how many people working 9 hours per day, 4 days per week can complete the same job in 2 years?   **Answer:** 8 people

3. A weight watcher exercising 3 hours per day, 5 times a week lost 10 pounds in 4 months. How many months will it take the weight watcher to lose 10 pounds exercising 2 hours per day, 6 times a week?   **Answer:** 5 months

4. A car traveling at 48 mph and for 6 hours per day covered a certain distance in 10 weeks. How many weeks will it take the car traveling at 60 mph and for 4 hours per day, to cover the same distance?   **Answer:** 12 weeks

5   If 10 people working 6 hours per day, 7 days per week earned a certain amount of income each week, how many people working 4 hours per day, 5 days per week would earn the same income in one week?   **Answer:** 21 people

6   Six men working 2 days a week can build a fence in 8 weeks. How many men working 4 days a week will build the same fence in 12 weeks.   **Answer:** 2 men

7. Six men working 2 days a week can build a fence in 8 weeks. How many weeks will it take 3 men working 4 days a week to build the same fence.   **Answer:** 8 weeks

8. If 8 men can make 20 tables in 6 days, how many men would be required to make 60 tables in 9 days?   **Answer:**   16 men

Lesson 26 Exercises B

1. Z varies jointly as x and y. If when $x = 2$, $y = 3$, $Z = 16$, what is Z when $x = 5$ and $y = 4$?

2. V is indirectly proportional to P and directly proportional to T. When $V = 30$, $T = 25$, $P = 15$. Find V when $T = 40$, and $P = 3$.

3. F is directly proportional to G and varies inversely as L. When $F = 256$, $G = 36$, and $L = 30$. What is G, when $F = 128$ and $L = 15$?

4. The kinetic energy E of a moving object varies as the mass M. of the object and the square of the velocity V. If when $E = 50$, $M = 4$, $V = 5$, find E when $M = 16$ and $V = 3$.

5. If 12 people working 2 days a week can complete a piece of work in 4 weeks, how many people working 3 days per week can complete the same piece of work in 2 weeks? (Assume that all the people involved work at the same rate as one another)

Answers: **1.** $53\frac{1}{3}$; **2.** 240; **3.** 9;   **4.** 72; **5.** 16 people;

EXTRA

Determining the type of relationship, given an equation

Determine how the first variable is related to each of the other variables in each equation.

(a) $\dfrac{P_1 V_1}{T_1} = \dfrac{P_2 V_2}{T_2}$;   (b) $\dfrac{A_1 B_1}{C_1} = \dfrac{A_2 B_2}{C_2}$;   (c) $\dfrac{B_1}{D_1} = \dfrac{B_2}{D_2}$;   (d) $\dfrac{A_1 B_1 C_1}{D_1 E_1} = \dfrac{A_2 B_2 C_2}{D_2 E_2}$

Solution
(a) P is inversely proportional to V and directly proportional to T
(b) A is inversely proportional to B and directly proportional to C
(c) B is directly proportional to D.
(d) A is inversely proportional to B, inversely proportional C, directly proportional to D, and directly proportional E.

Comparison of Sample Formulations of Proportion Problems

Proportion Formulation (subscript notation)	Variation Formulation	Statement of relationship
1. $\dfrac{V_1}{T_1} = \dfrac{V_2}{T_2}$	$\dfrac{V}{T} = k$ or $V = kT$	V is directly proportional to T
2. $P_1 V_1 = P_2 V_2$	$PV = k$ or $P = \dfrac{k}{V}$	P is inversely proportional to V
3. $\dfrac{P_1 V_1}{T_1} = \dfrac{P_2 V_2}{T_2}$	$\dfrac{PV}{T} = k$ or $PV = kT$	P is inversely proportional to V and directly proportional to T.
4. $I_1 R_1 = I_2 R_2$	$IR = k$ or $I = \dfrac{k}{R}$	I is inversely proportional to R
5. $E_1 I_1 = E_2 I_2$	$EI = k$ or $E = \dfrac{k}{I}$	E is inversely proportional to I
6. $\dfrac{E_1}{N_1} = \dfrac{E_2}{N_2}$	$\dfrac{E}{N} = k$ or $E = kN$	E is directly proportional to N
7. $I_1 N_1 = I_2 N_2$	$IN = k$ or $I = \dfrac{k}{N}$	I is inversely proportional to N
8.		
9.		

CHAPTER 9A
Factoring Polynomials

Lesson 27: **Greatest Common Factor and Common Monomial Factoring**
Lesson 28: **Difference between two Squares; Factoring by Grouping**
Lesson 29: **Factoring Monic Quadratic Trinomials** (Coefficient of the x^2-term is 1)
Lesson 30: **Non-monic Quadratic Trinomials** (Coefficient of the x^2-term is **not** 1)
Lesson 31: **Factoring trinomials quadratic in form (Substitution method)**
Lesson 32: **Solving Quadratic Equations by Factoring and Applications**

You are responsible for three main types of factoring:

1. Common monomial factoring. Example: $2x + 8 = \mathbf{2(x+ 4)}$
2. Factoring quadratic trinomials. Example: $x^2 + 7x + 12 = \mathbf{(x + 3)(x + 4)}$
3. Factoring the difference between two squares. Example: $x^2 - y^2 = \mathbf{(x + y)(x - y)}.$

Factoring Completely:

Note: Step 1. In any factoring, perform common monomial factoring first, if this is possible. (That is, factor out the greatest common divisor.)

Step 2. Look out for the difference between squares or quadratic trinomials and factor accordingly.

Lesson 27

Greatest Common Factor and Common Monomial Factoring

The Greatest Common Factor (GCF)

Definition: The **greatest common factor** of two or more given numbers is the largest **common** divisor of the given numbers.

We will cover two methods which are similar to the methods used in finding the LCM of numbers. Note that the greatest common factor is also the greatest common divisor (GCD) or the highest common factor (HCF).

Example   Find the greatest common factor (GCF) of 36 and 48.

Method 1

We will use a direct method which is similar to prime factorization techniques. We will use **prime numbers** successively as divisors. At each step of the division process, each prime number used **must divide all** the dividends. The division process ends when there are no more common divisors to be used. The GCF is the product of the divisors used.

```
    divisors
       |
       |   dividends
       |    |    |
       ↓    ↓    ↓
     2 | 36 | 48
     2 | 18 | 24
     3 |  9 | 12
          3    4
```

Answer: The GCF of 36 and 48 = $2 \times 2 \times 3$
$$= 12.$$

Method 2

Solution   Step 1 : Factor each number into primes, and write repeated factors in power form.

$$36 = 2 \times 2 \times 3 \times 3 = 2^2 \times 3^2$$
$$48 = 2 \times 2 \times 2 \times 2 \times 3 = 2^4 \times 3$$

Step 2 : Consider each prime factor which is common to both 36 and 48
(i.e., consider 2 and 3)
and choose the lowest power of each prime factor.

Step 3: The GCF = the product of the **lowest powers** of the prime factors considered.

$$= 2^2 \times 3$$
$$= 4 \times 3$$

The GCF of 36 and 48 = 12

Note : 2×2 or 2^2 is common to both 36 and 48; and also 3 is common to both 36 and 48.

Also, if a prime factor is **not common** to all the given numbers, we do not "consider" it.

Common Monomial Factoring

Example   Factor completely: $9x^3 - 6x$

$$9x^3 - 6x$$

Step 1: $= 3x\left(\dfrac{9x^3}{3x} - \dfrac{6x}{3x}\right)$ <---You may skip Step 1. ($3x$ is the greatest common factor (divisor) of $9x^3$ and $-6x$)

↓   ↓   ↓   Scrapwork:

Step 2: $= 3x(3x^2 - 2)$ $\left(\mathbf{1.}\ \dfrac{9x^3}{3x} = 3x^2;\ \ \mathbf{2.}\ \dfrac{-6x}{3x} = -2\right)$

Step 3: We perform two checks (The checking may be done by mere inspection)

Check # 1:   Multiply the factors:
$$3x\,(3x^2 - 2)$$
$$= 9x^3 - 6x \text{ <------------ This must agree with the original expression.}$$

Check # 2: Inspect $3x^2 - 2$ (the expression within the parentheses to make
sure the terms do not have anymore common factors (divisors).
This checking is usually done mentally (by inspection).

$\therefore\ 9x^3 - 6x\ =\ \boxed{3x(3x^2 - 2)}$ <----**answer**

Incomplete Factorization:

Example: $9x^3 - 6x$
$$= 3(3x^3 - 2x)\ \ \text{<------------This is incomplete factorization because}$$
$$3x^3 \text{ and } -2x\ \text{ still have common divisors (factors).}$$

To complete the factorization:
$$= 3(3x^3 - 2x)$$
$$= 3x(3x^2 - 2)$$
and then , $9x^3 - 6x = 3x(3x^2 - 2)$ as obtained previously.

Lesson 27 Exercises

A.
Find the GCF of the following: **1.** 24 and 30   **2.** 56, 70 and 105   **3.** 12, 64 and 72.

Answers: **1.** 6 ;   **2.** 7 ;   **3.** 4

B.
Factor completely: **1.** $4x^4y^6 - 20xy^2$ **2.** $2a^5b^3c^7 + a^4b^2c^6 - a^2b^2c^5$

3. $4x^4y^2 - 2x^2y$;   **4.** $6a^3bc - 2x^2y$;   **5.** $7x^5y^2z - 21x^4y$;   **6.** $8a^2b^2 - ab$; **7.** $4x^2 - 3xy$

Answers   **1.** $4xy^2(x^3y^4 - 5)$;   **2.** $a^2b^2c^5(2a^3bc^2 + a^2c - 1)$; **3.** $2x^2y(2x^2y - 1)$;
     4. $2(3a^3bc - x^2y)$;   **5.** $7x^4y(xyz - 3)$;   **6.** $ab(8ab - 1)$;   **7.** $x(4x - 3y)$

Lesson 28
Difference between two Squares; Factoring by Grouping

Factoring the Difference Between Two Squares

Examples **1.** $a^2 - b^2 = (a + b)(a - b)$

2. $c^2 - d^2 = (c + d)(c - d)$

3. $16 - x^2 = (4 + x)(4 - x)$

4. $9x^2 - 1 = (3x + 1)(3x - 1)$

5. $6x^4 - 6 = 6(x^4 - 1) = 6(x^2 + 1)(x^2 - 1) = 6(x^2 + 1)(x + 1)(x - 1)$

6. $a^4 - b^4 = (a^2 + b^2)(a^2 - b^2) = (a^2 + b^2)(a + b)(a - b)$

7. $3x^2 - 75 = 3(x^2 - 25) = 3(x + 5)(x - 5)$

8. $9y^2 - 16 = (3y + 4)(3y - 4)$

Factoring by Grouping

Sometimes, a seemingly non-factorable expression can be factored after a proper grouping of the terms of the expression. Look out for this type of factoring when there are four, six, or an even number of terms.

Example 1:   Factor by grouping: $ab + cd + ac + bd$

Step 1: Pick the first term and then pick another term (of the expression)
which has a common factor with the first term picked.
  $\underline{ab + ac}$ $+ \underline{cd + bd}$

Step 2: Perform common monomial factoring on the first two terms, and then on the other terms.
  $a(b + c) + d(c + b)$
  $= a(b + c) + d(b + c)$ (Note that $c + b = b + c$)

Step 3:   Perform "common binomial" factoring. (This is similar to common monomial factoring)
Since $(b + c)$ is common , we pick it first.
Then we obtain $(b + c)(a + d)$

Note : Step 3 is a common monomial factoring if we consider $(b + c)$ as a single quantity.

Lesson 28: Difference between two Squares; Factoring by Grouping

Example 2: Factor by grouping (without adding any like terms).

$$6x^2 + 15x - 4x - 10$$

Step 1: Perform common monomial factoring on the first two terms and then on the other terms.

$$3x(2x + 5) - 2(2x + 5) \quad \text{<------- Adjust the signs here so that we obtain}$$
$$- 4x - 10x \text{ when the parentheses are removed.}$$

Step 2: Pick the common factor $2x + 5$ and factor as done in common monomial factoring.

$$(2x + 5)(3x - 2)$$

You can check this factoring by multiplying the factors.
Note: We will obtain the same factors if we group the first and the third terms.
i.e. $6x^2 - 4x + 15x - 10$
$$2x(3x - 2) + 5(3x - 2)$$

Remove or factor out the common factor $(3x - 2)$, and then we obtain

$$(3x - 2)(2x + 5)$$
Again, we obtain the same binomial factors.

Lesson 28 Exercises

A
Factor completely: **1.** $y^2 - 16$; **2.** $4x^2 - 36$; **3.** $4x^2 - 9$; **4.** $25 - 4y^2$

Answers: **1.** $(y + 4)(y - 4)$; **2.** $4(x + 3)(x - 3)$; **3.** $(2x + 3)(2x - 3)$; **4.** $(5 + 2y)(5 - 2y)$ or $- (2y + 5)(2y - 5)$

B.
Factor completely : **1.** $x^2 + 2y + xy + 2x$; **2.** $15x^2 - 9x + 20x - 12$
3. $ac - bc + ad - bd$ **4.** $ax + by + bx + ay$ **5.** $8x^2 + 12x + 2x + 3$

Answers: **1.** $(x + y)(x + 2)$; **2.** $(3x + 4)(5x - 3)$; **3.** $(a - b)(c + d)$; **4.** $(a + b)(x + y)$; **5.** $(2x + 3)(4x + 1)$

Lesson 29
*Factoring Monic Quadratic Trinomials

(Coefficient of the x^2-term is 1)

Case 1: All the terms are positive

Factor completely:

Example 1: $x^2 + 7x + 12$

Step 1: $(x\quad)(x\quad)$

Step 2: $(x + 3)(x + 4)$

Scrapwork:
Find two numbers whose product is 12 and whose sum is 7
$12 = 6 \times 2$
$12 = \boxed{4 \times 3}$

Always check:
$(x + 3)(x + 4) = x^2 + 4x + 3x + 12$ (multiplying the factors)

$\qquad\qquad\qquad = x^2 + 7x + 12$ <------------------ This agrees with the original expression.

$\therefore x^2 + 7x + 12 = (x + 3)\ (x + 4)$

Case 2: The last term is positive and the middle term is negative

Example 2:   Factor completely: $x^2 - 5x + 6$

Step 1: $(x\quad)(x\quad)$

Step 2: $(x - 2)(x - 3)$

These two terms must have the same sign because this is positive.

Scrapwork : $\quad 6 \times 1 = 6$
$\qquad\qquad\quad \boxed{2 \times 3} = 6$

Always check by multiplying.
$\qquad\qquad (x - 2)(x - 3)$
$\qquad\qquad = x^2 - 3x - 2x + 6$
$\qquad\qquad = x^2 - 5x + 6$
$\therefore\quad x^2 - 5x + 6 = (x - 2)(x - 3)$

(We want two numbers whose
   product is +6, but whose sum is -5)

*The coefficient of the x^2-term of the quadratic trinomial is also called the leading coefficient of the quadratic trinomial. Also, a quadratic trinomial in which the coefficient of the x^2-term is 1 is called a **monic** quadratic trinomial.

Lesson 29: Factoring Monic Quadratic Trinomials (Coefficient of the x^2-term is 1)

Case 3: The last term is negative and the middle term is negative or positive

134

Example 3 Factor completely:

$$x^2 - 7x - 18$$

Step 1: $(x \quad)(x \quad)$ | These two terms | must have different signs because | this term | is negative

Step 2: $(x + 2)(x - 9)$

Scrapwork:
$9 \times 2 = 18$
$3 \times 6 = 18$

We want two numbers whose product is -18 and whose sum is -7
(Give the sign of the middle term to the larger of 2 and 9 (absolute values))
After checking, $x^2 - 7x - 18 = (x + 2)(x - 9)$.

Note that $x^2 + 7x - 18$ would be factored as $(x - 2)(x + 9)$

Example 4   Factor completely: $2x^2 - 12x + 16$

$2x^2 - 12x + 16$

Step 1: $2(x^2 - 6x + 8)$ (Factoring out the greatest common divisor)

Step 2: $2(x - 2)(x - 4)$ (Factoring the quadratic trinomial)

After checking, $2x^2 - 12x + 16 = 2(x - 2)(x - 4)$

Lesson 29 Exercises

A. Factor completely: **1.** $x^2 + 9x + 20$;   **2.** $x^2 + 16x + 28$;   **3.** $x^2 + 11x + 30$

Answers **1.** $(x + 4)(x + 5)$; **3.** $(x + 2)(x + 14)$; **4.** $(x + 5)(x + 6)$

B. Factor completely: **1.** $x^2 - 9x + 18$;   **2.** $x^2 - 15x + 36$;   **3.** $x^2 - 13x + 40$.

Answers **1.** $(x - 3)(x - 6)$ **2.** $(x - 3)(x - 12)$ **3.** $(x - 5)(x - 8)$

C. Factor completely: **1.** $x^2 + 5x - 24$;   **2.** $x^2 - 5x - 24$;   **3.** $x^2 - 13x - 48$

Answers: **1.** $(x - 3)(x + 8)$; **2.** $(x + 3)(x - 8)$; **3.** $(x + 3)(x - 16)$

D. Factor completely: **1.** $3x^2 - 24x + 36$;   **2.** $2x^2 + 2x - 84$

Answers **1.** $3(x - 2)(x - 6)$; **2.** $2(x - 6)(x + 7)$

Lesson 29: Factoring Monic Quadratic Trinomials (Coefficient of the x^2-term is 1)

135

E. Factor completely

1. $x^2 + 9x + 18$ **2**. $x^2 + 8x + 15$ **3**. $x^2 + 17x + 72$

4. $x^2 + 3x + 2$ **5**. $x^2 + 14x + 33$ **6**. $x^2 + 14x + 45$

7. $x^2 + 15x + 36$ **8**. $x^2 - 4x - 21$ **9**. $x^2 + 4x - 21$

10. $x^2 - 11x + 18$ **11**. $x^2 - 10x + 21$ **12**. $x^2 - 9x + 18$

13. $x^2 + 8x - 33$ **14**. $4x^2 + 44x + 72$ **15**. $2x^2 + 28x + 90$

Answers: 1. $(x + 3)(x + 6)$; **2**. $(x + 3)(x + 5)$; **3**. $(x + 8)(x + 9)$; **4**. $(x + 1)(x + 2)$; **5**. $(x + 3)(x + 11)$;
6. $(x + 5)(x + 9)$; **7**. $(x + 3)(x + 12)$; **8**. $(x + 3)(x - 7)$; **9**. $(x - 3)(x + 7)$; **10**. $(x - 2)(x - 9)$;
11. $(x - 3)(x - 7)$; **12**. $(x - 3)(x - 6)$; **13**. $(x - 3)(x + 11)$; **14**. $4(x + 2)(x + 9)$; **15**. $2(x + 5)(x + 9)$.

F. Factor completely:

1. $x^2 - 9x + 20$;   **2**. $x^2 + 3x - 18$ **3**. $6x^3y - 14x^2y$;   **4**. $x^2 - 11x + 24$;

5. $2x^2 - 12x - 54$;   **6**. $a^2 + 3a - 28$;   **7**. $b^2 - 2b - 48$;   **8**. $x^2 - 49$;   **9**. $2y^2 - 32$;

10. $4x^2y^2 - 9b^2$;   **11**. $b^4 - a^4$;   **12**. $9y^2 - 36$;   **13**. $9 - x^2$ **14**. $-x^2 - 7x - 12$.

Answers: 1. $(x - 5)(x - 4)$;   **2**. $(x + 6)(x - 3)$;   **3**. $2x^2y(3x - 7)$ **4**. $(x - 8)(x - 3)$;   **5**. $2(x + 3)(x - 9)$; **6**. $(a + 7)(a - 4)$;

     7. $(b + 6)(b - 8)$;   **8**. $(x + 7)(x - 7)$;   **9**. $2(y + 4)(y - 4)$;   **10**. $(2xy + 3b)(2xy - 3b)$; **11**. $(b^2 + a^2)(b + a)(b - a)$

     12. $9(y + 2)(y - 2)$; **13**. $(3 + x)(3 - x)$ or $-(x + 3)(x - 3)$;   **14**. $-(x + 3)(x + 4)$

Lesson 29: Factoring Monic Quadratic Trinomials (Coefficient of the x^2-term is 1)

Test # 8 - Student's Self-Test (Always, Test yourself before you are tested)

Attempt all questions on clean sheets of paper, **Do not write in the book** Show all necessary work.

---

1. Find the GCF of the following:

(a). 32 and 36;   (b) 14, 70 and 182 ;

(c). 12, 56 and 72.

2. Factor completely: (a) $x^2 + 11x + 18$;

(b) $x^2 - 11x + 18$.

---

3. Factor completely: (a) $x^2 + 9x + 20$;

(b) $x^2 + 13x + 42$.

4. Factor completely: (a) $x^2 + 12x + 20$;

(b) $x^2 - 17x + 72$.

---

5. Factor completely: (a) $x^2 - 10x + 24$;

(b) $2x^2 + 22x + 60$.

6. Factor completely: (a) $x^2 - 15x + 36$;

(b) $4x^2 + 44x + 72$.

---

7. Factor completely: (a) $x^2 - 12x + 32$;
(b) $x^2 + 4x - 77$.

8. Factor completely: (a) $x^2 + 5x - 24$;
(b) $x^2 - 9x + 18$.

---

9. Factor completely: (a) $x^2 - 5x - 24$;

(b) $x^2 - 10x + 9$.

10. Factor completely: (a) $x^2 - 13x - 48$;

(b) $x^2 + 4x - 21$.

---

11. Factor completely: (a) $10x^2 + 4x - 6$;

(b) $x^2 + 14x + 33$

12. Factor completely: (a) $2x^2 - 4x - 96$;

(b) $x^2 + 14x + 45$.

---

13. Factor completely: (a) $x^2 - 5x - 36$;

(b) $x^2 + 3x + 2$.

14. Factor completely: (a) $x^2 + 9x + 18$;

(b) $x^2 + 8x + 15$.

---

15. Factor completely: (a) $4x^2 - 16$;

(b) $4 - 9x^2$.

16. Factor completely: (a) $y^2 - 16$;

(b) $25 - 4y^2$.

17. Factor completely: (a) $9y^2 - 36$;

(b) $-x^2 - 7x - 12$.

18. Factor completely: (a) $4x^2y^2 - 9b^2$;

(b) $b^4 - a^4$.

19. Factor completely: (a) $x^2 - 49$;

(b) $2y^2 - 32$.

20. Factor completely: (a) $a^2 + 3a - 28$;

(b) $b^2 - 2b - 48$.

21. Factor completely: (a) $x^2 - 11x + 24$;

(b) $2x^2 - 12x - 54$.

22. Factor completely: (a) $-x^2 - 3x + 18$

(b) $x^2 + 3xy + 2y^2$

23. Factor completely: (a) $x^2 - 17x + 72$;

(b) $2x^2 + x - 15$

Factor completely:

24. **(a)** $ad - bc - ac + bd$

(b) $ay + bx + by + ax$

25. Factor completely: $a^2 + 2x + ax + 2a$

Bonus: Write down your first and last names, and place a plus sign between these names. Assuming that each letter in your name represents a real number, factor completely, if possible, the expression obtained.

Answers: 1 (a) 4; (b) 14 ; (c) 4; **2.** (a)$(x+2)(x+9)$; (b) $(x-2)(x-9)$; **3.** (a) $(x+4)(x+5)$; (b) $(x+6)(x+7)$; **4.** (a) $(x+2)(x+10)$; (b) $(x-8)(x-9)$; **5.** (a) $(x-6)(x-4)$; (b) $2(x+5)(x+6)$; **6.** (a) $(x-3)(x-12)$; (b) $4(x+2)(x+9)$; **7.** (a) $(x-4)(x-8)$; (b) $(x-7)(x+11)$; **8.** (a) $(x-3)(x+8)$; (b) $(x-3)(x-6)$; **9.**(a) $(x-8)(x+3)$; (b) $(x-1)(x-9)$; **10.** (a) $(x+3)(x-16)$; (b) $(x+7)(x-3)$; **11.** (a) $2(5x-3)(x+1)$; (b) $(x+3)(x+11)$; **12.** (a) $2(x-8)(x+6)$; (b) $(x+9)(x+5)$; **13.** (a) $(x-9)(x+4)$; (b) $(x+1)(x+2)$; **14.** (a) $(x+3)(x+6)$; (b) $(x+3)(x+5)$; **15.** (a) $4(x+2)(x-2)$; (b) $(2+3x)(2-3x)$; **16.** (a) $(y+4)(y-4)$; (b) $(5+2y)(5-2y)$; **17.** (a) $9(y+2)(y-2)$; (b) $-(x+3)(x+4)$; **18.** (a) $(2xy+3b)(2xy-3b)$; (b) $(b^2+a^2)(b+a)(b-a)$; **19.** (a) $(x+7)(x-7)$; (b) $2(y+4)(y-4)$; **20.** (a) $(a-4)(a+7)$; (b) $(b+6)(b-8)$; **21.** (a) $(x-3)(x-8)$; (b) $2(x+3)(x-9)$; **22.** (a) $-(x-3)(x+6)$; (b) $(x+y)(x+2y)$; **23.** (a) $(x-8)(x-9)$; (b) $(2x-5)(x+3)$; **24** (a) $(d-c)(a+b)$; (b) $(a+b)(x+y)$; **25.** $(a+x)(a+2)$.
Bonus: Answer varies.

Lesson 30

Factoring Non-monic Quadratic Trinomials

(Coefficient of the x^2-term is **not** 1)

Example   Factor completely: $6x^2 + 11x - 10$

Solution

There are a number of methods for factoring non-monic (the coefficient of the x^2-term is not 1) quadratic trinomials, namely
  1. A general method (Method 1)
  2. A Substitution Method (Method 2) and
  3. A third method (Method 3, the **ac**-method), which involves factoring by grouping.

Method 1   The General Method

This method involves considering the products of the various possible binomial factors to determine which factors yield the original quadratic trinomial.

We will now factor $6x^2 + 11x - 10$
We consider the possible binomials
(a) $(3x + 2)(2x - 5)$
(b) $(3x - 2)(2x + 5)$
(c) $(3x + 5)(2x - 2)$
(d) $(3x - 5)(2x + 2)$
(e) $(6x + 5)(x - 2)$
(f) $(6x - 5)(x + 2)$
(g) $(6x + 2)(x - 5)$
(h) $(6x - 2)(x + 5)$
Multiply the above factors. The pair of binomial factors which yield the original quadratic expression are the correct factors. **Note** that since the original quadratic trinomial did not have any common factors, each possible binomial factor will not have any common factors; and therefore, we could have eliminated the factors in (c),(d) and (h) without having to multiply them out, and then check only the products for cases (a), (b), (e) and (f).
In the above example, the correct factors are $(3x - 2)$ and $(2x + 5)$.

$\therefore$ $6x^2 + 11x - 10 = (3x - 2)(2x + 5)$

Factorability of a quadratic trinomial: $ax^2 + bx + c$

A quadratic trinomial is factorable if the discriminant $b^2 - 4ac$ is a perfect square.

Example Is $6x^2 + 11x - 10$ factorable?
        Check: $a = 6, b = 11, c = -10$
        $b^2 - 4ac = 11^2 - 4(6)(-10)$
                $= 121 + 240$
                $= 361$

        361 is a perfect square because $\sqrt{361} = 19$, a rational number.
Answer: Yes. $6x^2 + 11x - 10$ is factorable.

Method 2:   The Substitution Method

In this method, we change the quadratic trinomial to a monic trinomial which is then easily factored.

Example   Factor $6x^2 + 11x - 10$

Step 1: Multiply the expression by the coefficient of the x^2-term.

$6(6x^2) + 6(11x) - 6(10)$

Step 2: Write the first term as a square and interchange the 6 and the 11 in the second term.

$36x^2 + 11(6x) - 60$

$(6x)^2 + 11(6x) - 60$ ....(A)

Step 3: Let $6x = s$ (That is, replace $6x$ by s in expression (A))

Then, we obtain $s^2 + 11s - 60$

Now, since this is a monic trinomial (coefficient of s^2 is 1), factor easily.

$(s - 4)(s + 15)$ ....(B)

Scrapwork

$60 \times 1 = 60$
$30 \times 2 = 60$
$\boxed{15 \times 4 = 60}$

Step 4:   Replace s by $6x$,and then, expression (B) becomes

$(6x - 4)(6x + 15)$.............................(C)

Since we multiplied the original trinomial by 6, we must divide expression (C) by 6 (that is we must undo the "6" we introduced in Step 1.)

In order to divide (C) by 6, we perform common monomial factoring on the two binomial factors (in some cases, this factoring is performed only on one of the binomial factors).

$(6x - 4)(6x + 15)$
$2(3x - 2) \ 3(2x + 5)$
$2(3)(3x - 2)(2x + 5)$
$6(3x - 2)(2x + 5)$

Now, we divide by 6: $\dfrac{6(3x - 2)(2x + 5)}{6}$

and then the complete factorization of

$6x^2 + 11x - 10$ is $(3x - 2)(2x + 5)$

Method 3 Factoring by the ac-Method

The "a" is the coefficient of the x^2-term, and the "c" is the constant term of the general quadratic trinomial $ax^2 + bx + c.$

Example   Factor completely. $6x^2 + 11x - 10$
$a = 6,\ c = -10.$

Step 1: Multiply the coefficient of the x^2-term by the last term (the constant term).

$$6x^2 + 11x - 10 \ ....(A)$$

$$6(-10) = -60$$

Step 2: Replace the $+11x$ (the middle-term) of expression (A) by two terms whose coefficients add up to 11 and whose product is -60.

Then, we obtain

$$6x^2 + 15x - 4x - 10$$

Scrapwork

$60 \times 1 = 60$
$30 \times 2 = 60$
$\boxed{15 \times (-4) = -60}$
$20 \times 3 = 60$

Step 3: Factor the expression by grouping. (Break up the expression into two pairs and factor)

$$6x^2 + 15x - 4x - 10$$
$$3x(2x + 5) - 2(2x + 5)$$
$$(2x + 5)(3x - 2)$$

Therefore, $6x^2 + 11x - 10 = (2x + 5)(3x - 2)$

Factoring by any method

Example   Factor by any method

$$3x^2 - 7x - 10$$

We will use Method 2 and Method 3.

Method 2

Step 1:   Multiply every term by 3 (the coefficient of the x^2-term)

$$3(3x^2) - 7x(3) - 30$$
$$9x^2 - 7(3x) - 30$$
$$(3x)^2 - 7(3x) - 30 \ ....(A)$$

Step 2: Let $3x = s$ in expression (A) and factor the resulting monic trinomial.

Then $s^2 - 7s - 30$
$$(s + 3)(s - 10)$$

Step 3:   Replace s by $3x$.
$$(3x + 3)(3x - 10)$$

Step 4:   Factor the first binomial and divide by 3

$$\frac{3(x + 1)(3x - 10)}{3}$$ This "3" was used as a multiplier in Step 1

$$\therefore\ 3x^2 - 7x - 10 = (x + 1)(3x - 10)$$

Method 3   Factor $3x^2 - 7x - 10$.

Step 1: Multiply the coefficient of the x^2-term by the last term.

$$3x^2 - 7x - 10$$

$$3(-10) = -30$$

Step 2: Replace $-7x$ by two terms whose coefficients add up to -7 and whose product is -30.
$$3x^2 - 10x + 3x - 10$$

Step 3:   Factor by grouping
$$3x^2 - 10x + 3x - 10$$
$$= x(3x - 10) + 1(3x - 10)$$
$$= (3x - 10)(x + 1)$$
$$\therefore 3x^2 - 7x - 10 = (x + 1)(3x - 10)$$

Lesson 30 Exercises

A. Factor completely: **1.** $12x^2 - x - 6$;   **2.** $18x^2 - 3x - 10$;   **3.** $10x^2 - 29x + 21$

Answers: **1**. $(3x + 2)(4x - 3)$;   **2**. $(6x - 5)(3x + 2)$;   **3**. $(2x - 3)(5x - 7)$

B. Factor completely: **1.** $12x^2 - x - 6$;   **2.** $18x^2 - 3x - 10$;   **3.** $10x^2 - 29x + 21$

Answers: **1**. $(3x + 2)(4x - 3)$;   **2**. $(6x - 5)(3x + 2)$;   **3**. $(2x - 3)(5x - 7)$

C Factor completely:

1. $8x^2 - 2x - 15$ **2.** $4x^2 - 12x + 9$ **3.** $6x^2 - 19x + 10$

4. $15x^2 - 14x - 8$ **5.** $12x^2 + x - 6$

Answers: **1** . $(2x - 3)(4x + 5)$ **2.** $(2x - 3)(2x - 3)$ or $(2x - 3)^2$ **3.** $(2x - 5)(3x - 2)$
4. $(3x - 4)(5x + 2)$ **5** . $(3x - 2)(4x + 3)$

Lesson 31

Factoring trinomials quadratic in form (Substitution method)

Example (a) Factor completely: $x^4 + 7x^2 + 12$ ....(1)

Solution Step 1: Let $x^2 = u$. Then, expression (1) becomes $u^2 + 7u + 12$...............(2)

Step 2 : Factoring (2), we obtain $(u + 3)(u + 4)$................................(3)

Step 3: Replacing u by x^2 in expression (3), we obtain $(x^2 + 3)(x^2 + 4)$
$\therefore x^4 + 7x^2 + 12 = (x^2 + 3)(x^2 + 4)$.

Example (b)   Factor completely: $2a^4 - 5a^2 - 12$ ....(1)

Solution Let $a^2 = u$. Then, we obtain $2u^2 - 5u - 12$ ....(2)

Factoring (2), we obtain $(2u + 3)(u - 4)$ .... (3)

Replacing u by a^2 in expression (3), we obtain $(2a^2 + 3)(a^2 - 4) = (2a^2 + 3)(a + 2)(a - 2)$

$\therefore 2a^4 - 5a^2 - 12 = (2a^2 + 3)(a + 2)(a - 2)$.

Lesson 31 Exercises

Factor completely: **1**. $x^4 - 7x^2 + 12$;   **2**. $2a^4 + 5a^2 - 12$

Answers. **1.** $(x^2 - 3)(x + 2)(x - 2)$;   **2**. $(2a^2 - 3)(a^2 + 4)$

EXTRA
Factoring Perfect Square Trinomials. See Appendix. p.305

Lesson 32

Solving Quadratic Equations by Factoring and Applications

Standard form of the quadratic equation: $ax^2 + bx + c = 0$, where a, b, and c are constants and $a \neq 0$

Examples 1. $x^2 - 3x - 28 = 0$<---------in standard form

      2. $x^2 - 2x = 0$ <-------- in standard form

      3. $x^2 = 5x + 14$<---------**not** in standard form

      4. $5x^2 = 8x$ <---------**not** in standard form

Principle of zero products: If $ab = 0$ then either $a = 0$, or $b = 0$ (or both $= 0$).

To solve by factoring, the quadratic equation must be in standard form before factoring

Example 1   Solve by factoring
$$x^2 - 3x - 28 = 0$$

Step 1:   Factor the quadratic trinomial.
$$(x + 4)(x - 7) = 0$$

Step 2:   Set each factor equal to zero and solve for x.

$$x + 4 = 0 \qquad x - 7 = 0$$
$$\underline{-4 \ -4} \qquad \underline{+7 \ +7}$$
$$x = -4 \qquad x = 7$$
$$\therefore \ x = -4, x = 7$$

The solutions are -4 and 7.

Example 2   Solve for x by factoring.
$$3x^2 - 4x = 0$$

Step 1: Factor. $3x^2 - 4x = 0$

$$x(3x - 4) = 0 \qquad (\text{performing common monomial factoring})$$

Step 2:   Set each factor equal to zero and solve for x.   (We set only factors having a **variable** to zero)

$$x = 0 \qquad \text{or} \qquad 3x - 4 = 0$$
$$\underline{+4 \ +4}$$
$$\frac{3}{3}x = \frac{4}{3}$$
$$x = \frac{4}{3}$$
$$\therefore x = 0, \quad x = \frac{4}{3} \qquad\qquad \textbf{Note} \text{ that } 0 \text{ is also a solution .}$$

The solutions are 0 and $\frac{4}{3}$.

Note: A common **wrong** approach in the above problem:

$$3x^2 - 4x = 0$$
$$3x^2 = 4x$$

$3x = 4$ (Canceling, that is dividing out an x on both sides of the equation ; such a cancellation excludes
      $x = 0$ as a solution.)

$x = \frac{4}{3}$ (Of course, you still obtain one of the solutions but lose the other solution).

Example 3 Solve by factoring: $9x^2 = 36$

Solution $9x^2 = 36$

Step 1: $9x^2 - 36 = 0$

Step 2: Factor the left-hand side completely.

$9(x^2 - 4) = 0$

$9(x + 2)(x - 2) = 0$

Step 3: Set each factor equal zero and solve for x: (Note that $9 \neq 0$; only factors with a variable $= 0$)

$$x + 2 = 0 \quad \text{or} \quad x - 2 = 0$$
$$\underline{ - 2 \quad -2} \qquad \underline{+2 \quad +2}$$
$$x = -2 \qquad\qquad x = 2$$

The solutions are 2 and -2.

Another method (The square root method)

$$9x^2 = 36$$
$$\frac{9x^2}{9} = \frac{36}{9}$$
$$x^2 = 4$$
$$x = \pm\sqrt{4}$$
$$x = \pm 2 \ (+2 \text{ or } -2)$$

Again, the solutions are 2 and -2.

Applications of the quadratic equation

Example The width of a rectangle is 3 units less than the length. If the area of the rectangle is 28 sq. units, determine the dimensions of this rectangle.

Solution

Let the length of the rectangle $= x$.

Then, the width of this rectangle $= (x - 3)$

Area of a rectangle is given by LW (Where L is the length and W is the width).

The required equation: $x(x - 3) = 28$.

$$x^2 - 3x - 28 = 0$$
$$(x - 7)(x + 4) = 0$$
$$x = 7 \text{ or } x = -4 \qquad \text{(Solving by factoring)}$$

We reject the negative value, -4, since the dimension of a rectangle cannot be negative.

When the length is 7 units, the width is 7 - 3 = 4 units.

(**Check**: If the length is 7 and the width is 4, the area is 7(4) = 28 sq. units; and also the width, 4, is 3 less than the length, 7. Therefore, the conditions in the original word problem have been satisfied.)

The length is 7 units and the width is 4 units.

Lesson 32 Exercises

A. Solve for x by factoring: **1.** $x^2 - 3x - 10 = 0$; **2.** $2x^2 - 6x = 0$ **3.** $10x^2 = 15x$

Answers: **1.** -2 and 5; **2.** 0 and 3 ; **3.** 0 and $\frac{3}{2}$

B. 1. Solve for x: $2x^2 = 32$; **2.** Solve for x: $3x^2 - 75 = 0$.

3. The sum of two numbers is 20, and their product is 96. Find these numbers.

4. The length of a rectangle is 5 units more than the width. If the area of this rectangle is 126 sq, units, find the length and width of this rectangle.

Ans: **1.** 4 and -4; **2.** 5 and -5; **3.** 8 and 12; **4.** Length = 14 units, width = 9 units.

Test # 9 -Student's Self-Test (Always, Test yourself before you are tested)

Attempt all questions on clean sheets of paper, **Do not write in the book** Show all necessary work.

1 Factor completely: $6x^2 - 13x + 6$.

2. Factor completely: $10x^2 - 21x + 2$.

3. Factor completely: $3x^2 + 13x + 4$.

4. Factor completely: $10x^2 - 13x$.

5. Factor completely: $2x^2 + 20x + 18$.

6. Factor completely: $4x^2 + 44x + 72$.

7. Factor completely: $8x^2 - 2x - 1$.

8. Factor completely: $8x^2 - 26x + 15$.

9. Factor completely: $18x^2 - 3x - 10$.

10. Factor completely: $15x^2 - 14x - 8$.

11. Factor completely: $3x^2 - 24x + 36$.

12. Factor completely: $2x^2 + 2x - 84$.

13. Factor completely: $7x^2 - 23x + 6$.

14. Solve for x by factoring: $4x^2 - 9 = 0$.

15. Solve for x by factoring: $4x^2 - 36 = 0$.

16. Factor completely: $24x^2 + 2x - 12$.

17. Solve for x by factoring: $-x^2 - 7x - 12 = 0$

18. Factor completely: $6x^2 - 19x + 10$.

19. Solve for x by factoring: $3x^2 - 6x = 0$.

20. Factor completely: $10x^2 - 29x + 21$.

21. Solve for x by factoring: $4x^2 = 36$.

22. Solve for x by factoring: $3x^2 - 75 = 0$.

23. Solve for x by factoring:
$x^2 - 3x - 10 = 0$.

24. The sum of two numbers is 24, and their product is 108. Find these numbers.

25. The length of a rectangle is 6 units more than the width. If the area of this rectangle is 112 sq. units, find the length and width of this rectangle.

Bonus:

$a^4 - 7a^2 + 12$;

$2y^4 + 5y^2 - 12$.

Answers: 1. $(2x - 3)(3x - 2)$; **2.** $(x - 2)(10x - 1)$; **3.** $(x + 4)(3x + 1)$; **4.** $x(10x - 13)$;
5. $2(x + 1)(x + 9)$; **6.** $4(x + 2)(x + 9)$; **7.** $(2x - 1)(4x + 1)$; **8.** $(2x - 5)(4x - 3)$;
9. $(6x - 5)(3x + 2)$; **10.** $(3x - 4)(5x + 2)$; **11.** $3(x - 6)(x - 2)$; **12.** $2(x - 6)(x + 7)$;
13. $(x - 3)(7x - 2)$; **14.** $x = \pm\frac{3}{2}$; **15.** $x = \pm 3$; **16.** $2(3x - 2)(4x + 3)$; **17.** $x = -3$, $x = -4$;
18. $(2x - 5)(3x - 2)$; **19.** $x = 0$, $x = 2$; **20.** $(5x - 7)(2x - 3)$; **21.** $x = \pm 3$; **22** $x = \pm 5$;
23. $x = 5$, $x = -2$; **24.** The numbers are 6 and 18;
25. The length is 14 units and the width is 8 units.
Bonus: (a) $(a^2 - 3)(a + 2)(a - 2)$; (b) $(2y^2 - 3)(y^2 + 4)$.

CHAPTER 9B

More Factoring Polynomials
Lesson 33

Factoring the Sum and Difference between two Cube;

Factoring the Sum of Two Cubes

$$\boxed{a^3 + b^3 = (a + b)(a^2 - ab + b^2)} \dots\dots\dots\dots\dots(1)$$

Similarly, $x^3 + y^3 = (x + y)(x^2 - xy + y^2)$

Example Factor completely: $8x^3 + y^3$

$8x^3 + y^3$
$= (2x)^3 + y^3$
$= (2x + y)((2x)^2 - 2x(y) + y^2)$ (Replace a by $2x$; b by y in the factoring formula (1) above.
$= (2x + y)(4x^2 - 2xy + y^2)$

Factoring the Difference between (of) Two Cubes

$$\boxed{a^3 - b^3 = (a - b)(a^2 + ab + b^2)} \dots\dots\dots\dots(2)$$

Similarly, $x^3 - y^3 = (x - y)(x^2 + xy + y^2)$

Example Factor completely: $x^3 - 8y^3$

$x^3 - 8y^3$
$= x^3 - (2y)^3$
$= (x - 2y)(x^2 + x(2y) + (2y)^2)$ (Replace a by x; and b by $2y$ in the factoring formula (2) above)
$= (x - 2y)(x^2 + 2xy + 4y^2)$

It is good practice to check the factoring by multiplying the factors.
In the future, imitate the substitution (replacement) technique when factoring the sum of or the difference between two cubes.

Mnemonic device:
Find a way to remember and distinguish between the basic factoring formulas as in (1) and (2) above: You may observe that in the factors on the right-hand side, there is only one minus, "−", involved; and the other signs are all "+" signs.

$a^3 + b^3 = \boxed{(a + b)(a^2 - ab + b^2)}$. <-- the minus sign precedes the ab-term. Also there is only one minus sign.

$a^3 - b^3 = \boxed{(a - b)(a^2 + ab + b^2)}$ <--The minus sign precedes the b-term . Also there is only one minus sign

To make sure you have the correct factorization, check by multiplying the binomial and the trinomial.

Lesson 33 Exercises

A Factor completely: $y^3 + 8$;   **2.** $x^3 + 27$;   **3.** $x^3 - 27$;   **4.** $8x^3 - 27$;   **5.** $3x^3 - 81$

Answers: **1.** $(y + 2)(y^2 - 2y + 4)$; **2.** $(x + 3)(x^2 - 3x + 9)$; **3.** $(x - 3)(x^2 + 3x + 9)$; **4.** $(2x - 3)(4x^2 + 6x + 9)$
5. $3(x - 3)(x^2 + 3x + 9)$

Introductory Theme for Chapters 10A & 10B

(Next two chapters)

Straight Line
Theme: Two points

1. Why two points? Two points, because given or knowing two points, a straight line can be drawn by connecting the two points, using a straight edge and pencil.

2. Why two points? Two points, because given (knowing) two points; the slope, m, of the line segment connecting the two points $P_1(x_1, y_1)$ and $P_2(x_2, y_2)$ can be found by applying $m = \dfrac{y_2 - y_1}{x_2 - x_1}$

3a. Why two points? Two points, because if we know the **slope, m,** and the **y-intercept, b,** of the line, we can obtain two points and draw the graph of the line

3b Note: y–intercept, b implies the point $(0, b)$, By choosing a point (x, y) on a line, we have two points, and the slope (as well as an equation) of the line connecting the two pints $(0, b)$ and (x, y) is given by $m = \dfrac{y - b}{x - 0}$ <--**slope = slope**

$mx = y - b$ or $\boxed{y = mx + b}$ <------**slope-intercept form** of the equation of a line

4. Why two points? Two points, because given or knowing two points an equation of the line segment connecting the two points $P_1(x_1, y_1)$ and $P_2(x_2, y_2)$ can be found by

applying $\boxed{y - y_1 = \left(\dfrac{y_2 - y_1}{x_2 - x_1}\right)(x - x_1)}$ (from $\dfrac{y - y_1}{x - x_1} = \dfrac{y_2 - y_1}{x_2 - x_1}$ <-- **is slope = slope**

or $\boxed{y - y_1 = m(x - x_1)}$ <------**point-slope form,** where $m = \dfrac{y_2 - y_1}{x_2 - x_1}$

5. Why two points? Two points, because given or knowing the two-intercept points $(a, 0)$, $(0, b)$ an equation of the line segment connecting the two points $P_1(a, 0)$ and $P_2(0, b)$ can be found by

applying $y - y_1 = \left(\dfrac{y_2 - y_1}{x_2 - x_1}\right)(x - x_1)$ to obtain $y - 0 = \dfrac{b - 0}{0 - a}(x - a)$; from $\dfrac{y - 0}{x - a} = \dfrac{b - 0}{0 - a}$)

or $y = \dfrac{b}{-a}(x - a)$ or $y = \dfrac{b}{-a}x + (\dfrac{b}{-a})(-a)$ or $\boxed{y = -\dfrac{b}{a}x + b}$ also $\dfrac{x}{a} + \dfrac{y}{b} = 1$ (Two intercept form 6.

Why two points? Two points, because given the graph (picture) of a line , we are given infinitely many points from which we can read the coordinates of any two points on the line and write an equation of a line by applying **3, 4** or **5** above. A picture is worth a thousand words

7. Why two points? Two points, because given or knowing two points; the **midpoint** of the line segment connecting the points $P_1(x_1, y_1)$ and $P_2(x_2, y_2)$ is given by x-coordinate, $x_m = \dfrac{x_1 + x_2}{2}$, and the y-coordinate, $y_m = \dfrac{y_1 + y_2}{2}$

8. Why two points? Two points, because given or knowing) two points; $P_1(x_1, y_1)$, $P_2(x_2, y_2)$, the distance, d, between the two points on a line **in a plane** is given by
$$d = \sqrt{(x_2 - x_1)^2 + (y_2 - y_1)^2}$$

9. Why two points? Two points, because given or knowing two points, $P_1(x_1, y_1)$, $P_2(x_2, y_2)$ on each of two lines, the slopes m_1, m_2 as well as parallelism or perpendicularity can be determined. **The above theme** summarizes next two chapters.

CHAPTER 10A

Lesson 34A: **Rectangular Coordinate System**
Lesson 34B: **Drawing the Graph of a Straight Line**
Lesson 35: **Points on a Line; Slopes of lines; Intercepts**
Lesson 36: **Equations of Straight Lines**

Lesson 34A

Graphing in the Rectangular Coordinate System of Axes

We begin by means of an example. Let us consider the equation $y = 3x + 2$ (1)

If we let $x = 1$ in equation (1) then $y = 3(1) + 2$
$$= 5$$

Thus, when $x = 1$, $y = 5$

If we let $x = 2$ in equation (1) m then $y = 3(2) + 2$
$$= 8$$

Similarly, when $x = 2$, $y = 8$.

From the above example, we can say that for each value of x we choose, there is a corresponding value of y. By agreement, we can write the $x-$ and $y-$values values as an ordered pair (x, y), where x is the first component and y is the second component.

Then for the solutions to $y = 3x + 2$, when $x = 1$, $y = 5$, we can represent the solution as the ordered pair $(1, 5)$.

For $x = 2$, $y = 8$, we can write the ordered pair $(2, 8)$

Similarly, when $x = 4$, $y = 14$, giving us the ordered pair $(4, 14)$.

The equation $y = 3x + 2$ has infinitely many solutions. The solution set consists of all ordered pairs. If an ordered pair is a solution of a given equation, then these numbers when substituted in the equation for the unknowns, should make both sides of the equation equal. We should note that we could have chosen values for y and then calculate the corresponding values of x. In fact, in some instances, it may be more convenient to choose y and calculate x

Rectangular Coordinate System of Axes (Cartesian Coordinate System)

Consider a horizontal number line (scale) labeled as in Figure. On this line, all numbers to the right of the origin are positive, and all numbers to the left of the origin are negative.

Figure 1: Horizontal real number line

Consider also a vertical line (**Figure 2**). On this line, all numbers above the origin are positive, and all numbers below the origin are negative.

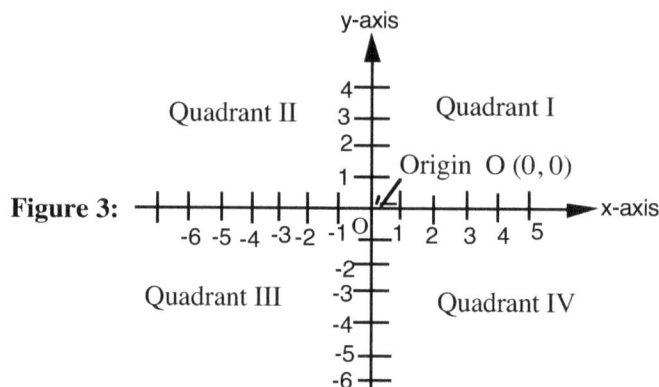

Figure 2:
Vertical line

Figure 3:

If we combine the horizontal and vertical lines at right angles so that the zero points (the origins) coincide, then we obtain what is called the rectangular coordinate system of axes. (**Figure 3**). We call the intersection of these lines, the origin and we label it $O(0,0)$. We call the horizontal number scale the x–axis, and we call the vertical number scale the y–axis. The two axes divide the plane (area) into four quadrants. These quadrants are numbered counterclockwise as quadrants I, II, III, and IV (i.e., first quadrant, second quadrant, third quadrant and fourth quadrant).

Graphing Ordered Pairs in a Rectangular Coordinate Plane

Geometrically, the first component (element) of each ordered pair is called the x-coordinate or the or ordinate of the point. The x-coordinate is the directed (positive or negative) distance from the y–axis and the y-coordinate is the directed distance from the x–axis. In quadrant I, both the x- and y-coordinates are positive. In quadrant II, the x–coordinate is negative but the y–coordinate is positive, In quadrant III, both coordinates are negative, In quadrant IV, the x–coordinate is positive, but the y–coordinate is negative.

Geometrically, each ordered pair represents a **point** in an x–y coordinate system of axes.

Plotting Points

In plotting points, we will be guided as follows:

1. We always count from the origin $O(0,0)$.

2. If the x–coordinate is positive, we count horizontally to the right from the origin, but if the x–coordinate is negative, we count horizontally to the left from the origin.

3. If the y–coordinate is positive, we count vertically upwards from the origin, but if the y–coordinate is negative, we count vertically downwards from the origin. The counting is done visually, and we do not make any marks on the graph paper as we count.

Example .Graph the following ordered pairs (points) in a rectangular coordinate system of axes. 1 5 1
$A(1, 5)$. $B(-2, 4)$, $P_1(-3, -4)$. $P_2(1, -2)$.

Solution

For $A(1, 5)$: From the origin, count horizontally one unit to the right ($x =1$) and stop.
From this point, count 5 units ($y =5$) vertically upwards and stop. Place a dot
here and label this point as $A(1, 5)$.

For $B(-2, 4)$: From the origin, count horizontally 2 units to the left ($x =$-2) and stop,
From this point, count 4 units ($y =4$) vertically upwards and stop. Place a
dot here and label this point as $B(-2, 4)$.

For $P_1(-3, -4)$: From the origin, count horizontally 3 units to the left ($x =$-3) and
stop. From this point, count 4 units vertically downwards ($y =$-4) and
stop. Place a dot here and label this point as $P_1(-3, -4)$.

Similarly, for $P_2(1, -2)$, count horizontally 1 unit to the right ($x = 1$) and 2 units vertically
downwards ($y =$-2) and stop. Place a dot here and label this point as $P_1(-3, -4)$.

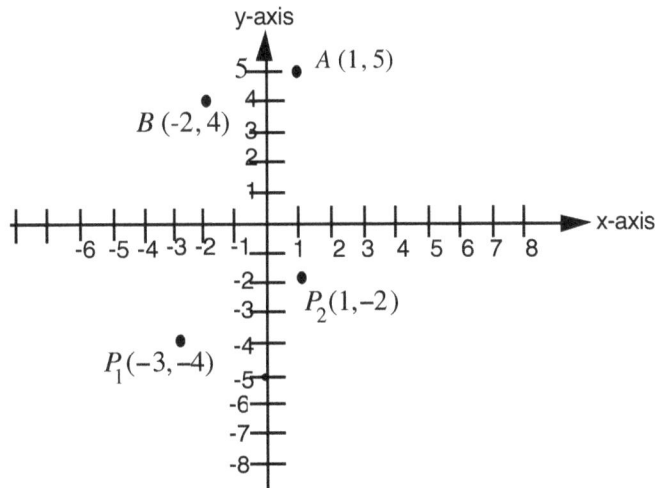

Figure 4

Lesson 34A Exercises

Graph the following ordered pairs in a system of rectangular axes.
$A(2, 3)$, $B(-2, -3)$, $C(-4, 5)$, $D(5, -4)$

Lesson 34B
Drawing the Graph of a Straight Line

Example Draw the graph of the line whose equation is given by : $y = 3x - 5$

We will cover three methods, namely a general method; the x- and y-intercepts method; and the intercept-slope method.

Method 1: General method

Step 1: Choose three convenient x-values and calculate the corresponding y-values to obtain ordered pairs.
(Actually, two ordered pairs will be sufficient; the third pair is used as a check: all **three** points must be in line)

choosing $x = 1$, $y = 3(1) - 5$ ordered pairs

$$\left. \begin{array}{l} y = 3 - 5 \\ y = -2 \end{array} \right\} \text{------->} (1, -2)$$

choosing $x = 0$, $y = 3(0) - 5$
$$\left. \begin{array}{l} \\ y = -5 \end{array} \right\} \text{------->} (0, -5)$$

choosing $x = -2$, $y = 3(-2) - 5$

$$\left. \begin{array}{l} y = -6 - 5 \\ y = -11 \end{array} \right\} \text{------->} (-2, -11)$$

Step 2: Plot the points $(1, -2), (0, -5)$ and $(-2, -11)$ on a rectangular coordinate system of axes (on graph paper) and connect the points by a straight line.

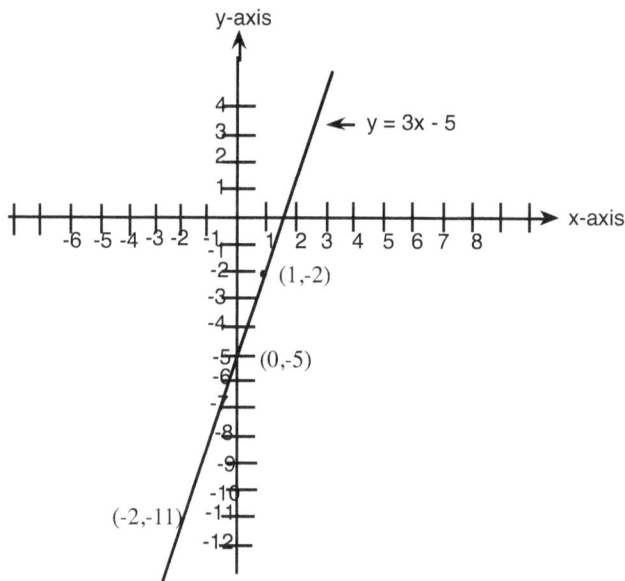

Note that from the given equation, the line has a positive slope and therefore the direction (see also page 164) of the line (it **leans to the right**) in the figure is appropriate.

Note also that the above general method can be extended to draw the graphs of nonlinear equations such as $y = x^2$.

Method 2: The *x*- and *y*-intercepts method

We need two points to plot and connect by a straight line. The coordinates of the *x*-intercept yield one point and the coordinates of the *y*-intercept yield a second point..

Step 1: Find the *x*-intercept by letting $y = 0$ in the equation, $y = 3x - 5$, and solving for *x*.

Then $0 = 3x - 5$ and from which $x = \frac{5}{3}$. This step yields the point $(\frac{5}{3}, 0)$.

Step 2: Find the *y*-intercept by letting $x = 0$ in the equation and solving for *y*.

$$y = 3(0) - 5$$
$$y = -5$$

This yields the point $(0, -5)$

Step 3: Plot these two points from Steps 1 and 2 and connect them by a straight line.

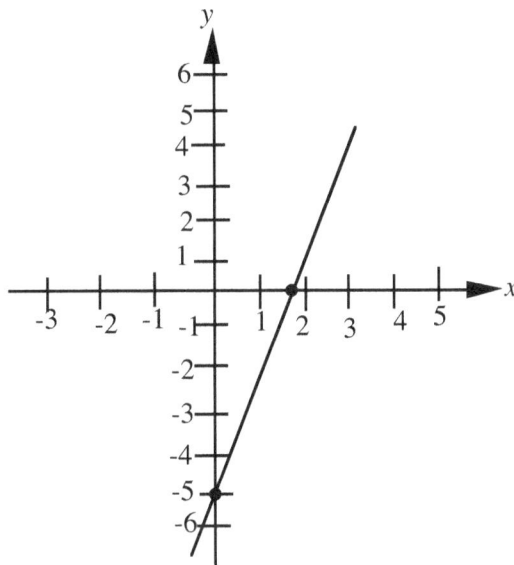

Figure: Graph of $y = 3x - 5$

Note that in Method 2, if the *y*-intercept is 0 (i.e., the equation is of the form $y = mx$), then we need to obtain another point by choosing another *x*-value (other than 0), calculate the corresponding *y*-value to obtain an ordered pair which we plot to locate a second point. An example of such a case is given by the line $y = 3x$. Note also that if the *y*-intercept is 0, the line passes through the origin.

Method 3: Slope-Intercept method

Here also, we need two points to plot and connect by a straight line. The y-intercept locates one point, and we use the slope to locate a second point.

Step 1: Solve the equation for y, and read the y-intercept.
   Since the equation has already been solved for y, we read the intercept, -5.
   This step locates the point $(0,-5)$.

Step 2: Graph the point $(0, -5)$.

Step 3: Since the slope $= +3 = \frac{3}{1}$, the vertical change is +3 and the horizontal change is +1.

   From the point $(0, -5)$, the y-intercept, count 3 units vertically upwards and 1 unit horizontally to the right, stop and place a dot here; this is a second point.
   Connect the two points by a straight line.

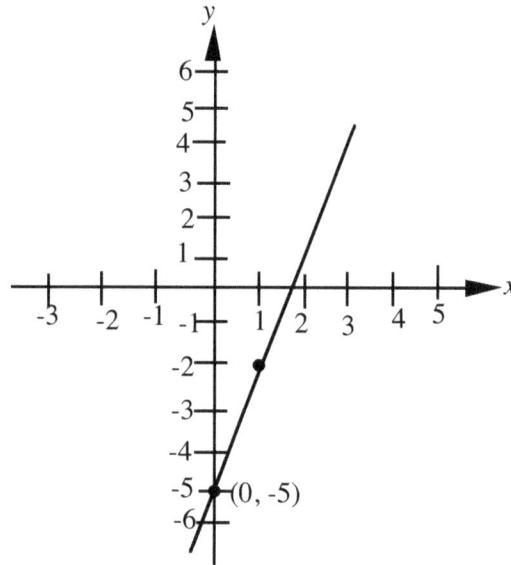

Figure: Graph of $y = 3x - 5$

Note In Step 3, we adopt the following convention.

1. Write the slope as a fraction if it is not a fraction. and if the slope is negative, give the minus sign to the numerator.

   For example, a slope of -3 is expressed as $\frac{-3}{1}$; $-\frac{4}{5}$ is expressed as $\frac{-4}{5}$; $-\frac{1}{2}$ is expressed as $\frac{-1}{2}$

2. In counting from the y-intercept, if the numerator (change in y) is positive, we count vertically upwards, but if it is negative, we count vertically downwards.
   For the denominator (change in x) we always count to the right since we have agreed to give the minus sign to the numerator.

   Examples: For a slope of $\frac{-2}{3}$, starting from the y-intercept, we count 2 units vertically downwards (negative numerator) and 3 units horizontally to the right. For a slope of $\frac{-1}{2}$, starting from the y-intercept, we count

   1 unit vertically downwards (negative numerator) and 2 units horizontally to the right. For a slope of $\frac{4}{5}$,

   starting from the y-intercept, we count 4 units upwards (positive numerator) and 5 units to the right

Lesson 34B Exercises

Draw the graphs of the following lines whose equations are given:

1. $y = -3x + 5$ **2.** $6y = 18 - 3x$ **3.** $2y - 6x = -3$

Answers: Graphs of the lines **1.** $y = -3x + 5$ **2.** $6y = 18 - 3x$ **3.** $2y - 6x = -3$

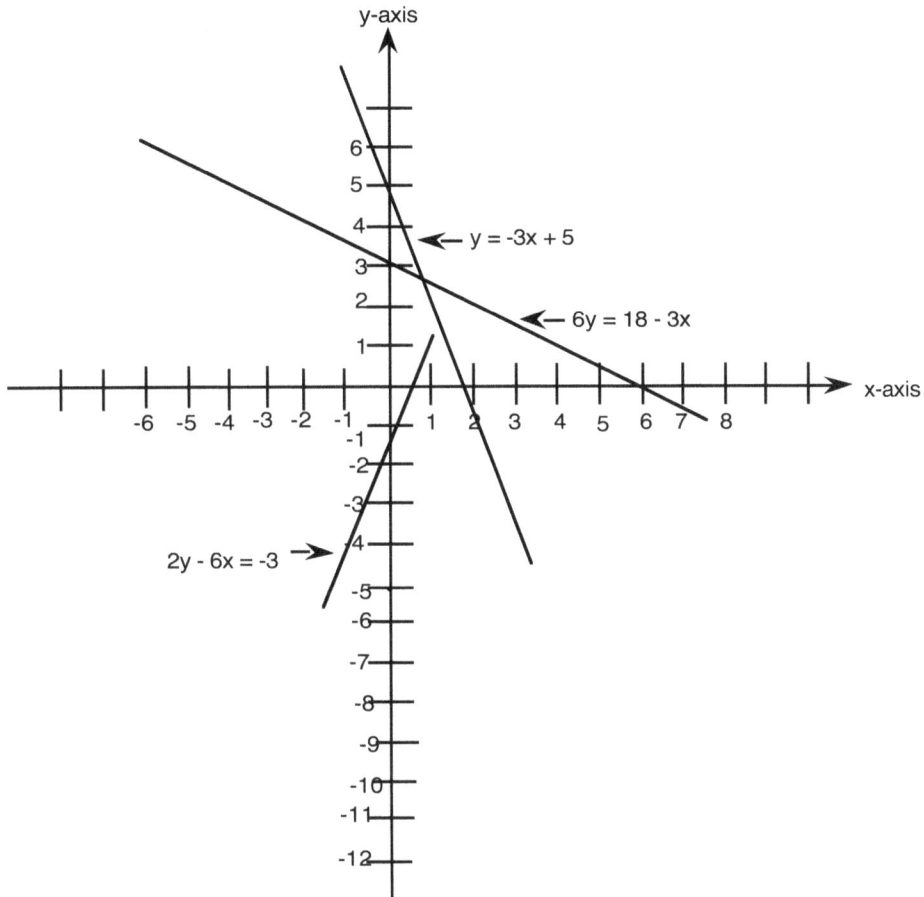

Lesson 35

Points on a Line; Slopes of Lines; *x*- and *y*-intercepts

Determining if a point is on a line whose equation is given (See also p.49)

Example   Which of the points $(5, 2)$ and $(3, 11)$ is on the line whose equation
is given by $y = 3x + 2$.

We will substitute each ordered pair in turn in the given equation. The ordered pair whose
substitution makes the left-hand side of the equation **equal** to the right-hand side **is on** the line.

Step 1:   Checking for the point $(5,2)$

Substitute the ordered pair $(5,2)$ in

$y = 3x + 2$ $(x = 5, y = 2)$

The question mark "?" above the equality symbol

then, $2 \overset{?}{=} 3(5) + 2$ is there to show that at this step, we do not yet know
if the left-hand side and the right-hand side of the

$2 \overset{?}{=} 15 + 2$ equation are equal. In checking solutions, it is a good
practice to use the question mark.

$2 \overset{?}{=} 17$ No   (or $2 = 17$ is False)

The left-hand side is **not** equal to the right-hand side of the equation and therefore the point $(5,2)$
is **not** on the given line.

We also say (algebraically) that the ordered pair $(5,2)$ is **not** a solution of the
the equation $y = 3x + 2$

Step 2: Checking for the point $(3,11)$

Substitute $(3, 11)$ in $y = 3x + 2$

then, $11 \overset{?}{=} 3(3) + 2$

$11 \overset{?}{=} 9 + 2$

$11 = 11$ True   (We no longer use the question mark since it is obvious that
the left- hand side equals the right-hand side of the equation)

Since the left-hand side **is equal to** the right-hand side of the equation,
the point $(3,11)$ **is on** the line $y = 3x + 2$.

Algebraically, we also say that
the ordered pair $(3,11)$ **is a** **solution** of the equation $y = 3x + 2$.

In fact, the last example (question) could have been posed as : Which of the following
ordered pairs is a solution of the equation $y = 3x + 2$? (a) $(5,2)$, (b) $(3,11)$

Solution: Proceed exactly as steps 1 and 2 of example above.

In calculations, the following have the same interchangeable implications:

1. A line passes through a given point.

2. A line contains a given point.

3. A given point is on a line.

4. The coordinates of a given point satisfy an equation of a line.

5. The coordinate pair of a given point is a solution of an equation of a line.

Example: The problem " find an equation of the line **passing through** the points.$(2, 3)$ and
$(6, 8)$" is **equivalent to** "find an equation of the line **containing** the points $(2, 3)$ and $(6, 8)$.

Finding the slope of a line given the coordinates of two points on the line 57

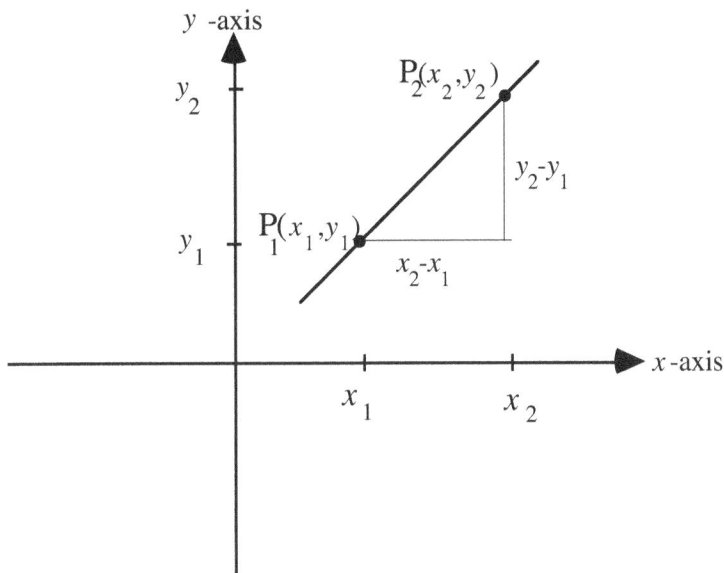

The slope, m, of the line segment connecting the points $P_1(x_1, y_1)$ and $P_2(x_2, y_2)$ is given by

$$m = \frac{y_2 - y_1}{x_2 - x_1}$$

Example   Find the slope of the line passing through the points (2,3) and (6,8).

Solution :   Identify the first point as $P_1(2,3)$ and the second point as $P_2(6,8)$

Then $x_1 = 2, y_1 = 3;$ and $x_2 = 6, \ y_2 = 8$

Applying the slope formula, $m = \frac{y_2 - y_1}{x_2 - x_1}$

$$m = \frac{8 - 3}{6 - 2}$$

$$m = \frac{5}{4}$$

The slope is $\frac{5}{4}$.

Special Cases of the Slopes of Lines
The special cases are for horizontal and vertical lines

Slope of a horizontal Line
The **slope** of a **horizontal** line is **zero**, since the vertical change is zero. For example, the slope,
m, of the horizontal line in Figure **1**, below, by the slope formula is $m = \frac{3-3}{5-2} = \frac{0}{3} = 0.$

$\left(\text{Note that } m = \frac{\text{vertical change}}{\text{horizontal change}} = \frac{\text{change in } y}{\text{change in } x}\right)$

Slope of a Vertical Line

The **slope** of a **vertical** line is **undefined** since the horizontal change is zero. For example, the slope, m, of the vertical line in Figure **2**, below, by the slope formula is $m = \frac{2-(-3)}{4-4} = \frac{5}{0}$ is undefined.

Figure 1

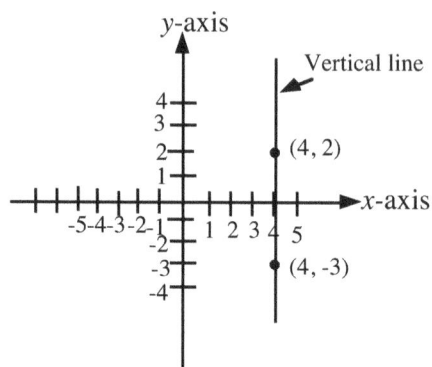

Figure 2

The *x*- and *y*-intercepts of a line

The **x-intercept** of a line is the *x*-coordinate of the point where the line crosses or meets the *x*-axis. At this point, $y = 0$. (Geometrically, we should also say that the *x*-intercept is the point where the line intersects the *x*-axis, and in which case, we must specify two coordinates, with the y-coordinate always being zero.) See Figure 1 below.

The y-intercept of a line is the *y*-coordinate of the point where the line crosses or meets the *y*-axis. At this point, $x = 0$. (Geometrically, we should also say that the *y*-intercept is the point where the line intersects the *y*-axis, and in which case, we must specify two coordinates, with the *x*-coordinate always being zero.) See Figure 1 below.

Example

In Figure 1 below, the *x*-intercept is -3; and the *y*-intercept is 2. The *x*- and *y*--intercepts are at (-3,0) and (0,2) respectively.

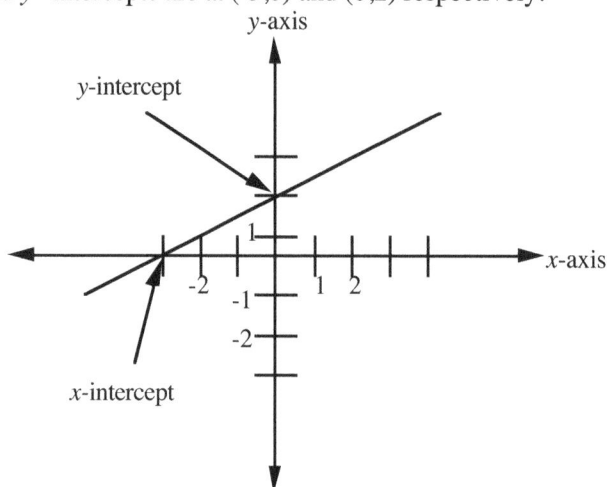

Figure 1

Example   Find the x- and y-intercepts for the graph of $3x + 2y = 12$

For the x-intercept: Let $y = 0$ in $3x + 2y = 12$.

Then $3x + 2(0) = 12$; $3x = 12$; and from which $x = 4$.

The x-intercept is 4.

For the y-intercept: Let $x = 0$ in $3x + 2y = 12$.

Then $3(0) + 2y = 12$; $2y = 12$; and from which $y = 6$.

The y-intercept is 6.

Finding the slope and the y-intercept of a line, given the equation of the line

Example 1 Find the slope and the y-intercept of the line whose equation is given by $y = -5x + 6$.

The slope-intercept form of the equation of a straight line is given by $y = mx + b$, where m is the slope and b is the y-intercept.

Compare $y = -5x + 6$ to   **Note:** (The slope is the coefficient of the x- term
$y = mx + b$,   if the equation is in slope-intercept form.)

then, the slope, $m = -5$,
and the y-intercept $= 6$

Example 2   Find the slope and the y-intercept of the line whose equation is given by
$$5y + 4x = 20$$

Step 1: Solve the equation for y (that is, write the equation in slope-intercept form).
$$5y + 4x = 20$$
$$\underline{\quad -4x \qquad\qquad -4x\quad}$$
$$5y = -4x + 20$$
$$\frac{5}{5}y = \frac{-4}{5}x + \frac{20}{5}$$
$$y = -\frac{4}{5}x + 4$$

Step 2: By comparison with $y = mx + b$, the slope is $-\frac{4}{5}$ (the coefficient of the x-term).

and the y-intercept is +4 or 4.

Lesson 35 Exercises   160

A Determine which of the points (2,4), (-2,5), (-2,-16) are on the line whose equation is given by
$$y = 5x - 6$$

Answer: The points (2,4), and (-2,-16) are on the given line

B Find the slope of each line passing through the given points:

1. (2, 3) and (6, 4)   **2.** (3, -2) and (2, -4)   **3.** (5, 2) and (1, -1)
4. (2, 2) and (-3, 3)   **5.** (-4, 2) and (5, 2)   **6.** (-3, -5) and (-3, -6)
7. Find c so that the slope of the line passing through the points $(c, 2c)$ and (-6, 8) is 1.

Answers: 1. $\frac{1}{4}$; **2.** 2 ; **3.** $\frac{3}{4}$; **4.** -$\frac{1}{5}$; **5.** 0 ; **6.** undefined; **7.** $c = 14$

C 1.. Find the slope of the line passing through the points (3,2) and (5,9).

2. The points (2,2) and (-5,6) are on a line whose slope we want to determine. What is the slope of this line?
3. Find the slope of the line containing the points (1,6) and (2,8).
4. Find the slope of the line passing through the points (1,-1), (3,3) and (-2,-7).

Answers: 1. slope $= \frac{7}{2}$ **2.** slope $= -\frac{4}{7}$ **3.** slope = 2 ;   **4.** slope =2

D Find the slope and the y-intercept of the following:

1.. $y = -6x + 4$; **2.** $-7 + 5x = y$; **3.** $y = -\frac{x}{2} + 8$

Answers: 1. slope = - 6, y-intercept = 4; **2.** slope = 5, y-intercept = -7;   **3.** slope = -$\frac{1}{2}$, y-intercept = 8

E Find the slope and y-intercept: **1.** $3x + 8y = 32$ **2.** $2y - 6x = 21$;   **3.** $y = -3x + 4$
4. $y = \frac{1}{2}x - 3$;   **5.** $3y = 6x + 12$;   **6.** $y = \frac{x}{3} + 5$;   **7.** $2y + 5x = 8$;   **8.** $4x + 3y = 12$

Ans 1. slope = -$\frac{3}{8}$, y -intercept = 4;   **2.** slope =3, y -intercept $= \frac{21}{2}$; **3.** slope = - 3 ; y -intercept = 4;

4. slope = $\frac{1}{2}$, y -intercept = -3 ;   **5.** slope = 2, y -intercept = 4;   **6.** slope = $\frac{1}{3}$, y -intercept = 5;

7. slope = -$\frac{5}{2}$, y -intercept = 4;   **8.** slope = -$\frac{4}{3}$, y -intercept = 4;

Lesson 36
Equations of Straight Lines

There are a number of approaches that we can use in **finding equations** of straight lines. Each approach depends upon what we are given.

Generally, we can easily write down an equation of a line if we know any of the pairs of properties or characteristics of (information about) a line in the formulas presented below.

These formulas are based on finding two different expressions for the slope of a line and equating these expressions to each other.

1. If we know the **slope, m**, and the **y-intercept, b,** of the line,
   we can apply the slope-intercept form, $y = mx + b$.

2. If we know the slope, m, and the coordinates (x_1, y_1) of one point on the line,
   we can apply the point-slope form, $(y - y_1) = m(x - x_1)$.

3. If we know the coordinates (x_1, y_1) and (x_2, y_2) of two points on the line,

   we can apply $(y - y_1) = \dfrac{y_2 - y_1}{x_2 - x_1}(x - x_1)$.

4. If we are given the graph (picture), we can determine any of the above pairs of
   properties (information) from the graph, and then apply the formulas in the above cases;
   however, if we want to memorize one more formula, we can apply the equation

   $y = -\dfrac{b}{a}x + b$, where a is the x-intercept and b is the y-intercept. (If $b = 0$ or $a = 0$ use the other methods.)

(This equation is another form of the two-intercept form: $\dfrac{x}{a} + \dfrac{y}{b} = 1$)

   We will now cover the above **four** cases in detail with examples.

Case 1: Finding an equation of a line given the slope and the y-intercept of the line

Example 1 Find an equation of the line with slope 3 and y-intercept of 4.

Solution We will apply the slope-intercept form of the equation
of a straight line, $y = mx + b$.
Substituting the slope, $m = 3$, and the y-intercept, $b = 4$ in this equation,
$$y = 3x + 4$$

Example 2 Find an equation of the line with slope - 2 and y-intercept of - 5.

Solution: Substituting $m = -2$, $b = -5$ in $y = mx + b$, we obtain
the equation $y = -2x - 5$

Case 2: **Finding an equation of a line given the slope of the line and the coordinates of one point on the line**

Example 1 Find an equation of the line passing through the point $(3, -2)$ and having a slope 4.

Solution. We will cover two methods.

Method 1

Step 1: Find the y-intercept, b, by substituting $x_1 = 3$, $y_1 = -2$, $m = 4$ in

$$y = mx + b$$
$$\text{then, } -2 = 4(3) + b$$
$$-2 = 12 + b$$
$$-14 = b$$

Step 2: Now, since we know that $m = 4$, $b = -14$, we can apply $y = mx + b$ (as in Case 1, above) and then, $y = 4x - 14$

Method 2

We will use the point-slope form of the equation of a straight line which is given by

$$y - y_1 = m(x - x_1) \quad \text{<------point-slope form.} \tag{1}$$

Substituting the coordinates, $x_1 = 3$, $y_1 = -2$ and slope, $m = 4$ in equation (1), we obtain

$$y - (-2) = 4(x - 3)$$

$$y + 2 = 4(x - 3) \tag{2}$$

Equation (2) is the point-slope form of the required equation.

By solving equation (2) for y, we obtain

$$y = 4x - 14 \quad \text{<----------- slope-intercept form} \tag{3}$$

Equation (3) is the slope-intercept form of the required equation.
In this form, we can, by inspection, determine the slope and the y-intercept.

The author recommends the slope-intercept form (for this course), unless otherwise specified.

Case 3: **Finding an equation of a line given the coordinates of two different points on the line**

Example Find an equation of the line passing through the points (2,1) and (-3,-4).

Solution

Step 1: Find the slope, m, with $x_1 = 2, y_1 = 1, x_2 = -3$ $y_2 = -4$

$$m = \frac{y_2 - y_1}{x_2 - x_1}$$

Scrapwork

$$m = \frac{-4 - 1}{-3 - 2}$$

$$\frac{-4 - 1}{-3 - 2} = \frac{-5}{-5} = 1$$

$$m = 1$$

Now, $m = 1, x_1 = 2, y_1 = 1, x_2 = -3, y_2 = -4$, and we can apply the procedure in Case 2.

Step 2: Applying Method 2 of Case 2 and
Substituting $m = 1, x_1 = 2, y_1 = 1$ in $y - y_1 = m(x - x_1)$, we obtain

$$y - 1 = 1(x - 2)$$

$$y = x - 1$$

Also, If we substitute $m = 1, x_2 = -3, y_2 = -4$ in $y - y_2 = m(x - x_2)$, we obtain

$$y - (-4) = 1(x - (-3))$$

$$y + 4 = x + 3$$
$$y = x - 1$$

Again, we obtain the same equation. An equation of the line is $y = x - 1$.
We conclude also that any of the two given points can be used in finding an equation of the straight line. From the above solution, we can state a general formula for Case 3 as

$$(y - y_1) = \frac{y_2 - y_1}{x_2 - x_1}(x - x_1)$$

Case 4: **Finding an equation of a line given the graph (picture) of the line**

If we are given the graph (picture) , we can determine any of the above pairs of properties (information) from the graph (see p.161, case 4).

It is also useful to be able to tell immediately from the graph if the line has a positive slope, a negative slope, a zero slope, or an undefined slope.

The signs of the slopes of lines

The lines in Figure 1 have positive slopes.   The lines in Figure 2 have negative slopes.

Figure 1

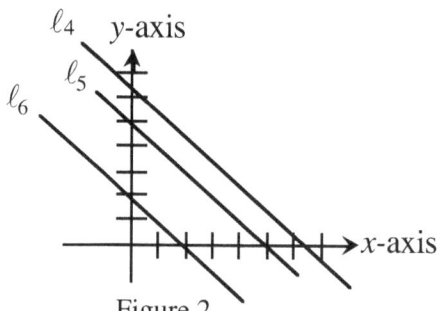

Figure 2

Lines ℓ_1, ℓ_2, and ℓ_3 have positive slopes. (Simply, these lines **lean** to the **right** in the page; or these lines rise as one moves ones head from the left to the right in the page.)

Lines ℓ_4, ℓ_5 and ℓ_6 have negative slopes. (Simply, these lines **lean** to the **left** in the page; or these lines fall as one moves ones head from the left to the right in the page.)

Note: The slope of a **horizontal line** is zero. The slope of a **vertical line** is undefined. (For the equations of horizontal and vertical lines, see p.167-168)

Example 1 Find an equation of the line whose graph is given below.

Method 1: Apply $y = -\dfrac{b}{a}x + b,$

where a is the x-intercept and b is the y-intercept. (see also p.158)

Step 1: From the graph, we read the values of a and b: $a = 3, b = -4$

Step 2: Substitute $a = 3, b = -4$ in

$$y = -\dfrac{b}{a}x + b.$$

Then $y = -\dfrac{-4}{3}x + (-4)$

(Make sure you take into account the **minus sign** that comes with the formula.)

$y = +\dfrac{4}{3}x - 4$ (Two minus signs make the x-term positive)

An equation of the line is $y = \dfrac{4}{3}x - 4$.

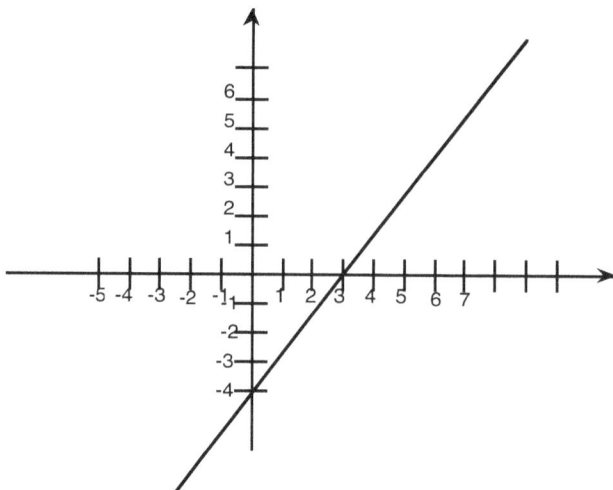

Method 2

Step 1: By inspection, the line has a positive slope. (See figure 1 above.)

Step 2: Slope, $m = \dfrac{\text{vertical change}}{\text{horizontal change}}$

$= \dfrac{\text{change in } y}{\text{change in } x}$

Pick any two convenient points on the line, say, $P_1(0,-4)$ and $P_2(3,0)$ (see figure 2, below).

Finding the vertical and horizontal changes by counting.

We will assume that we want to go from the point P_1 to the point P_2, and count the number of equal intervals (units) we will travel vertically, and the number of equal intervals (units) we will travel horizontally. Then, we will go up vertically 4 equal intervals and horizontally to the right 3 equal intervals. Divide the vertical change, 4, by the horizontal change, 3 , to obtain the slope $\dfrac{4}{3}$

Note that the slope is positive. (see p.164, Figure 1).

Step 3: From the graph , the y-intercept, $b = -4$

Step 4: With $m = \dfrac{4}{3}$, $b = -4$, apply $y = mx + b$

then, $y = \dfrac{4}{3}x - 4$

Therefore, an equation of the line is

$y = \dfrac{4}{3}x - 4$

In fact, once we have been able to specify the points $P_1(0,-4)$, $P_2(3,0)$, we

can apply m $= \dfrac{y_2 - y_1}{x_2 - x_1} = \dfrac{0 - (-4)}{3 - 0} = \dfrac{+4}{3}$

to find the slope $m = \dfrac{4}{3}$,

and then apply $y - y_1 = m(x - x_1)$ to obtain

$y + 4 = \dfrac{4}{3}(x - 0)$ $(x_1 = 0, y_1 = -4)$

$y + 4 = \dfrac{4}{3}x - 0$

$y = \dfrac{4}{3}x - 4$ < -------slope-intercept form of the equation of a straight line.

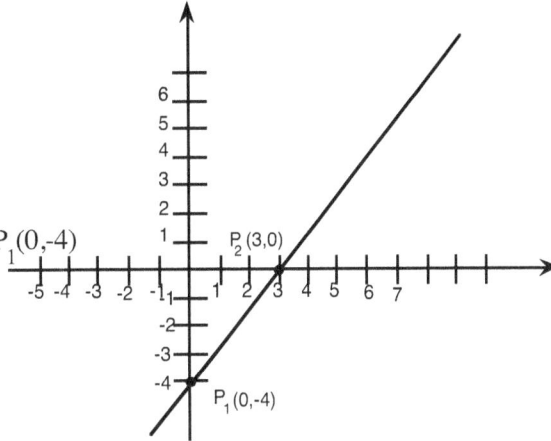

Figure 2

Example 2 Find an equation of the line whose graph is given below.

Method 1 Apply $y = -\dfrac{b}{a}x + b$ where a is the x-intercept and b is the y-intercept.

Step 1: From the graph, $a = 3, b = 6$

Step 2: Substitute $a = 3, b = 6$ in $y = -\dfrac{b}{a}x + b$

Then $y = -\dfrac{6}{3}x + 6$

$y = -2x + 6$ < ---slope-intercept form of the equation of a straight line.

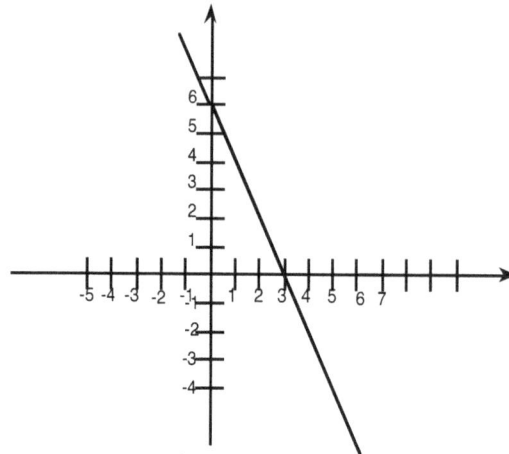

Figure

Method 2

Step 1: Note that the slope is negative.
(See p.164, Figure 2.)

Step 2: Pick any two convenient points on the line, say P_1 and P_2

Step 3: Assume that we want to travel (vertically and horizontally only) from P_1 to P_2 .

Then, we will go down 6 units and then to the right 3 units.

The slope, $m = -\dfrac{6}{3} = -2$

(The slope is negative since the line leans to the left. See page 164, Fig. 2)

Step 4: Read the y-intercept, $b = 6$

Step 5: Apply $y = mx + b$, with $m = -2, b = 6$

Then, an equation of the given graph is
$y = -2x + 6$

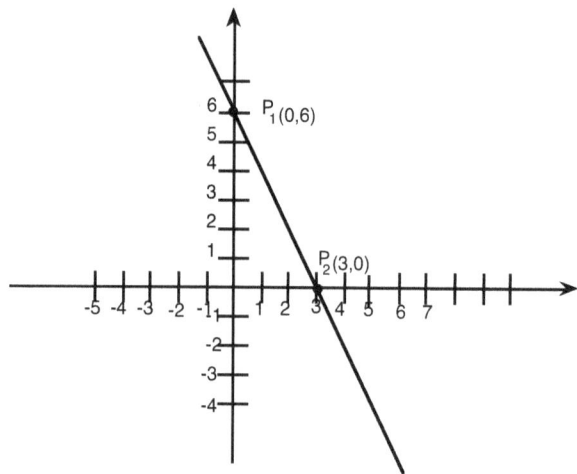

Figure

Method 3 From $P_1 (0,6)$ and $P_2(3,0)$

Step 1: $m = \dfrac{y_2 - y_1}{x_2 - x_1}$ $\qquad (x_1 = 0, y_1 = 6)$

$= \dfrac{0 - 6}{3 - 0} = \dfrac{-6}{3} = -2$

$m = -2$

Step 2: Apply $y - y_1 = m (x - x_1)$

$y - 6 = -2(x - 0)$

$y - 6 = -2x + 0$

$y = -2x + 6$

An equation of the line is $y = -2x + 6$

Note that, in **Method 3**, the negative sign results solely from the calculation and we do not have to know in advance the sign of the slope)

In **Case 4**, the author recommends **Method 1.**

Special Cases of the Equations of Straight Lines and their Graphs

The equation $y = mx + b$, (with m defined and, $m \neq 0$) geometrically represents oblique lines (i.e., lines which are neither vertical nor horizontal). The special cases of the equation of a straight line are for horizontal and vertical lines. In these cases, either the x- or the y-term is missing.

The Equation and Graph of a Horizontal Line

The equation of a horizontal line is of the form $y = b$, where b is the y-intercept.

   The slope, $m = 0$, since the vertical change is zero.

Substituting $m = 0$, in $y = mx + b$,

$$y = 0x + b$$

$$y = b \qquad\qquad (1)$$

Equation (1) means that as x varies, y remains unchanged.

Examples: Sketch the graphs of the following lines:

   1. $y = 3$,   **2.** $y = -4$.   **3.** $y = 0$.

Solution: The line $y = 3$ is the horizontal line passing through (0,3). See Figure 1, below.

The line $y = -4$ is the horizontal line passing through (0,-4).

The line $y = 0$ is the horizontal line along the x-axis (i.e., the line $y = 0$ is the x-axis)

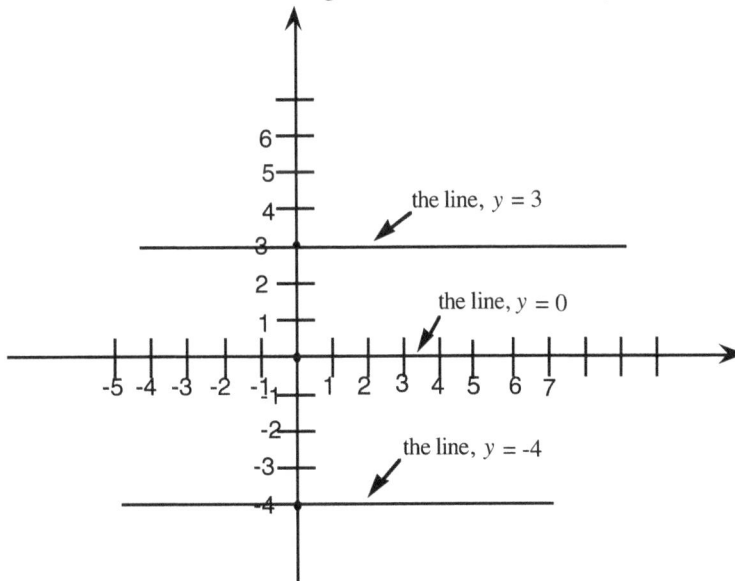

Figure 1

The Equation and Graph of a Vertical Line

The slope of a vertical line is undefined since the horizontal change is zero.
The equation of a vertical line is of the form $x = a$, where a is the x-intercept. This form of the equation means that as y varies, x remains unchanged.

Examples   Sketch the lines with the following equations:

         1. $x = 2$,   **2.** $x = 0$,   **3.** $x = -4$

Solution: See Figure 2, below.

   1. The line $x = 2$ is the vertical line passing the point (2,0).

   2. The line $x = 0$ is the vertical line along the y-axis (i.e., the line $x = 0$ is the y-axis).

   3. The line $x = -4$ is the vertical line passing through the point (-4,0).

   4. The line $x = -2$ is the vertical line passing through the point (-2,0).

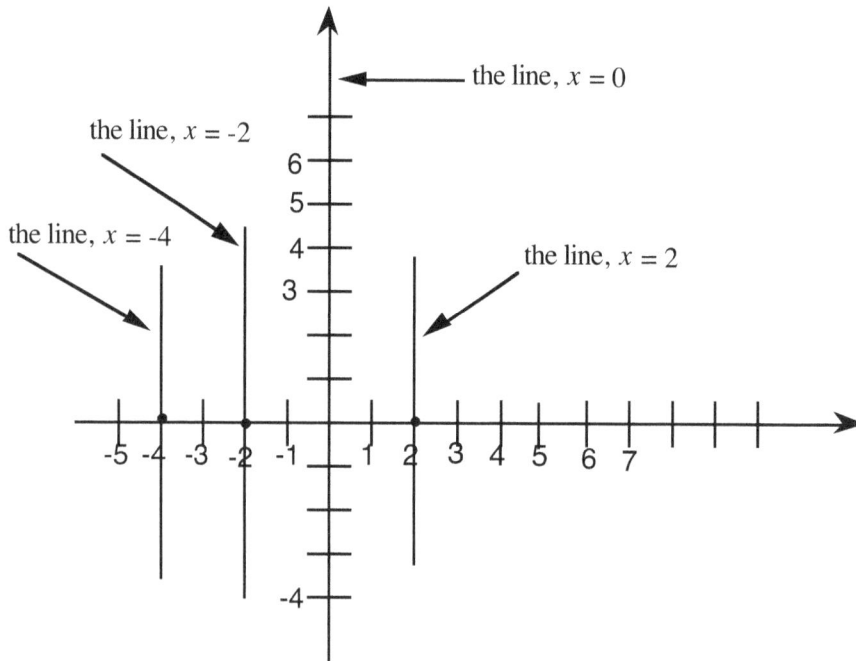

Figure 2

Lesson 36 Exercises

A 1. Find an equation of the line with slope 7 and y-intercept -2.

2. Find an equation of the line with slope $\frac{2}{3}$ and y-intercept 5.

3. A line has a y-intercept -3 and a slope of 6. Find an equation for this line.

Answers: 1. $y = 7x - 2$;　　**2.** $y = \frac{2}{3}x + 5$;　　**3.** $y = 6x - 3$

B 1. Find an equation of the line with slope -4 and passing through the point (3,-2).

2. A line passes through the point (-1,-7) and has a slope of 5. Find an equation for this line.

Answers: **1.** $y = -4x + 10$;　　**2.** $y = 5x - 2$

C　**1.** Find an equation of the line passing through the points (2,2) and (-5,6)

　2. If the points (2,-5) and (-3,1) are on a certain line, find an equation for this line.

　3. Find an equation of the line with slope -3 and y-intercept 8.

　4. Find an equation of the line with slope 2 and passing through the point (1,6)

Answers:　**1.** $y = -\frac{4}{7}x + \frac{22}{7}$;　**2.** $y = -\frac{6}{5}x - \frac{13}{5}$;　**3.** $y = -3x + 8$;　**4.** $y = 2x + 4$.

D Find the equations (slope-intercept forms) of the lines ℓ_1, ℓ_2, ℓ_3, ℓ_4

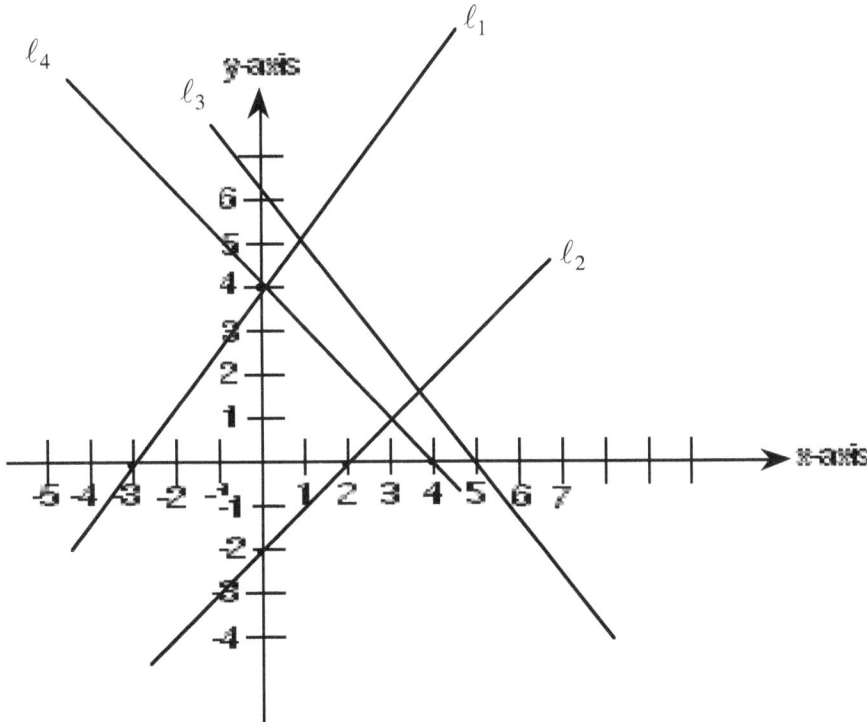

Answers: For ℓ_1: $y = \frac{4}{3}x + 4$.　For ℓ_2: $y = x - 2$.　For ℓ_3: $y = -\frac{5}{4}x + \frac{25}{4}$　For ℓ_4: $y = -x + 4$

E 1. Find an equation of the horizontal line passing through the point (3,-4).

  2. Sketch the graph the line in Problem 1.

  3. Does the line in Problem 1 have a y-intercept ? If yes, find it.

  4. Sketch the graph of the line $y = $ -5.

  5. Sketch the graph of the line $y = $ 2.

Answers: **1.** $y = - 4$; **2.** See Fig 1, above and imitate. ; **3.** Yes. y-intercept = - 4 ; **4.** & **5.** Imitate Fig. 1 above

F 1. Find an equation of the vertical line passing through the point (2,-3).

  2. Sketch the graph the line in Problem **1**.

  3. Does the line in Problem **1** have a y-intercept ? If yes, find it.

Answers: **1.** $x = 2$; **3.** No.

G 1. Draw the graph of the line $x = 4$.   **2.** Draw the graph of the line $x = $ -1.

H   **1.** Find an equation of the line with slope -5 and y-intercept 2.

  2. Find an equation of the line whose y-intercept is -3 and whose slope is $\frac{1}{2}$.

  3. Find an equation of the line whose slope is -2 and which passes through the point (4, -5).

  4. A line passes through the point (-3, 2) and has slope 4. Find an equation for this line.

  5. Find an equation of the line passing through the points (5, 4) and (8, 6).

Answers: **1.** $y = - 5x + 2$; **2.** $y = \frac{1}{2}x - 3$;   **3.** $y = - 2x + 3$;   **4.** $y = 4x + 14$;   **5.** $y = \frac{2}{3}x + \frac{2}{3}$

CHAPTER 10B

Lesson 37A: **Equations of Parallel lines and Perpendicular Lines**
Lesson 37B: **Midpoint of a Line; Distance between Points;** Distance Formula
Lesson 37C: **Applications: 1. Distance from a point to a line whose equation is given**
                                   2. Perpendicular Bisector of a Line

Lesson 37A

Parallel Lines: Slopes and Equations

Note: If two lines are parallel, then they have the same slope.

If the slope of one line is m_1, then the slope of the other parallel line is also m_1.

Example Find an equation of the line passing through the point (2,-5), and parallel to the
              line $y = -3x + 2$.

Solution
The slope of the line whose equation is given = -3. (comparing $y = -3x + 2$. with $y = mx + b$. $m = -3$)
The slope of the line whose equation we want to find = -3 (since two parallel lines have the same slope).
Applying $y - y_1 = m(x - x_1)$; with $m = -3, x_1 = 2, y_1 = -5$

$$y - (-5) = -3(x - 2)$$
$$y + 5 = -3x + 6$$
$$y = -3x + 1 \;\; \text{<------slope-intercept form.}$$

Perpendicular Lines: Slopes and Equations

If two lines are perpendicular, then their slopes are negative reciprocals of each other; or simply, if

m_1 and m_2 are their slopes, then $m_1 = -\dfrac{1}{m_2}$ or $m_2 = -\dfrac{1}{m_1}$ or $m_1 \cdot m_2 = -1$

Example   Find an equation of the line passing through the point (-2,4) and
                perpendicular to the line $y = 3x - 5$.

Solution

Step 1: Determine the slope of the line whose equation we want to find.

The slope of $y = 3x - 5$ is 3

The slope of the line whose equation we are to find is $-\dfrac{1}{3}$ (Since the two lines are perpendicular).

Step 2:   Apply $y - y_1 = m (x - x_1)$

               with $m = -\dfrac{1}{3}$ and $x_1 = -2, y_1 = 4$

               Then, $y - 4 = -\dfrac{1}{3} (x - (-2))$

$$y - 4 = -\frac{1}{3} (x + 2)$$
$$y - 4 = -\frac{1}{3} x - \frac{2}{3}$$
$$y = -\frac{1}{3} x - \frac{2}{3} + 4$$
$$y = -\frac{1}{3} x + \frac{10}{3} \;\; \text{<---------slope-intercept form.}$$

A **note** about finding the slope in the above problem:

The slope of the given line is 3

To find the slope of the other (perpendicular line) line,

Step 1. Invert 3 to obtain $\frac{1}{3}$

Step 2. Change the sign. Since $\frac{1}{3}$ has a plus sign, after the change, we obtain $-\frac{1}{3}$

Similarly, if the slope of one line were -5, the slope of the other perpendicular line would be $+\frac{1}{5}$.

If the slope of one line were $-\frac{1}{4}$, the slope of the other perpendicular line would be +4.

Summary: Invert and change the sign, or change the sign and invert.

Distinction between the point $x = a$ and the line $x = a$

We must distinguish between, for example, the point $x = 2$ and the line $x = 2$. We use a dot to locate the point $x = 2$ on a number line.

For the graph of the line $x = 2$, we draw a vertical line through the point $(2, 0)$

Figure: The graph of the point $x = 2$.

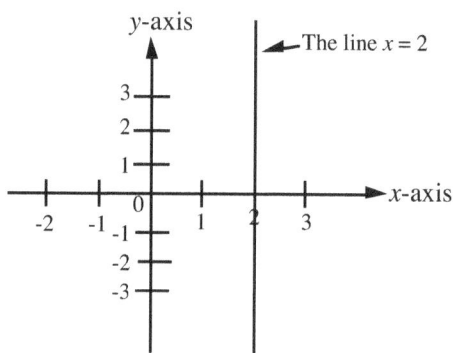

Figure: The graph of the line $x = 2$.

Similarly, the point $y = 2$ and the line (horizontal line) $y = 2$ are shown in Figures .. and respectively.

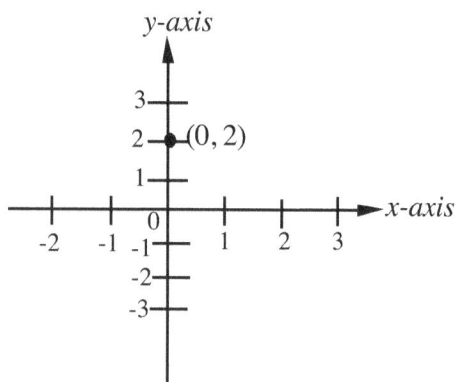

Figure: The graph of the point $y = 2$.

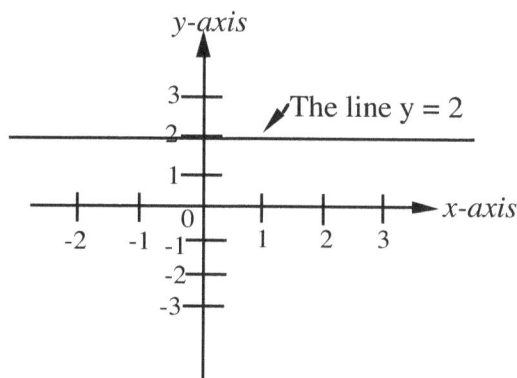

Figure: The graph of the line $y = 2$.

Lesson 37A Exercises

A 1.Find an equation of the line through the point $(3,-4)$ and perpendicular to the line $y = 2x + 5$.

2. Find an equation of the line with the same y-intercept as the line $2y = 3x + 8$ and parallel to the line $y = -4x + 2$.

Answers: **1.** $y = -\frac{1}{2}x - \frac{5}{2}$; **2.** $y = -4x + 4$

B **1. By** graphing, distinguish between the point $x = 2$ on the number line and the line $x = 2$.

2. By graphing, distinguish between the point $y = 2$ on the y-axis and the line $y = 2$.

1.

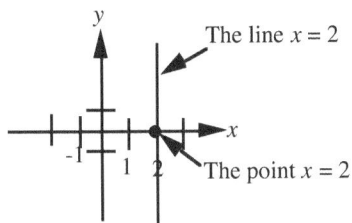

The line $x = 2$

The point $x = 2$

2.

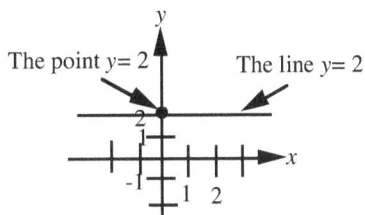

The point $y = 2$

The line $y = 2$

Lesson 37B
Midpoint of a Line; Distance between Points; Distance Formula

Midpoint of a Line

The **midpoint** of (Figure below), the line, P_1P_2, connecting the points $P_1(x_1,y_1)$ and $P_2(x_2,y_2)$ has the coordinates given by the following formulas:

The x-coordinate, x_m, of the midpoint is given by $x_m = \dfrac{x_1 + x_2}{2}$

The y-coordinate, y_m of the midpoint is given by $y_m = \dfrac{y_1 + y_2}{2}$

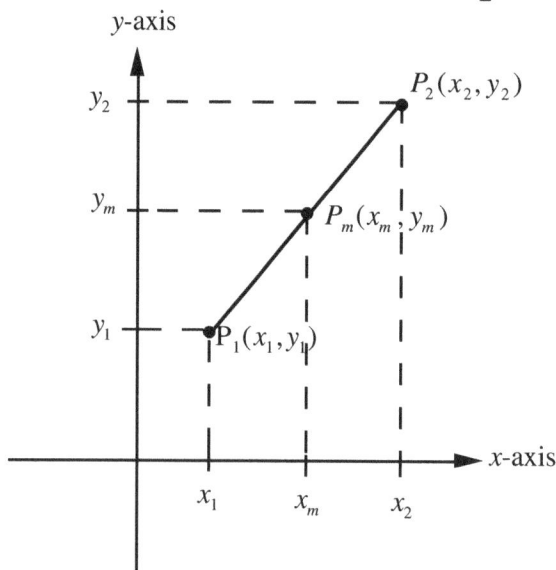

Example Find the coordinates of the mid-point of the line connecting the points $(3, 2)$ and $(-9, 4)$.

Solution

The x-coordinate of the mid-point $= \dfrac{x_1 + x_2}{2} = \dfrac{3 + (-9)}{2}$ $\qquad$ ($x_1 = 3, x_2 = -9$)

$$= \dfrac{-6}{2}$$

$$= -3$$

The y-coordinate of the mid-point $= \dfrac{y_1 + y_2}{2} = \dfrac{2 + 4}{2}$ $\qquad$ ($y_1 = 2, y_2 = 4$)

$$= \dfrac{6}{2}$$

$$= 3$$

Therefore $(x_m, y_m) = (-3, 3)$ where x_m and y_m are the coordinates of the mid-point. The mid-point is at $(-3, 3)$.

Distance Between two Points on a Line

The distance between two points A and B on a line is equal to the number of units (equal intervals) between A and B. If we denote the distance between A and B by $d(A, B)$. Then, we can write

$$d(A, B) = |A - B|$$

that is, the distance between A and B is the absolute value of the difference between A and B.

Figure: Distance on a horizontal line

Figure: Distance on a vertical line

Example

$$d(\text{-}3, 1) = |\text{-}3 - 1|$$
$$= |\text{-}4|$$
$$= 4$$

We conclude also that the distance between two points on the same **horizontal line** or the same **vertical line** can be found algebraically.

However, the distance between two points in a plane cannot be found algebraically, but must be found geometrically as discussed in the next section.

Distance Between two Points on a Line in a Plane: Distance Formula 176

The distance, d, between the points $P_1(x_1, y_1)$ and $P_2(x_2, y_2)$ on a line **in a plane** (Fig.1) is given by

$d = \sqrt{(x_2 - x_1)^2 + (y_2 - y_1)^2}$ (By applying the Pythagorean theorem to the right triangle, and solving for d)

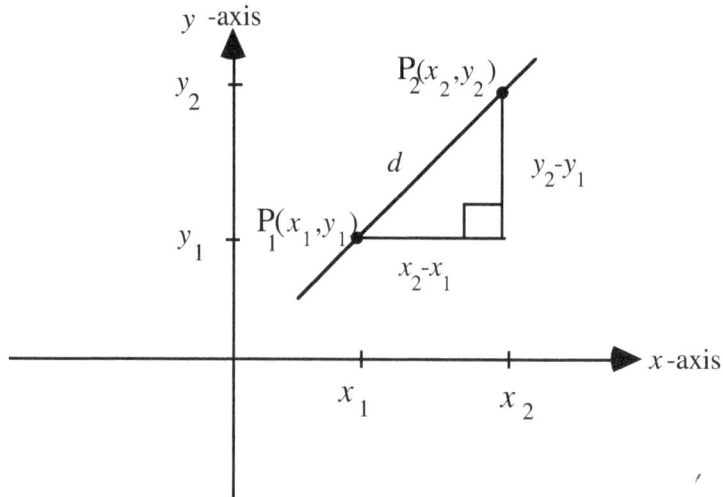

Figure 1

Example Find the distance between the points (-2,3) and (4, -5) .

Solution Apply the distance formula:

$$d = \sqrt{(x_2 - x_1)^2 + (y_2 - y_1)^2} \qquad (1)$$

(where d is the distance between the points $P_1(x_1, y_1)$ and $P_2(x_2, y_2)$).

Substituting $x_1 = -2, y_1 = 3, x_2 = 4, y_2 = -5$ in equation (1) above,

$d = \sqrt{(4 - (-2))^2 + (-5 - 3)^2}$

$d = \sqrt{(4 + 2)^2 + (-8)^2}$

$\quad = \sqrt{(6)^2 + (-8)^2}$

$\quad = \sqrt{36 + 64}$

$\quad = \sqrt{100}$

$\quad = 10$

∴ the distance between the given points is 10 units.

Lesson 37B Exercises

A   **1.** Find the distance between the points $(3,4)$ and $(5,-1)$.

   2. Find the distance between the points $(-4, 2)$ and $(-6, -3)$

Answers: **1.** $\sqrt{29}$; **2.** $\sqrt{29}$

B   Find the distance between each given pair of points.

   1. $(2, 3)$ and $(-7, 4)$;   **2.** $(-4, 3)$ and $(6, 2)$;   **3.** $(-3, 5)$ and $(7, 10)$

   4. (a, b), and (c, d);   **5.** $(-3, 5)$, and $(4, 7)$

   6. (a) Find the directed distance from $(4,-5)$ to $(4,-7)$;

      (b) Find the distance from $(4,-5)$ to $(4,-7)$

   Find the coordinates of the mid–point of each of the following points :

   7. $(2,-4)$, and $(4, 0)$;   **8.** $(3, 2)$, and $(4, 6)$.

   9. $(-4, -4)$, and $(5, 3)$;   **10.** $(5, 12)$, and $(0, 0)$.

Answers: **1.** $\sqrt{82}$; **2.** $\sqrt{101}$; **3.** $5\sqrt{5}$; **4.** $\sqrt{a^2 + b^2 + c^2 + d^2 - 2ac - 2bd}$; **5.** $\sqrt{53}$; **6.** (a) -2; (b) 2.
7. $(3, -2)$; **8.** $(\frac{7}{2}, 4)$; **9.** $(\frac{1}{2}, -\frac{1}{2})$; **10.** $(\frac{5}{2}, 6)$.

Lesson 37C
Applications
1. Distance from a point to a line whose equation is given
2. Perpendicular Bisector of a Line

Distance from a point to a line whose equation is given

Find a formula for the distance, d, from the point $A(x_0, y_0)$ to the line $y = mx + b$.

Solution
Method 1

Given: The point $A(x_0, y_0)$ and the line $y = mx + b$.

Required: To find a formula for the distance, d, between the point $A(x_0, y_0)$ and the line $y = mx + b$.

Construction: Draw the perpendicular line from $A(x_0, y_0)$ to meet the line $y = mx + b$

(line ℓ_2) at E. Also, draw $\overline{AC} \perp$ to the x-axis.

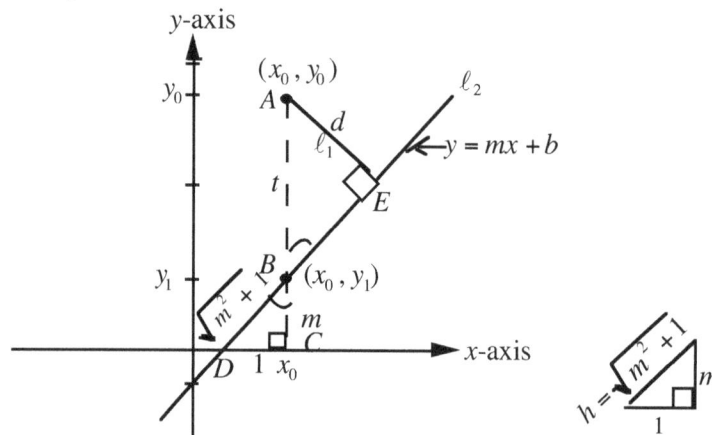

Step 1: Let $h = DB$. If the slope of $y = mx + b$ is $m = \frac{m}{1}$, then $BC = m$, $DC = 1$, and $h = \sqrt{m^2 + 1}$.

ΔABE and ΔDBC are similar: (Two angles of ΔABE are congruent to two angles of ΔDBC)

$$\therefore \frac{AE}{DC} = \frac{AB}{DB} \quad \text{(Corresponding sides of similar triangles are in proportion)}$$

$$\frac{d}{1} = \frac{t}{\sqrt{m^2 + 1}} \quad (1) \quad (AE = d, \ AB = t, \ DC = 1, \ DB = \sqrt{m^2 + 1})$$

Step 2: $t = y_0 - y_1$

$\qquad t = y_0 - (mx_0 + b) \qquad (y_1 = mx_0 + b$ is obtained by substituting x_0 in $y = mx + b$)

$\qquad t = y_0 - mx_0 - b$

Step 3: Substitute for t in equation (1)

$$\frac{d}{1} = \frac{y_0 - mx_0 - b}{\sqrt{1 + m^2}}$$

$$d = \frac{y_0 - mx_0 - b}{\sqrt{1 + m^2}}$$

$$\therefore d = \frac{|y_0 - mx_0 - b|}{\sqrt{m^2 + 1}}$$

Note: If given $Ax + By + C = 0$, $y = -\frac{A}{B}x - \frac{C}{B}$

Then $d = \dfrac{|Ax_0 + By_0 + C|}{\sqrt{A^2 + B^2}}$ by substitution.

(We take the absolute value of the numerator in case the numerator is negative)

Method 2

Given: The point $A(x_0, y_0)$ and the line $y = mx + b$.

Required: To find a formula for the distance, d, between the point $A(x_0, y_0)$ and the line $y = mx + b$.

Construction: Draw the perpendicular line from the point $A(x_0, y_0)$ to meet the line
$y = mx + b$ ℓ_2 at E.

Step 1: Let lines ℓ_1, and ℓ_2 meet at (r, s). (Figure below)

Let the distance between the points (x_0, y_0) and $(r, s) = d$.

Then $d^2 = (x_0 - r)^2 + (y_0 - r)^2$ or $d = \sqrt{(x_0 - r)^2 + (y_0 - r)^2}$

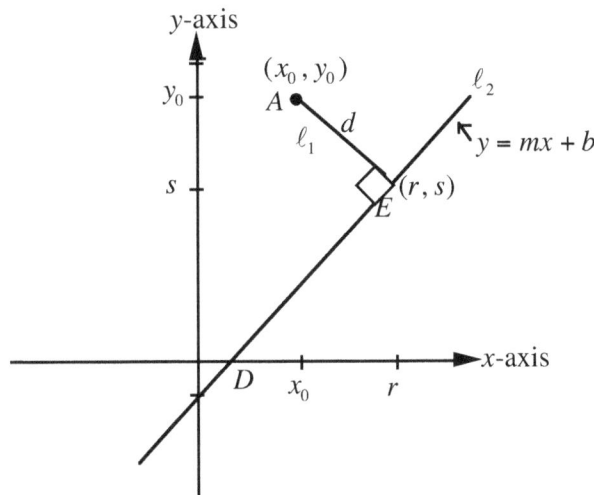

Step 2: We shall now express d in terms of x_0, y_0, m, and b only.

(That is we shall eliminate r, and s from the formula)

The slope of line $\ell_1 = \dfrac{y_0 - s}{x_0 - r}$.   (1)

The slope of line $\ell_2 = m$. (from $y = mx + b$)

Therefore slope of $\ell_1 = -\dfrac{1}{m}$ (2)   (S i n c e ℓ_1 and ℓ_2 meet at right angles at E, the slopes are negative reciprocals of each other)

Equating right-hand-sides of (1) and (2) to each other, we obtain

$\dfrac{y_0 - s}{x_0 - r} = -\dfrac{1}{m}$

$my_0 - ms = -x_0 + r$ (cross-multiplying)

$r = my_0 + x_0 - ms$ (3)   (solving for r)

Since (r, s) is on the line $y = mx + b$

$s = mr + b$ (4)   (substituting r for x and s for y.)

Step 3: We now solve equations (3) and (4) simultaneously for r and s.

Substitute for s from (4) in (3): $my_0 + x_0 - m(mr + b) = r$

Now, we solve for r: $my_0 + x_0 - m^2r - mb = r$

$my_0 + x_0 - mb = m^2r + r$

$my_0 + x_0 - mb = r(m^2 + 1)$

$$r = \frac{my_0 + x_0 - mb}{m^2 + 1} \qquad (5)$$

Substitute for r from (5) in (4). Then we obtain

$$s = m\left[\frac{my_0 + x_0 - mb}{m^2 + 1}\right] + b$$

$$= \frac{m^2 y_0 + mx_0 - m^2 b + m^2 b + b}{m^2 + 1}$$

$$s = \frac{m^2 y_0 + mx_0 + b}{m^2 + 1} \qquad (6)$$

Step 4: We now substitute right-hand sides of equations (5) and (6) for r and s respectively in the formula $d^2 = (x_0 - r)^2 + (y_0 - r)^2$. Then we obtain

$$d^2 = \left\{x_0 - \left[\frac{my_0 + x_0 - mb}{m^2 + 1}\right]\right\}^2 + \left\{y_0 - \left[\frac{m^2 y_0 + mx_0 + b}{m^2 + 1}\right]\right\}$$

$$= \frac{\{m^2 x_0 + x_0 - my_0 - x_0 + mb\}^2}{(m^2 + 1)^2} + \frac{\{m^2 y_0 + y_0 - m^2 y_0 - mx_0 - b\}^2}{(m^2 + 1)^2}$$

$$= \frac{\{m^2 x_0 - my_0 + mb\}^2}{(m^2 + 1)^2} + \frac{\{y_0 - mx_0 - b\}^2}{(m^2 + 1)^2}$$

$$= \frac{\{-m(-mx_0 + y_0 - b)\}^2}{(m^2 + 1)^2} + \frac{\{y_0 - mx_0 - b\}^2}{(m^2 + 1)^2}$$

$$= \frac{\{(-m)^2(-mx_0 + y_0 - b)\}^2}{(m^2 + 1)^2} + \frac{\{y_0 - mx_0 - b\}^2}{(m^2 + 1)^2}$$

$$= \frac{m^2(y_0 - mx_0 - b)^2}{(m^2 + 1)^2} + \frac{(y_0 - mx_0 - b)^2}{(m^2 + 1)^2}$$

$$= \frac{m^2(y_0 - mx_0 - b)^2 + (y_0 - mx_0 - b)^2}{(m^2 + 1)^2}$$

$$= \frac{(y_0 - mx_0 - b)^2(m^2 + 1)}{(m^2 + 1)^2}$$

$$= \frac{(y_0 - mx_0 - b)^2(m^2 + 1)}{(m^2 + 1)^2}$$

$$d^2 = \frac{(y_0 - mx_0 - b)^2}{(m^2 + 1)}$$

$$d = \sqrt{\frac{(y_0 - mx_0 - b)^2}{(m^2 + 1)}}$$

$$= \frac{\sqrt{(y_0 - mx_0 - b)^2}}{\sqrt{m^2 + 1}}$$

$$\therefore d = \frac{|y_0 - mx_0 - b|}{\sqrt{m^2 + 1}} \qquad \text{(noting that } \sqrt{a^2} = |a|\text{)}$$

and the derivation is complete.

Application 2: Perpendicular Bisector of a Line

Example

Find the slope-intercept form of the equation of line L_1 (Figure below) which passes through the mid-point C of line L_2 given that line L_2 passes through the two points A (1 ,4) and B (7, 8)

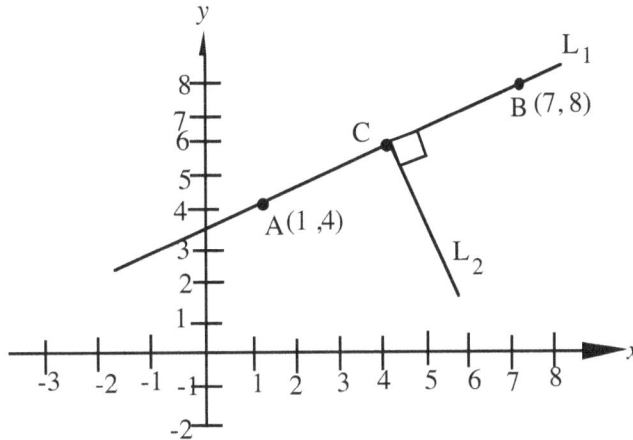

Solution

Step 1 Find the slope of line L_2

$$m = \frac{y_2 - y_1}{x_2 - x_1}$$

$$= \frac{8 - 4}{7 - 1}$$

$$= \frac{4}{6}$$

$$m = \frac{2}{3}$$

Step 2: Find the slope of L_1

Since L_1 is perpendicular to L_2, the slope of L_2 is $-\frac{3}{2}$ ($m_2 = -\frac{1}{m_1}$. The lines are perpendicular)

Step 3 : Find the coordinates of the midpoint C

$$x_m = \frac{1+7}{2} = 4 \quad y_m = \frac{4+8}{2} = 6$$

The mid-point is at (4, 6)

Step 4: now find the slope-intercept form of the equation of the line with slope $-\frac{3}{2}$ and passing through the point (4, 6).

Applying $y - y_1 = m(x - x_1)$, we obtain

$$y - 6 = -\frac{3}{2}(x - 4) \quad \text{<-------- point-slope form}$$

$$y - 6 = -\frac{3}{2}x + 6$$

$$y = -\frac{3}{2}x + 6 + 6$$

$$y = -\frac{3}{2}x + 12 \quad \text{<-----slope-intercept form}$$

Test # 10 -Student's Self-Test (**Always**, **Test yourself before you are tested**)

Attempt all questions on clean sheets of paper, **Do not write in the book** Show all necessary work.

1. Draw the graphs of the following lines whose equations are given:

    (a) $y = -2x + 5$

    (b) $3y = 18 - 3x$

    (c) $2y - 6x = -8$

2. Determine which of the points $(2, 3), (-2, 5)$, $(-2, -16)$ are on the line whose equation is given by $y = 5x - 6$.

3. Find the slope of each line passing through the given points:

    (a) $(5, 6)$ and $(3, 2)$;

    (b) $(2, -4)$ and $(3, -2)$.

4. Find the slope of each line passing through the given points:

    (a) $(1, -1)$; and $(5, 2)$.

    (b) $(-3, 3)$ and $(2, 2)$.

5. Find the slope of each line passing through the given points:

    (a) $(5, 2)$ and $(-4, 2)$;

    (b) $(-3, -6)$ and $(-3, -5)$.

6. Find the slope of the line passing through the points $(5, 9)$ and $(3, 2)$.

7. The points $(-5, 6)$ and $(2, 2)$ are on a line whose slope we want to determine. What is the slope of this line?

8. Find the slope of the line containing the points $(2, 8)$ and $(1, 6)$.

9. Find the slope of the line passing through the points $(1, -1)$, $(3, 3)$ and $(-2, -7)$.

10. Find the slope and the y-intercept of the following:
    (a) $y = -6x + 4$;   (b) $-7 + 5x = y$;
    (c) $y = -\frac{x}{2} + 8$

11. Find the slope and the y-intercept:

    (a) $4x + 8y = 32$

    (b) $5y - 6x = 21$

12. Find the slope and the y-intercept:

    (a) $y = -2x + 4$

    (b) $y = \frac{1}{5}x - 3$

13. Find the slope and the y-intercept:

    (a) $4y = 6x + 12$

    (b) $y = \frac{x}{3} + 5$

14. Find the slope and the y-intercept:

    (a) $3y + 5x = 8$

    (b) $5x + 4y = 12$

15. Find an equation of the line with slope 5 and y-intercept -3.

16. Find an equation of the line with slope $\frac{2}{3}$ and y-intercept 5.

Lesson 37C: 1. Distance from a point to a line; 2. Perpendicular Bisector of a Line

17. A line has a *y*-intercept -2 and a slope of 6. Find an equation for this line.	**18.** Find an equation of the line with slope -3 and passing through the point (5, -2).
19. A line passes through the point (-1, -7) and has a slope of 5. Find an equation for this line.	**20.** Find an equation of the line passing through the points (2, 2) and (-5, 6)
21 (a) If the points (2, -5) and (-3, 1) are on a certain line, find an equation for this line. (b) Find an equation of the line with slope -3 and *y*-intercept 8.	**22.** Find equations for lines *m* and l
23. (a) Find an equation of the line with slope 2 and passing through the point (1, 6) (b) Sketch the graph of the line $y = 2$ (c) Draw the graph of the line $x = 4$. (d) Draw the graph of the line $x = -1$.	**24.** (a) Find an equation of the horizontal line passing through the point (3, -4). (b) Find an equation of the vertical line passing through the point (2, -3). (c) Does the line in (b) have a *y*-intercept ? If yes, find it.
25. (a) Does the line in Problem 23 (b) have a *y*-intercept ? If yes, find it. **(b)** Sketch the graph of the line $y = -5$.	**Bonus:** (a) Sketch the graph of the line $y = -4$. (b) Sketch the graph of the line $y = 1$.

Answers: **2.** The point (-2, -16) ; **3.** (a) 2 ; (b) 2; **4.** (a) $\frac{3}{4}$; (b) $-\frac{1}{5}$; **5.** (a) 0 ;

(b) undefined; **6.** $\frac{7}{2}$; **7.** $-\frac{4}{7}$; **8.** 2 ; **9.** 2 ; **10.** (a) Slope:-6, *y*-intercept: 4;

(b) Slope: 5, *y*-intercept:: -7; (c) Slope: $-\frac{1}{2}$, *y*-intercept : 8 ; **11.** (a) slope: $-\frac{1}{2}$, *y*-intercept: 4;

(b) Slope: $\frac{6}{5}$, *y*-intercept: $\frac{21}{5}$; **12.** (a) Slope: -2, *y*-intercept : 4 ; (b) Slope: $\frac{1}{5}$, *y*-intercept: -3;

13. (a) Slope: $\frac{3}{2}$, *y*-intercept: 3 ; (b) Slope: $\frac{1}{3}$, *y*-intercept: 5; **14.** (a) Slope: $-\frac{5}{3}$, *y*-intercept: $\frac{8}{3}$;

(b) Slope: $-\frac{5}{4}$, *y*-intercept: 3; **15.** $y = 5x - 3$; **16.** $y = \frac{2}{3}x + 5$; **17.** $y = 6x - 2$;

18. $y = -3x + 13$; **19.** $y = 5x - 2$; **20.** $y = -\frac{4}{7}x + \frac{22}{7}$ or $4x + 7y = 22$;

21. (a) $y = -\frac{6}{5}x - \frac{13}{5}$ or $6x + 5y = -13$; (b) $y = -3x + 8$; **22.** For the line *m*: $y = -2x - 2$,

for the line l : $y = \frac{3}{2}x - 3$; **23.** (a) $y = 2x + 4$; **24.** (a) $y = -4$; (b) $x = 2$; (c) No;

25. (a) Yes. *y*-intercept: 2.

CHAPTER 11

Solving Systems of Linear Equations (Simultaneous Equations)

Lesson 38: **Addition Method; Substitution Method; Equality Method**
Lesson 39: **Graphical Method**

Lesson 38

Solving by the Addition (or Subtraction) Method

Example 1 Solve for x and y by the elimination method.

$$\begin{cases} 2x + 4y = 6 ....(A) \\ 3x + 5y = 7 ....(B) \end{cases}$$

You may read the following discussion first, or you may skip it and read it after studying the solution.

Discussion: The above system is an example of a general form of a linear system of two equations involving two variables.

We will combine (**add** or subtract) the two equations or their equivalent equations so that either the x-term or the y-term drops out (**is eliminated**). We will then have an equation of a single variable; this equation is then solved for this variable. We then substitute the value obtained in one of the original equations to find the value of the other variable. Finally, we will check the x- and y-values in the original system of equations to determine if they satisfy the system of equations. If they do, then, these values are accepted as solutions of the system of equations, otherwise they are rejected as solutions.

First, let us eliminate x (or the x-term) by adding the equivalent forms of the two equations. We would like the coefficients of the x-terms to be the same, but of opposite signs. More preferably, we want the coefficients of the x-terms to be the LCM of the coefficients of the x-terms of the original equations. We will multiply equation (A) by 3 and equation (B) by -2. Note that 3 is the coefficient of the x-term of (B) and 2 is the coefficient of the x-term of equation (A). We use a minus sign for the 2, so that the new coefficients obtained have different signs. We could also give the minus to the 3 and then, we would multiply equation (A) by -3 and equation (B) by 2. We use different signs for the "3" and "2" because the given x-terms have the same sign. If the x-terms had different signs, then the multipliers 3 and 2 must have the same sign, and then we would have multiplied equation (A) by 3 and equation (B) by 2.

Solution: Now, let us apply the above discussion to solve the given system of linear equations.

$$\begin{cases} 2x + 4y = 6 ....(A) \\ 3x + 5y = 7 ....(B) \end{cases}$$

(A) × 3: $\qquad\qquad 3(2x + 4y = 6)$
(B) × -2: $\qquad\qquad -2(3x + 5y = 7)$

$$6x + 12y = 18 .... (C)$$
$$\underline{-6x - 10y = -14 ....(D)}$$

adding (C) and (D) $\qquad 2y = 4$ $\qquad\qquad$ (The x-term has been **eliminated**)
$\qquad\qquad\qquad\qquad\qquad y = 2$

Next, we find the value of x:

Substitute $y = 2$ in (A) or (B) and solve for x.
We substitute in (A): $\qquad 2x + 4(2) = 6$
$\qquad\qquad\qquad\qquad\qquad 2x + 8 = 6$
$\qquad\qquad\qquad\qquad\quad \underline{\quad -8 \quad -8 \quad}$
$\qquad\qquad\qquad\qquad\qquad\quad 2x = -2$
$\qquad\qquad\qquad\qquad\qquad\quad x = -1$
$\qquad\qquad\quad \therefore \ x = -1, \ y = 2.$

We will check the values $x = -1$, $y = 2$ in equations (A) and (B) to determine if this ordered pair 185 (-1,2) is a solution of the system of equations.

Check #1:   For $x = -1$ and $y = 2$ in (A):

$$2(-1) + 4(2) \overset{?}{=} 6$$
$$-2 + 8 \overset{?}{=} 6$$
$$6 = 6 \quad \text{True}$$

.: The ordered pair (-1, 2) is a solution of (A).

Check #2:   For $x = -1$ and $y = 2$ in (B):

we have $3(-1) + 5(2) \overset{?}{=} 7$
$$-3 + 10 \overset{?}{=} 7$$
$$7 = 7 \quad \text{True}$$

.: The ordered pair (-1, 2) is a solution of (B).

Since the ordered pair (-1, 2) is a solution of both equation (A) and (B).
the ordered pair (-1, 2) is the solution of the given system of equations.

Let us solve the same system of equations by eliminating y first (instead of x).

$$\begin{cases} 2x + 4y = 6 .... (A) \\ 3x + 5y = 7 .... (B) \end{cases}$$

(A) × 5: $5(2x + 4y = 6)\text{---------}> 10x + 20y = 30 ....(C)$
(B) × –4: $-4(3x + 5y = 7).\text{--------}> \underline{-12x - 20y = -28} ....(D)$
(C) + (D): $-2x = 2$
$$\frac{2x}{2} = \frac{2}{-2}$$
$$x = -1$$

To find y, substitute $x = -1$ in (A)

$$2(-1) + 4y = 6$$
$$-2 + 4y = 6$$
$$\underline{+2 \qquad\quad +2}$$
$$4y = 8$$
$$y = 2$$
$$\therefore \ x = -1, \ y = 2.$$

Again, we obtain the same solution as in the case in which we eliminated x first.
The above problem was a most general example. Sometimes we may not have to multiply any of the given equations by any (nonzero constant) number, but rather, we may just add the two equations to eliminate one of the variables, as the next example shows.

Example 2   Solve for x and y: $\begin{cases} 2x + 5y = 4 .... (A) \\ 3x - 5y = 11 .... (B) \end{cases}$

We can see that the coefficients of the y-terms are the same (absolute values) but of opposite signs.

Step 1:   Add (A) and (B)   (The objective here is to combine the two equations so that
$$2x + 5y = 4 \qquad \text{either the } x\text{-term or the } y\text{-term drops out)}$$
$$\underline{3x - 5y = 11}$$
$$5x + 0 = 15$$

Step 2:   Solve for x:
$$5x = 15$$
$$x = 3$$

To find y, substitute $x = 3$ in (A) and solve for y

$$2(3) + 5y = 4$$
$$6 + 5y = 4$$
$$\underline{-6 \qquad -6}$$
$$5y = -2$$
$$\frac{5y}{5} = \frac{-2}{5}$$

Thus $x = 3$, $y = -\frac{2}{5}$

Checking in (A):   Substitute $x = 3$, $y = -\frac{2}{5}$

$$2(3) + 5\frac{(-2)}{5} \overset{?}{=} 4$$
$$6 - 2 \overset{?}{=} 4$$
$$4 = 4 \qquad \text{True}$$

Checking in (B):   Substitute $x = 3$, $y = -\frac{2}{5}$

$$3(3) - 5\frac{(-2)}{5} \overset{?}{=} 11$$
$$9 + 2 \overset{?}{=} 11$$
$$11 = 11 \qquad \text{True}$$

$\therefore$ The solution is $x = 3$ and $y = -\frac{2}{5}$

Example 3   If $2x + 5y = 4$ and $3x - 5y = 11$, then $x = ?$ and $y = ?$

The solution is the same as that of **Example 2**, above. $x = 3$ and $y = -\frac{2}{5}$.

Note: Each of the above systems of linear equations is called a consistent and independent system with a unique solution. Two other systems follow.

Special Cases of Systems of Linear Equations

System with infinitely many solutions. **Consistent but dependent system**	**System with no solution. Inconsistent System**
Example Solve for x and y $\begin{cases} x + 2y = 4 & (A) \\ 2x + 4y = 8 & (B) \end{cases}$	**Example** Solve for x and y $\begin{cases} 2x + 4y = 2 & (D) \\ x + 2y = -3 & (E) \end{cases}$
Solution $(A) \times -2: \quad -2x - 4y = -8 \qquad (C)$ $(C) + (B): \quad \underline{2x + 4y = 8} \qquad (B)$ $\qquad\qquad\qquad 0 = 0 \quad \text{True}$	**Solution** $(E) \times -2: \quad -2x - 4y = 6 \qquad (F)$ $\qquad\qquad\quad \underline{2x + 4y = 2} \qquad (D)$ $(F) + (D): \qquad 0 = 8 \quad \text{False}$
Such a result, $0 = 0$, an identity, implies that the system is consistent, and there are infinitely **many solutions**. For example, $x = 1$ in (A), $1 + 2y = 4$ and from which $y = \frac{3}{2}$. **Note**: If the graphs of (A) and (B) were drawn, they would coincide since they have the same slope **and** the same y–intercept.	Such a false statement, $0 = 8$, a contradiction, implies that the original system of equations inconsistent. and there are **no** solutions. **Note**: If the graphs of (D) and (E) were drawn, they would be parallel, since they have the same slope ($m = -\frac{1}{2}$, but different y–intercepts).

Solving by the Substitution Method

Example 1   Solve for x and y by the substitution method

$$\begin{cases} 3x + 4y = 4 ....\text{(A)} \\ x + 2y = 6 ....\text{(B)} \end{cases}$$

Step 1: Solve for one of the variables (either x or y) in equation (A) or (B).
It is easier to solve for x in equation (B) since the coefficient of x is 1.

$$x + 2y = 6$$
$$x = 6 - 2y$$

Step 2: **Substitute** $(6 - 2y)$ for x in equation (A).   (The **parentheses** enclosing the 6 - 2y are important)
(That is replace x by $(6 - 2y)$ in the other equation)
Then, $3(6 - 2y) + 4y = 4$

Step 3: Solve for y.

$$3(6 - 2y) + 4y = 4$$
$$18 - 6y + 4y = 4$$
$$18 - 2y = 4$$
$$-2y = -14$$
$$y = \frac{-14}{-2}$$
$$y = 7$$

Step 4:   To find the value of x
substitute $y = 7$ in equation (B) and solve for x.

$$x + 2y = 6$$
$$x + 2(7) = 6$$
$$x + 14 = 6$$
$$x = -8$$

Now, $x = -8, y = 7$

Step 5: To check for $x = -8, y = 7$ in equations (A) and (B).
Substitute $x = -8$ and $y = 7$ in these equations.

Check # 1:   Checking in (A)
$$3(-8) + 4(7) \overset{?}{=} 4$$
$$-24 + 28 \overset{?}{=} 4$$
$$4 = 4 \quad \text{True (left-hand side of equation equals right-hand side of equation)}$$

Therefore, the ordered pair (-8,7) is a solution of equation (A)
Check # 2: Checking in (B)

$$-8 + 2(7) \overset{?}{=} 6$$
$$-8 + 14 \overset{?}{=} 6$$
$$6 = 6 \quad \text{True}$$

Therefore, the ordered pair (-8,7) is a solution of equation (B)
The solution of the system of equations is the ordered pair (-8,7)

Note.   Solving by the substitution method: In Step 1, it is easier to solve for the variable whose coefficient is 1, **if** there is a variable with this coefficient.

Note also that if the problem does not specify which method (addition or substitution) to use, then we can use either method. Experience will dictate which method to use. It must also be remarked that there are relative merits for both methods.

Solving by the Equality Method

Example Solve for x and y:

$$\begin{cases} 2x + 5y = -11 & (1) \\ 3x - 4y = 18 & (2) \end{cases}$$

Step 1: Solve each equation for y or x.
We solve for y.

From equation (1) , $y = -\frac{2}{5}x - \frac{11}{5}$ (3)

From equation (2) , $y = \frac{3}{4}x - \frac{9}{2}$ (4)

Step 2: **Equate** the right-hand sides of equations (3) and (4) (since if $A = B$ and $A = C$ then $B = C$)

Then $-\frac{2}{5}x - \frac{11}{5} = \frac{3}{4}x - \frac{9}{2}$

Step 3: Solve for x:

$$-\frac{2(20)}{5}x - \frac{11(20)}{5} = \frac{3(20)}{4}x - \frac{9(20)}{2}$$
$$-8x - 44 = 15x - 90$$
$$46 = 23x$$
$$2 = x$$

Step 4: To find y, substitute $x = 2$ in equation (1):

Then $2(2) + 5y = -11$
$$4 + 5y = -11$$
$$5y = -15$$
$$y = -3$$

From Steps 3 and 4, the solution is the ordered pair (2, -3).

Lesson 38 Exercises

A Solve the systems of equations for x and y by the elimination method:

1. $\begin{cases} 5x + 2y = 24 \\ 15x + 3y = 6 \end{cases}$ **2..** $\begin{cases} 7x + 3y = 15 \\ 3x - 2y = 13 \end{cases}$ **3.** $\begin{cases} 2x - 2y = 9 \\ x + 2y = 3 \end{cases}$ **4.** $\begin{cases} 3x + 5y = 6 \\ 6x + 5y = 12 \end{cases}$

Answers: **1.** $x = -4, y = 22$; **2.** $x = 3, y = -2$; **3.** $x = 4, y = -\frac{1}{2}$; **4.** $x = 2, y = 0$

B Solve the systems of equations for x and y by the substitution method

1. $\begin{cases} y = 7 - 2x \\ x - 5y = 9 \end{cases}$ **2.** $\begin{cases} 4y + 3x = 4 \\ y + 3x = 4 \end{cases}$

Answers: **1.** $x = 4$, $y = -1$; **2.** $x = \frac{4}{3}$, $y = 0$

C Solve the systems for x and y:

1. $\begin{cases} 3x + 4y = 4 \\ 2x + y = 6 \end{cases}$ **2.** $\begin{cases} 2x + 5y = -11 \\ 3x - 4y = 18 \end{cases}$

Answers: **1.** (4, -2) ;   **2.** (2, -3)

Lesson 39

Graphical Method of Solving Systems of Equations

Solve for x and y graphically: $\begin{cases} y = x + 1 \\ y = -x + 5 \end{cases}$

Solution: (See Figure 1 below)

    Step 1: Refer to Lesson 53B and review how to draw the graph of a straight line.

    Step 2: Draw the graph of $y = x + 1$ in a rectangular coordinate system of axes (on graph paper).

    Step 3: On the same system of axes as in Step 2, draw the graph of $y = -x + 5$.

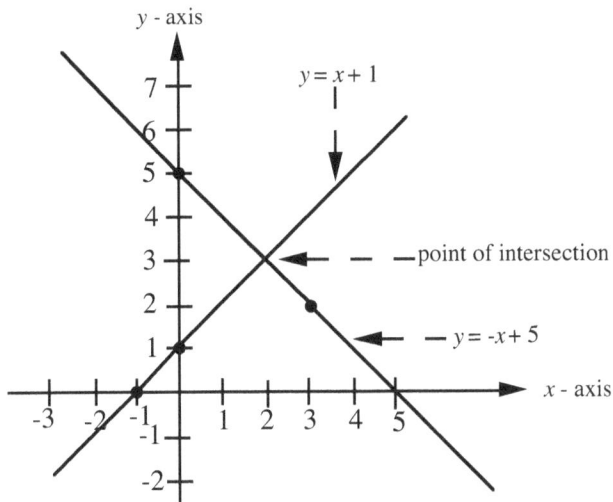

Figure 1

Step 4: From the graph, locate the point of intersection of the two lines drawn.
      Read the x- and y-coordinates of **the point of intersection** of the lines drawn.
Step 5: From the graph, the x-coordinate = 2, the y-coordinate = 3
      The solution is the ordered pair (2,3).

Note: The point of intersection is where the two lines meet. In drawing the lines, it is good practice to extend each line in both directions so that at least both the x- and y-axes are intercepted; and better, extend the lines to "fill" the graphing area.

Note also that : **1.** If the two lines are parallel (do not intersect), then there is no solution.
               2. If the two lines coincide, there are infinitely many solutions.

Lesson 39 Exercises

Solve by the graphical method

1. $\begin{cases} y = 3x - 5 \\ y = -x + 11 \end{cases}$ **2...** $\begin{cases} y = 7 - 2x \\ x - 5y = 9 \end{cases}$

Answer: 1. (4,7); 2. x=4. y = -1

CHAPTER 12

Applications of Solutions of Systems of Linear Equations

Lesson 40: **Coin Problems and Coin-Like Problems**
Lesson 41: **Distance-Time-Speed Problems: Travel by Water or by Air**

Lesson 40

Coin Problems and Coin-Like Problems

Example 1

Mary has 15 coins in dimes and nickels. The total value of the coins is $1.10. How many of each type of coin has she?

To solve the above problem, we can use either one variable (unknown) and one equation; or we can use two variables (unknowns) and two equations.
We will cover both methods.

Method 1:   Using two variables:

We need two equations. One of the equations relates or connects the number of each type of coin and the total number of coins. The other equation relates the sum of the values of the coins and the total value of the coins.

Now, back to the problem.
Let the number of dimes be d and let the number of nickels be n.

The value of d dimes is $10d$ cents. (Note that each dime is worth 10 cents)
The value of n nickels is $5n$ cents. (Each nickel is worth 5 cents)
(Examples: Value of 6 dimes is 6 x 10 = 60 cents. Value of 8 dimes is 8 x 10 = 80 cents).
Also, $1.10 x 100 = 110 cents.

We form the two equations now:
$$n + d = 15................. \quad \text{(A)}$$
$$5n + 10d = 110.................. \text{(B)}$$
Solve for n and d; by the elimination method or any method
We solve by the elimination method.
$$-5(n + d = 15)............\text{(C)}$$
$$-5n - 5d = -75...........\text{(C)}$$
$$\underline{5n + 10d = 110}...........\text{(B)}$$
Add (C) and (B): $5d = 35$
$$d = 7$$

To find n, substitute $d = 7$ in (A) and solve for n.
Then $n + 7 = 15$
$$\underline{\quad -7 \quad -7}$$
$$n = 8$$
$$d = 7$$

Check: 8 nickels are worth 8(5) = 40 cents
        7 dimes are worth 7(10) = 70 cents
                Total value = 110 cents.

Mary has 7 dimes and 8 nickels.

Method 2:   Using a single variable

Let the numbers of dimes be d
Then the number of nickels = $(15 - d)$
Value of d dimes = $10d$ cents
Value of $(15 - d)$ nickels = $5(15 - d)$ cents.

The equation: $10d + 5(15 - d) = 110$

Solve for d:
$$10d + 5(15 - d) = 110$$
$$10d + 75 - 5d = 110$$
$$5d + 75 = 110$$
$$\underline{\quad\quad -75 \quad -75}$$
$$5d = 35$$
$$d = 7$$

Number of dimes = 7
Number of nickels = $(15 - 7) = 8$ nickels
Again, we obtain the same solution as in Method 1.

Note: In using a single variable, the equation needed is a relationship between the value of each type of coin and the total value of the coins; and also, the number of one type of coin is expressed in terms of the number of the other type of coin. **In some coin problems, we are not given the total number of coins, but rather we are given how the number of one type of coin is related to (connected with) the number of another type of coin.** Example: James has 6 more nickels than dimes.

Example 2   A student has 6 more nickels than quarters. If the student has $1.20 altogether, how many of each type of coin has the student?

Method 1:   Using one variable

Let the number of quarters be q.
Then, the number of nickels is $(q + 6)$
(6 more than q means $q + 6$)
Value of the quarters = $25q$ cents.
Value of the nickels   = $5(q + 6)$

The equation: $25q + 5(q + 6) = 120$
Solve for q: $25q + 5q + 30 = 120$
$$30q + 30 = 120$$
$$\underline{\quad\quad -30 \quad -30}$$
$$30q = 90$$
$$\frac{30q}{30} = \frac{90}{30}$$
$$q = 3$$

Number of quarters = 3
Number of nickels = $(3 + 6) = 9$ nickels
Check: 3 quarters are worth $3(25) = 75$ cents
9 nickels are worth $5(9) = 45$ cents
Total value = 120 cents.
The student has 3 quarters and 9 dimes.

Method 2: **Using two variables**

Recall that we need two equations.

Let the number of quarters $= q$
Let the number of nickels $= n$

From the problem, "the number of nickels is 6 more that the number of quarters" translates to :

$n = q + 6$ . . . .(A)<------ This equation relates the number of coin types.

$5n + 25q = 120$. . . . . .(B)<------This equation relates the sum of the values
of the coins and the total value of the coins.

We solve for n and q by the substitution method.
Substitute for n from (A) in (B)

$$5(q + 6) + 25q = 120$$
$$5q + 30 + 25q = 120$$
$$30q + 30 = 120$$
$$\underline{\quad\quad\quad\quad -30 \quad\quad -30\quad}$$
$$30q = 90$$
$$q = 3$$

To find n, substitute $q = 3$ in (A)
Then $n = 3 + 6 = 9$
$$n = 9$$

Again we obtain the same solution as by Method 1 (using one variable).

Coin-Like Problems

There are problems which do not use coins as the types of items whose total values are given, but which can be solved in an identical manner as if they were coin problems.

Example   Let us call the following type " Stamp Problem". In fact, we will reword
Example 1 (Lesson 59) and replace coins by stamps.

Mary bought 15 stamps, some of which were worth 10 cents each, and the others 5 cents each. If she spent altogether $1.10 on these stamps, how many 10-cent and 5-cent stamps did she buy?

Solution   We solve this problem the same way as we did for Example 1, above,)

Let the number of 5-cent stamps $= x$
Let the number of 10-cent stamps $= y$
Then $x + y = 15$.................(A)
$5x + 10y = 110$.............(B)
Answer: $x = 8$, $y = 7$

$\therefore$ the number of 5-cent stamps Mary bought is 8 and
the number of 10-cent stamps Mary bought is 7

Another Coin-Like Problem

Example We can call this type " a ticket problem". Tickets to a concert are $5 for children and $10 for adults. If 21 tickets are sold and the total receipts are worth $165, how many children's tickets and how many adult's tickets were sold.

Let the number of children's tickets $= x$
Let the number adult's tickets $= y$.

Then, $x + y = 21$. . (A)<-----**This equation relates the number of each type of ticket to the total number of tickets.**
$5x + 10y = 165$. . .(B) <---**This equation relates the sum of the values of the tickets to the total value of the tickets.**

Solve for x and y:

Multiply (A) by -5 : $-5x - 5y = -105$. . . . . (C)
$\underline{5x + 10y = \ 165}$ . . . . (B)
Add (C) and (B) $5y = 60$
$y = 12$

To find x, substitute $y = 12$ in (A)

$x + 12 = \ 21$
$\underline{\ \ -12\ = -12}$
$x = \ 9$

Check: 9 children's tickets are worth 9(5) = $45
12 adult's tickets are worth 12(10) = $\underline{120}$
Total value $165
Therefore 9 children's tickets and 12 adult's tickets were sold.

Some Expressions for Coin and Coin-Like Problems

Example 1 James has x dimes and y quarters. How much money has James?

Solution: 1 dime is worth 10 cents
x dimes are worth $10x$ cents

1 quarter is worth 25 cents.
$\therefore$ y quarters are worth $25y$ cents.

The total amount of money James has is $(10x + 25y)$ cents

Example 2 Maria has p ten-dollar bills and r twenty-dollar bills. How much money has Maria?

Solution: Every ten-dollar bill is worth 10 dollars
$\therefore$ p ten -dollar bills are worth $10p$ dollars

Every twenty-dollar bill is worth 20 dollars
$\therefore$ r twenty-dollar bills are worth $20r$ dollars
The total amount of money Maria has is $(10p + 20r)$ dollars

Lesson 40 Exercises

A Betty has 15 coins, all in dimes and nickels, in her purse. The value of these coins is $1.05
How many of each type of coin has she?

Answer: 6 dimes and 9 nickels

B Lisa has four less dimes than nickels. The value of these coins is $1.40.
How many of each type of coin has she?

Answer: 12 nickels and 8 dimes

C 1. Peter has 9 coins in dimes and quarters. The total value of these coins is $1.35.
How many of each type of coin has he?

2. Nancy has two more dimes than nickels. If the total value of these coins is $1.10.
(*a*) How many of each type of coin has she?
(*b*) How many coins has she?

Answers: **1.** Ans: 6 dimes and 3 quarters ; **2.** (*a*) Ans: 8 dimes and 6 nickels ; (*b*) 14 coins

D 1. Tickets to a concert are $2 for children and $5 for adults. If 125 tickets are sold and the total
receipts are worth $505, how many children's tickets and how many adults' tickets were sold?

2. Maria bought 31 stamps, some of which were worth 29 cents each, and the others 40 cents each.
If she spent altogether $10.20 on these stamps, how many 29-cent and 40-cent stamps did she
buy?

3. The difference between two positive numbers is 92. The larger of the two numbers is 24 times the
smaller. Find these two numbers.

4. A ball thrown downwards from the top of a tower of height 240 feet. If the height, *s*, of this
tower is related to the time, *t*, in seconds the ball takes to strike the ground, according to the
equation $s = 32t + 16t^2$, how long does the ball take to strike the ground?

Answers:
1. # of children's tickets = 40; # of adult's tickets = 85.   **2** .# of 29-cent stamps = 20; # of 40-cent stamps = 11
3. The smaller number is 4, and the larger number is 96;   **4.** The ball takes 3 seconds to strike the ground.

Lesson 41

Distance-Time-Speed Problems: Travel by Water or by Air

The approach in solving problems involving travel on water (e.g., boat rowing or swimming) and air travel (e.g., aircraft or bird flight) is similar to the distance-time-speed problems covered already, except that we also have to consider the effect of the flow of water or air on the speed of the moving object. We introduce a new quantity, the **speed** of the **water** (river) **current** or the **wind current**. The equations derived below are based on **speed**. Also, see Lesson 17 techniques.

Boat rowing in a river (or any water body) downstream and upstream

Let S = the speed of boat (or ship or swimmer) in still (calm) water.

Let C = the speed of the river current.

Let S_N = the net speed of the boat.　　These quantities are related as follows:

$$S_N = S \pm C \qquad \text{(The net speed is either the sum or the difference of } S \text{ and } C.)$$

where we use the **plus** sign if the boat travels downstream or with the river current, and the **minus** sign if the boat travels upstream or against the river current. **Note** that traveling with the river current implies traveling in the direction of the current.

Example 1 A boat traveled 40 miles downstream a river in 4 hours and returned upstream to the starting point in 8 hours.
Find (a) the speed of the boat in still water ; (b) the speed of the river current.

Solution We will write two equations in terms of two variables and solve for the variables.
We can rewrite this word problem as two simple statements:
1. A boat traveled 40 miles downstream a river in 4 hours.
2. The boat returned upstream, traveling 40 miles to the starting point, in 8 hours.

Step 1:　Let S = the speed of boat in still (calm) water.
Let C = the speed of the river current.
Let S_N = the net speed of the boat.

Traveling downstream:　$S_N = \dfrac{40}{4}$　　$(\text{speed} = \dfrac{d}{t})$

$S_N = S + C$　(We use the **plus** sign since the boat is traveling with the river current).

$\dfrac{40}{4} = S + C$ 　　　　　(1)

Traveling upstream:　　$S_N = \dfrac{40}{8}$

$S_N = S - C$　We use the **minus** sign since the boat is traveling against the river current).

$\dfrac{40}{8} = S - C$ 　　　　　(2)

Step 2: Solve equation (1) and (2) simultaneously for S and C .

From equation (1):　$10 = S + C$　　　(3)
From equation (2):　$5 = S - C$　　　(4)
Adding (3) and (4):　$15 = 2S$

$$\dfrac{15}{2} = S$$

$$7\tfrac{1}{2} = S$$

To find C, substitute $S = 7\tfrac{1}{2}$ in equation (3) and solve for C.

Then $10 = 7\tfrac{1}{2} + C$ and from which $C = 2\tfrac{1}{2}$

(a) The speed of the boat in still water is $7\tfrac{1}{2}$ mph ; and (b) the speed of the river current is $2\tfrac{1}{2}$ mph.

Aircraft (or bird) flight with wind, and against wind

The net speed is similar to the case of boat rowing downstream or upstream.
Let S = the speed of aircraft or bird in still air.
Let C = the speed of the wind current .
Let S_N = the net speed of the aircraft or bird. These quantities are related as follows:

$$S_N = S \pm C$$

where we use the **plus** sign if the aircraft travels **with** the wind current, and the **minus** sign if the aircraft travels **against** the wind current. **Note** that traveling with the wind current implies traveling in the direction of the current.

Example 2 A bird flew 40 miles with the wind current in 4 hours and flew back against the wind current to the starting point in 8 hours.
Find (a) the speed of the bird in still air ; (b) the speed of the wind current.

Solution We will write two equations in terms of two variables and solve for the variables.
We can rewrite this word problem as two simple statements:
1. A bird flew 40 miles with the wind in 4 hours.
2. The bird flew back 40 miles against the wind to the starting point in 8 hours.

Step 1: Let S = the speed of bird in still (calm) air.
Let C = the speed of the wind current.
Let S_N = the net speed of the bird.

Flying with the wind: $S_N = \dfrac{40}{4}$ (speed $= \dfrac{d}{t}$)

$S_N = S + C$ (We use the **plus** sign since the bird is flying with the wind current).

$\dfrac{40}{4} = S + C$ (1)

Flying against the wind: $S_N = \dfrac{40}{8}$ (speed $= \dfrac{d}{t}$)

$S_N = S - C$ We use the **minus** sign since the bird is flying against the wind current).

$\dfrac{40}{8} = S - C$ (2)

Step 2: Solve equation (1) and (2) simultaneously for S and C .

From equation (1): $10 = S + C$ (3)
From equation (2): $5 = S - C$ (4)
Adding (3) and (4): $15 = 2S$

$\dfrac{15}{2} = S$

$7\frac{1}{2} = S$

To find C, substitute $S = 7\frac{1}{2}$ in equation (3) and solve for C.
Then $10 = 7\frac{1}{2} + C$ and
$C = 2\frac{1}{2}$

(a) The speed of the bird in still air is $7\frac{1}{2}$ mph ; and (b) the speed of the wind current is $2\frac{1}{2}$ mph.

Lesson 41: Distance-Time-Speed Problems: Travel by Water or by Air

Example 3 An aircraft can fly 1600 miles with the wind current in 4 hours, but against the wind current it can fly 1200 miles in 4 hours.
Find (a) the speed of the aircraft in still air ; (b) the speed of the wind current.

Solution

Step 1: Let $S =$ the speed of aircraft in still (calm) air.
Let $C =$ the speed of the wind current.
Let $S_N =$ the net speed of the aircraft.

Flying with the wind: $S_N = \dfrac{1600}{4}$ $(\text{speed} = \dfrac{d}{t})$

$\qquad\qquad S_N = S + C$ (We use the **plus** sign since the bird is flying with the wind current).

$\qquad\qquad \dfrac{1600}{4} = S + C \qquad\qquad (1)$

Flying against the wind: $S_N = \dfrac{1200}{4}$ $(\text{speed} = \dfrac{d}{t})$

$\qquad\qquad S_N = S + C$ (We use the **minus** sign since the bird is flying against the wind current).

$\qquad\qquad \dfrac{1200}{4} = S - C \qquad\qquad (2)$

Step 2: Solve equation (1) and (2) simultaneously for S and C .

$\qquad$ From equation (1): $\quad 400 = S + C \qquad (3)$
$\qquad$ From equation (2): $\quad 300 = S - C \qquad (4)$
$\qquad$ Adding (3) and (4): $\quad 700 = 2S$
$$\dfrac{700}{2} = S$$
$$350 = S$$

To find C, substitute $S = 350$ in equation (3) and solve for C
$\qquad\qquad$ Then $400 = 350 + C$ and from which
$$C = 50$$

(a) The speed of the aircraft in still air is 350 mph ; and (b) the speed of the wind current is 50 mph.

Example 4 An aircraft flew 1200 miles with the wind in the same time as it flew 800 miles against 198 the wind. If the speed of the aircraft in still air is 240 mph, find the speed of the wind.

Solution

Step 1: Let S = the speed of aircraft in still (calm) air.

Let C = the speed of the wind current.

Let S_N = the net speed of the aircraft.

Let t = the time taken to travel 1200 miles and also to travel 800 miles.

Also, $S = 240$.

Flying with the wind: $S_N = \dfrac{1200}{t}$

$S_N = S + C$ (We use the **plus** sign since the aircraft is flying with the wind current).

$$\dfrac{1200}{t} = 240 + C \qquad (1)$$

Flying against the wind: $S_N = \dfrac{800}{t}$

$S_N = S - C$. We use the **minus** sign since the aircraft is flying against the wind current).

$$\dfrac{800}{t} = 240 - C \qquad (2)$$

Step 2: Solve equation (1) and (2) simultaneously for C . (We are required to find C only and not t)

We can eliminate t by dividing equation (1) by equation (2):

$$\dfrac{1200}{t} \bullet \dfrac{t}{800} = \dfrac{240 + C}{240 - C}$$

$$\dfrac{3}{2} = \dfrac{240 + C}{240 - C} \qquad \text{(simplifying the left-hand side)}$$

$$2(240 + C) = 3(240 - C)$$

$$480 + 2C = 720 - 3C$$

$$5C = 240$$

$$C = 48$$

The speed of the wind current is 48 mph.

Note: We could have begun Step 2 by solving for t from each of equations (1) and (2) followed by equating the resulting expressions to each other.

A note about other units Example: 60 mph = 88 ft/sec (obtained from $\dfrac{60(5280)\,\text{ft}}{(60)(60)\,\text{sec}}$)

Lesson 41 Exercises

1. A boat traveled 60 miles downstream a river in 4 hours and returned upstream to the starting point in 6 hours. Find (a) the speed of the boat in still water ; (b) the speed of the river current.

2. A bird flew 60 miles with the wind current in 4 hours and flew back against the wind current to the starting point in 6 hours. Find (a) the speed of the bird in still air ; (b) the speed of the wind current.

3. An aircraft can fly 1600 miles with the wind current in 4 hours, but against the wind current it can fly 1200 miles in 4 hours. Find (a) the speed of the aircraft in still air ; (b) the speed of the wind current.

Answers: 1. (a) $12\frac{1}{2}$ mph; (b) $2\frac{1}{2}$ mph; **2.** (a) $12\frac{1}{2}$ mph; (b) $2\frac{1}{2}$ mph; **3.** (a) 350 mph; (b) 50 mph.

Test # 11 -Student's Self-Test (Always, Test yourself before you are tested)

Attempt all questions on clean sheets of paper, **Do not write in the book** Show all necessary work.

1. Solve the systems of equations for x and y by the elimination method:

$$\begin{cases} 3x + 2y = 29 \\ 8x - 3y = 19 \end{cases}$$

2. Solve the systems of equations for x and y by the elimination method:

$$\begin{cases} 3x + 5y = 6 \\ 6x + 5y = 12 \end{cases}$$

3. Solve the systems of equations for x and y by the elimination method:

$$\begin{cases} 2x - 2y = 9 \\ x + 2y = 3 \end{cases}$$

4. Solve the systems of equations for x and y

$$\begin{cases} 3x - 4 = -4y \\ 2x + y = 6 \end{cases}$$

5. Solve the systems of equations for x and y by the substitution method:

$$\begin{cases} y = 7 - 2x \\ x - 5y = 9 \end{cases}$$

6. Solve the systems of equations for x and y by the substitution method

$$\begin{cases} 4y + 3x = 4 \\ y + 3x = 4 \end{cases}$$

7. Solve the systems for x and y:

$$\begin{cases} 3x + 4y = 4 \\ 2x + y = 6 \end{cases}$$

8. Solve the systems for x and y:

$$\begin{cases} 2x - 3y = 6 \\ 5x - 7y = 11 \end{cases}$$

9. Solve by the graphical method

$$\begin{cases} y = 3x - 5 \\ y = -x + 11 \end{cases}$$

10. Betty has 10 coins, all in dimes and nickels, in her purse. The value of these coins is 60 cents. How many of each type of coin has she?

11. Lisa has three more dimes than nickels. The value of these coins is $1.05. How many of each type of coin has she?

12. Peter has 17 coins in dimes and quarters. The total value of these coins is $2.60 How many of each type of coin has he?

13. Nancy has four more dimes than nickels. If the total value of these coins is $2.80. (a) How many of each type of coin has she? (b) How many coins has she?

14. Tickets to a concert are $3 for children and $7 for adults. If 80 tickets are sold and the total receipts are worth $480, how many children's tickets and how many adults' tickets were sold?

15. Maria bought 12 stamps, some of which
were worth 35 cents each, and the others
20 cents each. If she spent altogether $3.45
on these stamps, how many 35-cent and
20-cent stamps did she buy?

16. The difference between two positive
numbers is 92. The larger of the two
numbers is 24 times the smaller.
Find these two numbers.

17. If John gives $8 to James, they would have
the same amount. If James gives John $8,
John would have three times as much as
James. How much does each have now?

18.

19.

20.

Answers: **1**. $x = 5$, $y = 7$; **2**. $x = 2$, $y = 0$; **3**. $x = 4$, $y = -\frac{1}{2}$; **4**. $x = 4$, $y = -2$;

5. $x = 4$, $y = -1$; **6**. $x = \frac{4}{3}$, $y = 0$; **7**. $x = 4$, $y = -2$; **8**. $x = -9$, $y = -8$;

9. $x = 4$, $y = 7$; **10**. Betty has 8 nickels and 2 dimes;

11. Lisa has 8 dimes and 5 nickels; **12**. Peter has 11 dimes and 6 quarters ;

13. Nancy has 20 dimes and 16 nickels. She has 36 coins ;

14. 60 adults' tickets and 20 children's tickets were sold;

15. The number of 35-cent stamps Maria bought is 7 and
the number of 20-cent stamps she bought is 5.

16. The smaller number is 4 and the larger number is 96.

17. John has $40; James has $24.

CHAPTER 13A

Exponents and Radicals

Lesson 42: **Review of Exponents and Radicals**
Lesson 43: **Definitions and Simplification of Radicals**
Lesson 44: **Addition and Multiplication of Radicals**

Lesson 42

Review of Exponents and Radicals (see also Chapter 3)

Mnemonic Examples **for the Rules of Exponents and Radicals**

1. $x^2 x^3 = x^5$

2. $\dfrac{x^7}{x^4} = x^{7-4} = x^3$

3. $\left(x^4\right)^2 = x^8$

4. $(xy)^5 = x^5 y^5$

5. $\left(\dfrac{x}{y}\right)^6 = \dfrac{x^6}{y^6}$

6. $x^{-2} = \dfrac{1}{x^2}$

7. $x^0 = 1$

8. $\dfrac{x^4}{x^7} = x^{4-7} = x^{-3} = \dfrac{1}{x^3}$

9. $9^{1/2} = \sqrt{9}$

10. $\sqrt[3]{xy} = \sqrt[3]{x}\,\sqrt[3]{y}$

11. $8^{1/3} = \sqrt[3]{8} = 2$

12. $8^{2/3} = \left(\sqrt[3]{8}\right)^2 = \sqrt[3]{8^2} = 4$

13. $\left(\sqrt[4]{8}\right)^4 = \left(\sqrt[4]{8^4}\right) = 8$

14. $\sqrt[3]{\dfrac{x}{y}} = \dfrac{\sqrt[3]{x}}{\sqrt[3]{y}}$

Extra: Show that $x^{-2} = \dfrac{1}{x^2} \cdot \dfrac{x^7}{x^3} = x^{7-4} = x^3$

Note: $x^{-2} = \dfrac{x^{-2}}{1} \cdot \dfrac{x^2}{x^2} = \dfrac{x^{-2+2}}{x^2} = \dfrac{x^0}{x^2} = \dfrac{1}{x^2}$.

Example 1 Evaluate $\quad 7^{-2}$

Step 1: Express with positive exponents.

$$7^{-2} = \dfrac{1}{7^2}$$

Step 2: Simplify. $\quad \dfrac{1}{7^2} = \dfrac{1}{49}$

Example 2 Simplify, leaving answer with only positive exponents

$$\dfrac{x^{-3} y^2 z^6}{x^2 y^{-3} z}$$

Lesson 42: Review of Exponents and Radicals

Method 1 (Strictly, using the rules of exponents)

$$\frac{x^{-3}y^2z^6}{x^2y^{-3}z} = x^{-3-2}y^{2+3}z^{6-1}$$

$$= x^{-5}y^5z^5$$

$$= \frac{y^5z^5}{x^5}$$

Method 2 Taking powers across the division bar, and changing the signs of exponents, followed by cancellation

$$\frac{x^{-3}y^2z^6}{x^2y^{-3}z} = \frac{y^2y^3\cancel{z^6}\;z^5}{x^2x^3\cancel{z}}$$

(change x^{-3} to x^3 and write it in the denominator. Change y^{-3} to y^3 and write it in the numerator)
If the exponent is positive, leave the power where it is. The changes are for the powers with negative exponents).

$$= \frac{y^5z^5}{x^5}$$

A note about cancellation in Method 2 above:
In order to apply cancellation (as done in arithmetic), the exponents of the powers involved in the cancellation must have the same sign; preferably, the exponents must be positive (or made positive by following the instructions outlined above) .

Example 3 Simplify $(4x)^0 - 4\,x^0$

Solution

$$(4x)^0 - 4x^0 = 4^0x^0 - 4x^0 \qquad\qquad \textbf{Note: } x^0 = 1,\; (4)^0 = 1$$

$$= (1)(1) - 4(1)$$

$$= 1 - 4$$
$$= -3$$

Example 4 Multiply $\left(\frac{-8a^2b}{5}\right)\left(\frac{9c}{3abc}\right)\left(\frac{-15}{27c}\right)$

Solution $\dfrac{-8(9)(-15)a^2bc}{5(3)(27)abc^2}$

$$= +\frac{8(9)(15)a^2bc}{5(3)(27)abc^2}$$

$$= \frac{8a}{3c} \qquad\qquad \text{(After canceling the common factors in the numerator and the denominator)}$$

Lesson 42 Exercises

A   Simplify the following:

1. $16^{1/2}$

2. $\sqrt[3]{x^6 y^{12}}$

3. $64^{1/3}$

4. $27^{2/3}$

5. $(\sqrt[4]{16})^3$

6. $\sqrt[3]{\dfrac{x^6}{y^{15}}}$

Answers: 1. 4 ;   **2.** $x^2 y^4$;   **3.** 4 ;   **4.** 9;   **5.** 8;   **6.** $\dfrac{x^2}{y^5}$

B Simplify the following:

1. 3^{-3}

3. $\left(\dfrac{4}{9}\right)^{-3/2}$

4. $\dfrac{x^{-4} y^3 z^2}{x^4 y^{-2} z}$

5. $(6x)^0 - 6x^0$

2. $100^{-1/2}$

6. $\left(\dfrac{-4a^3 b}{3b}\right)\left(\dfrac{9c}{2bc}\right)\left(\dfrac{-12}{18c}\right)$

7. $81^{-1/4}$

Answers: 1. $\dfrac{1}{27}$;   **2.** $\dfrac{1}{10}$;   **3.** $\dfrac{27}{8}$;   **4.** $\dfrac{y^5 z}{x^8}$;   **5.** -5;   **6.** $\dfrac{4a^3}{bc}$;   **7.** $\dfrac{1}{3}$

Lesson 43
Definitions and Simplification of Radicals

Definitions of **Rational and Irrational Numbers**

Rational number : A rational number (a fraction) is a real number which **can** be written as the ratio of two integers. The word **rational** pertains to the word **ratio.**

Examples are (a) $\dfrac{2}{3}$;   (b) $\dfrac{1}{5}$;   (c) 4   (since $4 = \dfrac{4}{1}$)

(d) 0   (since $0 = \dfrac{0}{7} = \dfrac{0}{3}$ ... or $0 = \dfrac{0}{b}$, where b is an integer and b ≠ 0)

(e) $\sqrt{4}$ (because $\sqrt{4} = 2 = \dfrac{2}{1}$)

Irrational number: An irrational number is a real number which **cannot** be written as the ratio of two integers. However, we can approximate irrational numbers as closely as we wish by rational numbers or decimals.

Examples of irrational numbers are $\sqrt{2}, \sqrt[3]{4}$, and π (pi). We can for example, approximate $\sqrt{2}$ by 1.414, and π by $\dfrac{22}{7}$ or 3.142. Except otherwise instructed, in simplifying radical expressions, we prefer to leave the irrational numbers in **radical** forms.

Radicals and Roots

Consider two real numbers r and A. We denote the **principal nth root** of A (n being a positive integer)

by $\sqrt[n]{A}$. Also $\sqrt[n]{A}$ is called a **radical**. We define the principal nth root of A as follows:

$\sqrt[n]{A} = r$ if $r^n = A$ (i.e., the nth root of A = r if r^n = A) with the following qualifications:

1. Any root of zero is zero (i.e. $\sqrt[n]{0} = 0$)
2. If A is a positive number, then the principal nth root of A is the positive nth root of A.
3. If A is a negative number, then the principal nth root of A is the negative nth root of A.

Note from above that a **radical** is an expression used for indicating the root of a number.

We call n the index of the radical (n indicating the type of root being considered); A is called the radicand (the number of which a root is being taken). The radicand is thus the expression under the symbol " $\sqrt[n]{}$ " . Thus, the radical consists of the index n with root symbol " $\sqrt[n]{}$ " and the radicand.

If $n = 2$, we obtain $\sqrt[2]{A}$, the square root of A, We usually omit the "2" and write $\sqrt{A}$.

Therefore, $\sqrt[2]{A}$ = $\sqrt{A}$. However, if explicit indication helps you to understand a problem, write the 2. When $n = 3$, we obtain $\sqrt[3]{A}$, the cube root A, but in this case, we have to write the "3". Similarly, for $n = 4$ and higher indices, we must always write the index. Some higher orders of the root of A are $\sqrt[4]{A}, \sqrt[5]{A}$, and $\sqrt[8]{A}$.

Some specific radicals are $\sqrt{2}, \sqrt[3]{4}$,$\sqrt{10}$, and $\sqrt{5}$.

Example 1 Name the parts of the radical $\sqrt{32}$

Solution   The index is understood to be 2, since, $\sqrt{32} = \sqrt[2]{32}$
           The radicand is 32..

Example 2   Name the parts of the radical $\sqrt[3]{16}$

Solution   The index is 3
             The radicand is 16

Square Roots

Example $\sqrt{9}$ or $\sqrt[2]{9}$

This radical is read as the square root of 9. The square root of 9 is 3 (because $3^2 = 9$).

The following definition is useful in finding the **square root** of a number:

The **principal square root** of a number (nonzero number) is one of the two equal positive factors of that number.

Thus, if we can "break up" a number into two equal positive factors, then, one of the positive factors is the square root.(Note that we exclude negative roots).

Square root of zero: $\sqrt{0} = 0$ (that is, the square root of zero is zero).
                        because $0^2 = 0$

Examples   (a) $\sqrt{9} = \sqrt{(3)(3)} = 3$;   (b) $\sqrt{64} = \sqrt{(8)(8)} = 8$ (one of (8)(8) is 8)

Cube Roots

Example $\sqrt[3]{8}$

The above radical consists of the index 3, the radical sign and the radicand 8.

This radical is read " the cube root of 8". The cube root of 8 is 2. (because $2^3 = 8$)

The following definition is also useful in finding the cube root of a number:

The **cube root** of a number is **one of the three equal** factors of that number. Here, we do not specify positive root, since the cube root may be positive or negative, depending on whether the given number is positive or negative.

Examples (a) $\sqrt[3]{8} = \sqrt[3]{(2)(2)(2)} = 2$ **(b)** $\sqrt[3]{-8} = \sqrt[3]{(-2)(-2)(-2)} = -2$

                Check: $2^3 = 8$ Check: $(-2)^3 = -8$

Note: For **even roots** such as the square root of a , $(\sqrt{a})$, the fourth root of a, $(\sqrt[4]{a})$, the sixth

root of a, $(\sqrt[6]{a})$, $\sqrt[n]{a}$ implies the positive root.

For **odd roots** such as the cube root $(\sqrt[3]{a})$, the fifth root $(\sqrt[5]{a})$, or the seventh root $(\sqrt[7]{a})$, the root may be positive or negative according to whether the given number is positive or negative.

More Examples

(a) $\sqrt[4]{16} = \textbf{2}$ or $\sqrt[4]{16} = \sqrt[4]{(2)(2)(2)(2)} = \textbf{2}$
      because $2^4 = 16$

(b) $\sqrt[5]{32} = 2$ or $\sqrt[5]{32} = \sqrt[5]{(2)(2)(2)(2)(2)} = \textbf{2}$
     because $2^5 = 32$

(c) $\sqrt[5]{-32} = -2$ or $\sqrt[5]{-32} = \sqrt[5]{(-2)(-2)(-2)(-2)(-2)} = \textbf{-2}$
     because $(-2)^5 = -32$

Simplifying Radicals

Two approaches are considered: One approach depends on guessing correctly a perfect power which is a factor of the radicand. The other approach is the application of prime factorization, and the grouping of the factors into n equal factors: for example, for square root, we would like to have two equal **positive** factors. For cube root, we would like to have three equal factors.

Example Simplify the following: **(a)** $\sqrt{32}$; **(b)** $\sqrt{\frac{49}{64}}$; **(c)** $\sqrt[3]{80}$.

Solutions

(a) Method 1

$$\sqrt{32} = \sqrt{(16)(2)}$$
$$= \sqrt{16}\sqrt{2}$$
$$= 4\sqrt{2} \quad (\sqrt{16} \text{ is } \textbf{rational} \text{ but } \sqrt{2} \text{ is } \textbf{irrational})$$

Method 2

$$\sqrt{32} = \sqrt{(2)(2)(2)(2)(2)}$$
$$= \sqrt{(4)(4)(2)} \quad (\textbf{Rule}: \sqrt[n]{ab} = \sqrt{a}\sqrt{b})$$
$$= 4\sqrt{2}$$

(b) $\sqrt{\frac{49}{64}} = \frac{\sqrt{49}}{\sqrt{64}} = \frac{7}{8}$ $\quad \left(\textbf{Rule:} \sqrt[n]{\frac{a}{b}} = \frac{\sqrt[n]{a}}{\sqrt[n]{b}}\right)$

(c)
$$\sqrt[3]{80} = \sqrt[3]{(2)(2)(2)(2)(5)}$$
$$= \sqrt[3]{(2)(2)(2)}\,\sqrt[3]{10}$$
$$= \sqrt[3]{8}\,\sqrt[3]{10}$$
$$= 2\sqrt[3]{10} \quad \text{(We leave the product of the "extra" 2 and the 5 under the radical sign as 10)}$$

Scrapwork:

$80 = 2 \times 2 \times 2 \times 2 \times 5$

Simplify the following: **1.** $\sqrt{x^3y^2}$; **2.** $\sqrt{27}$; **3.** $\sqrt{18x^3y^4}$; **4.** $\sqrt{64x^7y^{14}z^3}$; **5.** $\sqrt{\frac{8}{9}}$

Solutions

1. $\sqrt{x^3y^2} = \sqrt{x^2y^2x} = xy\sqrt{x}$

2. $\sqrt{27} = \sqrt{(9)(3)} = \sqrt{9}\sqrt{3} = 3\sqrt{3}$

3. $\sqrt{18x^3y^4} = \sqrt{(9)(2)x^2xy^4} = \sqrt{9x^2y^4 \cdot 2x} = 3xy^2\sqrt{2x}$

4. $\sqrt{64x^7y^{14}z^3} = \sqrt{64x^6y^{14}z^2 \cdot xz} = 8x^3y^7z\sqrt{xz}$;

5. $\sqrt{\frac{8}{9}} = \frac{\sqrt{8}}{\sqrt{9}} = \frac{2\sqrt{2}}{3}$

Lesson 43 Exercises

A Simplify:

1. $\sqrt{18}$;   2. $\sqrt{50}$;   3. $\sqrt{24}$;   4. $\sqrt{72}$;   5. $\sqrt{288}$;   6. $\sqrt{48}$;   7. $\sqrt{\dfrac{81}{100}}$

Answers: **1.** $3\sqrt{2}$;   **2.** $5\sqrt{2}$;   **3.** $2\sqrt{6}$;   **4.** $6\sqrt{2}$;   **5.** $12\sqrt{2}$;   **6.** $4\sqrt{3}$;   **7.** $\dfrac{9}{10}$

B Simplify the following: 1. $\sqrt{16x^6}$ 2. $\sqrt{9x^8y^2}$ 3. $\sqrt{27x^9y^4}$

4. $\sqrt{18x^3y^5}$;   5. $\sqrt{32x^8y}$;   6. $\sqrt{8x^4y^2z^3}$

Answers: **1.** $4x^3$;   **2.** $3x^4y$;   **3.** $3x^4y^2\sqrt{3x}$;   **4.** $3xy^2\sqrt{2xy}$;   **5.** $4x^4\sqrt{2y}$;   **6.** $2x^2yz\sqrt{2z}$

Lesson 44
Addition and Multiplication of Radicals

Addition of Radicals (Like Radicals or Similar Radicals)

Like radicals have the same index **and** the same radicand.

Examples: **1.** $4\sqrt{3}$ and $\sqrt{3}$ are like radicals
(index = 2 , the square root index; radicand = 3).

   2. $5\sqrt[3]{4}$ and $2\sqrt[3]{4}$ are like radicals
(index = 3, radicand = 4).

Unlike radicals: Unlike radicals have different indices **or** radicands or both.

Examples   **1.** $\sqrt{3}$ and $\sqrt{2}$ are unlike radicals. (They have different radicands.)

   2. $\sqrt[3]{2}$ and $\sqrt[4]{2}$ are unlike radicals (They have different indices.)

Like radicals can be combined into a single radical (i.e., added or subtracted in much the same
way as like terms of polynomials are added or subtracted).

Unlike radicals cannot be combined into a single radical (i.e., cannot be added). However, in
some cases, by simplifying the given radical(s), like radicals may be obtained, and which then may be added.

To **add like radicals**, add the non-radical parts (coefficients) and keep the radical part.

Example 1   Add: $3\sqrt{11} + 5\sqrt{11}$ (same index = 2; same radicand =11)

Solution $3\sqrt{11} + 5\sqrt{11}$
$= (3 + 5)\sqrt{11}$
$= 8\sqrt{11}$

Note that $3\sqrt{11}$ means 3 times $\sqrt{11}$; but $\sqrt[3]{11}$ means the cube root of 11.

Example 2   Simplify : $3\sqrt{19} + 8\sqrt{10} + 6\sqrt{19} - 2\sqrt{10}$

Solution $3\sqrt{19} + 8\sqrt{10} + 6\sqrt{19} - 2\sqrt{10}$

$= 3\sqrt{19} + 6\sqrt{19} + 8\sqrt{10} - 2\sqrt{10}$ <-------- You may skip this step.

$= (3 + 6)\sqrt{19} + (8 - 2)\sqrt{10}$
$= 9\sqrt{19} + 6\sqrt{10}$

Example 3 Add : $\sqrt{50}   + \sqrt{72}$

Solution Step 1: Simplify the radicals first to see if there are any like radicals

$\sqrt{50} = \sqrt{(25)(2)} =$ or $\sqrt{50} = \sqrt{(5)(5)(2)}$
$\sqrt{50} = 5\sqrt{2}$ $= 5\sqrt{2}$

$\sqrt{72} = \sqrt{(36)(2)}$ or $=\sqrt{(3)(3)(2)(2)(2)}$
$\sqrt{72} = 6\sqrt{2}$ $= (3)(2)\sqrt{2} = 6\sqrt{2}$

Step 2: Now, we have like radicals and therefore, we can add.

$\sqrt{50} + \sqrt{72}$
$= 5\sqrt{2} + 6\sqrt{2}$
$= (5 + 6)\sqrt{2}$
$= 11\sqrt{2}$

In the future, Step 1 could be considered as scrapwork and show only Step 2.

Example 4   Simplify: $2\sqrt{75} - 4\sqrt{27}$

Solution

$$= 2\sqrt{75} - 4\sqrt{27}$$
$$= 2\sqrt{(25)(3)} - 4\sqrt{(9)(3)}$$
$$= 2(5)\sqrt{3} - 4(3)\sqrt{3}$$
$$= 10\sqrt{3} - 12\sqrt{3}$$
$$= (10 - 12)\sqrt{3}$$
$$= -2\sqrt{3}$$

Example 5 Simplify: $4x\sqrt{25y} - 6x\sqrt{16y}$.

Solution

$$4x\sqrt{25y} - 6x\sqrt{16y}$$
$$= 4x(5)\sqrt{y} - 6x(4)\sqrt{y} \qquad \text{Scrapwork: } \sqrt{25} = 5; \sqrt{16} = 4$$
$$= 20x\sqrt{y} - 24x\sqrt{y}$$
$$= (20x - 24x)\sqrt{y} \qquad \text{(Adding the coefficients of } \sqrt{y}.)$$
$$= -4x\sqrt{y}$$

Multiplication of Radicals

Compared to the conditions in the addition of radicals, the only condition here is that the radicals to be multiplied must have the **same index**. Thus, the radicands may be different or the same, but the index must be the same. Note however that radicals with different indices can always be changed to radicals with a common index by using fractional exponents (see p. 216, Examples 6 & 7, bottom of page) and then multiplying the resulting radicals (The common index will be the LCM of the different indices.)

Example 1   Multiply $\sqrt{3}$ and $\sqrt{5}$

Solution

$$(\sqrt{3})(\sqrt{5}$$
$$= \sqrt{(3)(5)}$$
$$= \sqrt{15}$$

Example 2 Multiply $7\sqrt{3}$ and $9\sqrt{5}$

Solution

Procedure: First multiply the non-radical parts (coefficients) and then multiply the radical parts and simplify if possible.

$$(7\sqrt{3})(9\sqrt{5})$$
$$= (7)(9)\sqrt{(3)(5)}$$
$$= 63\sqrt{15}$$

Example 3   Multiply $4\sqrt{3}$ and $5\sqrt{3}$

Solution

$$(4\sqrt{3})(5\sqrt{3}) = (4)(5)\sqrt{(3)(3)}$$
$$= 20\sqrt{9}$$
$$= 20(3)$$
$$= 60$$

Example 4 Find the product of $2\sqrt{3}$ and $5\sqrt{2} - 4\sqrt{3}$

Solution $2\sqrt{3}\ (5\sqrt{2} - 4\sqrt{3})$<----------The parentheses are important.

$= (2\sqrt{3})(5\sqrt{2}) - (2\sqrt{3})(4\sqrt{3})$ <-------Application of the distributive rule.

$= (2)(5)\sqrt{(3)(2)} - (2)(4)\sqrt{(3)(3)}$

$= 10\sqrt{6} - 8\sqrt{9}$

$= 10\sqrt{6} - 8(3)$

$= 10\sqrt{6} - 24$

$= -24 + 10\sqrt{6}$

Note: In Example 4, you may skip lines 2 and 3.

Example 5 Simplify: $\sqrt{8}\sqrt{24}$
Method 1

$\sqrt{8}\sqrt{24} = \sqrt{(8)(24)}$

$= \sqrt{(2)(2)(2)(3)(2)(2)(2)}$

$= \sqrt{(2)(2)(2)\ (2)(2)(2)\ (3)}$

$= \sqrt{(8)(8)3}$

$= 8\sqrt{3}$

Method 2 $\sqrt{8}\sqrt{24}$

Step 1:   Simplify each radical first.

$\sqrt{8}\sqrt{24}$

$= (2\sqrt{2})(2\sqrt{6})$

Scrapwork:
1. $\sqrt{8} = 2\sqrt{2}$
2. $\sqrt{24} = 2\sqrt{6}$

Step 2: Multiply the radicals.

$= (2)(2)(\sqrt{2})(\sqrt{6})$

$= 4\sqrt{12}$

$= 4\sqrt{4}\sqrt{3}$

$= 4(2)\sqrt{3}$

$= 8\sqrt{3}$

Example 6 $(\sqrt{5})(\sqrt{5}) = \sqrt{25} = 5$

Example 7 $(\sqrt{6})(\sqrt{6}) = \sqrt{36} = 6$

Lesson 44 Exercises

A Add or subtract:

1. $2\sqrt{3} + 4\sqrt{3}$ **2.** $3\sqrt{5} + \sqrt{98} + \sqrt{20} + 8\sqrt{2}$;   **3.** $4\sqrt{48} - \sqrt{12} + \sqrt{75}$

4. $2x\sqrt{9y} - 5x\sqrt{4y}$

Answers: **1.** $6\sqrt{3}$;   **2.** $5\sqrt{5} + 15\sqrt{2}$;   **3.** $19\sqrt{3}$;   **4.** $-4x\sqrt{y}$

B Multiply and simplify:

1. $2\sqrt{5}$ and $3\sqrt{2}$;   **2.** $3\sqrt{5}$ and $4\sqrt{10}$;   **3.** $4\sqrt{2}$ and $5\sqrt{2}$

4. $4\sqrt{3}$ and $4\sqrt{3}$;   **5.** $2x\sqrt{3y}$ and $\sqrt{6y}$

Answers: **1.** $6\sqrt{10}$;   **2.** $60\sqrt{2}$;   **3.** 40;   **4.** 48;   **5.** $6xy\sqrt{2}$

CHAPTER 13B
More Radicals

Lesson 45: **Division of Radicals; Rationalization of Denominators**
Lesson 46: **Fractional Exponents and Reduction of Indices**

Lesson 45
Division of Radicals; Rationalization of Denominators,

Division of a Radical by a Rational Number

Example   Simplify: $\dfrac{6 - \sqrt{12}}{2}$

Solution $\dfrac{6 - \sqrt{12}}{2}$

$= \dfrac{6}{2} - \dfrac{\sqrt{12}}{2}$

$= 3 - \dfrac{\overset{1}{\cancel{2}}\sqrt{3}}{\underset{1}{\cancel{2}}}$

$= 3 - \sqrt{3}$

Scrapwork: $\sqrt{12} = \sqrt{(2)(2)(3)} = = 2\sqrt{3}$

Rationalization of Denominators

To rationalize a denominator, we change a given "fraction" to an "equivalent fraction" so that there are no radicals in the denominator. The equivalent fraction may have radicals in the numerator (but not in the denominator).

Example 1 Rationalize the denominator: $\dfrac{3}{\sqrt{6}}$

We will multiply both the denominator and the numerator by a radical such that the radicand in the denominator becomes a perfect square.

We will multiply both the denominator and the numerator by $\sqrt{6}$

$\dfrac{3}{\sqrt{6}} = \dfrac{3}{\sqrt{6}} \dfrac{\sqrt{6}}{\sqrt{6}}$

$= \dfrac{3\sqrt{6}}{\sqrt{36}}$

$= \dfrac{3\sqrt{6}}{6}$

$= \dfrac{\overset{1}{\cancel{3}}\sqrt{6}}{\underset{2}{\cancel{6}}}$

$= \dfrac{\sqrt{6}}{2}$

A motivation for rationalizing denominators:

Note above that in practical applications, $\dfrac{\sqrt{6}}{2}$ is more convenient to use than $\dfrac{3}{\sqrt{6}}$; for example, it is easier to divide the decimal approximation of $\sqrt{6}$ by 2 than to divide 3 by the decimal approximation of $\sqrt{6}$.

Example 2   Rationalize the denominator: $\sqrt{\dfrac{7}{3}}$ (one-term denominator)

Solution   We shall multiply both the denominator and the numerator by $\sqrt{3}$.

$$\sqrt{\frac{7}{3}} = \sqrt{\frac{7\cdot3}{3\cdot3}} \quad \text{or} \quad \frac{\sqrt{7}}{\sqrt{3}}\cdot\frac{\sqrt{3}}{\sqrt{3}}$$

$$= \frac{\sqrt{21}}{\sqrt{9}} \quad \text{or} \quad \frac{\sqrt{21}}{\sqrt{9}}$$

$$= \frac{\sqrt{21}}{3} \quad \text{or} \quad \frac{\sqrt{21}}{3}$$

Example 3   Rationalize the denominator : $\sqrt[3]{\dfrac{5}{4}}$

Solution

$$\sqrt[3]{\frac{5}{4}} = \sqrt[3]{\frac{5}{2\cdot2}}$$

$$= \sqrt[3]{\frac{5\cdot2}{2\cdot2\cdot2}} \;\text{<-------One more 2 will make the denominator a perfect cube.}$$

$$= \frac{\sqrt[3]{10}}{\sqrt[3]{8}}$$

$$= \frac{\sqrt[3]{10}}{2}.$$

Example 4 Rationalize the denominator: $\sqrt[3]{\dfrac{8}{9}}$

Method 1 $\sqrt[3]{\dfrac{8}{9}} = \sqrt[3]{\dfrac{(2)(2)(2)}{(3)(3)}}$

$$= \sqrt[3]{\frac{(2)(2)(2)(3)}{(3)(3)(3)}} \;\text{<--- one more 3 will make the denominator a perfect cube}$$

$$= \frac{2\sqrt[3]{3}}{3}$$

Method 2 $\sqrt[3]{\dfrac{8}{9}} = \dfrac{\sqrt[3]{8}}{\sqrt[3]{9}}$

$$= \frac{2}{\sqrt[3]{9}} \qquad (\sqrt[3]{8}=2)$$

$$= \frac{2\sqrt[3]{3}}{\sqrt[3]{9}\,\sqrt[3]{3}}$$

$$= \frac{2\sqrt[3]{3}}{\sqrt[3]{27}}$$

$$= \frac{2\sqrt[3]{3}}{3}$$

Lesson 45: Division of Radicals; Rationalization of Denominators

Method 3 In the following approach, simplifying the numerator becomes more involved than Methods 1 & 2: 2 1 4

$$\sqrt[3]{\frac{8}{9}} = \frac{\sqrt[3]{8}}{\sqrt[3]{9}} \cdot \frac{\sqrt[3]{9}(\sqrt[3]{9})}{\sqrt[3]{9}(\sqrt[3]{9})} = \frac{\sqrt[3]{8\cdot9\cdot9}}{\sqrt[3]{9\cdot9\cdot9}} = \frac{\sqrt[3]{8\cdot9\cdot9}}{9}$$

$$= \frac{\sqrt[3]{2\cdot2\cdot2\cdot3\cdot3\cdot3\cdot3}}{9} = \frac{2\cdot3\sqrt[3]{3}}{9} = \frac{2\sqrt[3]{3}}{3}$$

Example 5 Rationalize the denominator: $\dfrac{\sqrt{5}}{\sqrt[3]{2}}$

Solution $\dfrac{\sqrt{5}}{\sqrt[3]{2}} = \dfrac{\sqrt{5}}{\sqrt[3]{2}} \cdot \dfrac{\sqrt[3]{2}(\sqrt[3]{2})}{\sqrt[3]{2}(\sqrt[3]{2})} = \dfrac{\sqrt{5}\cdot\sqrt[3]{4}}{2}$ or $\dfrac{\sqrt[6]{2000}}{2}$ (See also page 216, Examples 6 & 7, bottom of page)

Definition : The **conjugate** of a given binomial is another binomial that differs from the given binomial only in the sign of one of the terms. The conjugate of $a + b$ is $a - b$; and the conjugate of $a - b$ is $a + b$.

Example 6 Rationalize the denominator $\dfrac{5}{\sqrt{3} + \sqrt{2}}$ (two-term denominator).

Procedure: Multiply both the denominator and the denominator by the conjugate of $\sqrt{3} + \sqrt{2}$.

The conjugate of $\sqrt{3} + \sqrt{2}$ is $\sqrt{3} - \sqrt{2}.$ (You may also multiply by $-\sqrt{3} + \sqrt{2}$; try it later on.)

Step 1: $\dfrac{5}{\sqrt{3} + \sqrt{2}} = \dfrac{5\ (\sqrt{3} - \sqrt{2})}{(\sqrt{3} + \sqrt{2})(\sqrt{3} - \sqrt{2})}$

Step 2: $= \dfrac{5(\sqrt{3} - \sqrt{2})}{(\sqrt{3})(\sqrt{3}) - (\sqrt{3})(\sqrt{2}) + (\sqrt{2})(\sqrt{3}) - (\sqrt{2})(\sqrt{2})}$

$= \dfrac{5(\sqrt{3} - \sqrt{2})}{\sqrt{9} + 0 - \sqrt{4}}.$ Note: $(-\sqrt{3})(\sqrt{2}) + (\sqrt{2})(\sqrt{3}) = 0$

$= \dfrac{5(\sqrt{3} - \sqrt{2})}{3 - 2}$

$= \dfrac{5(\sqrt{3} - \sqrt{2})}{1}$

$= 5(\sqrt{3} - \sqrt{2})$ or $5\sqrt{3} - 5\sqrt{2}$

Note above: As was suggested in the procedure, you could also have multiplied by $-\sqrt{3} + \sqrt{2}$ and obtained the same result. Therefore, it is not critical which of the terms differ in sign.

Example 7 Rationalize the denominator : $\dfrac{1}{\sqrt{8} - 2}$

Procedure: Multiply both the denominator and the numerator by the conjugate of $\sqrt{8}$ - 2.

The conjugate of $\sqrt{8}$ - 2 is $\sqrt{8} + 2$ (You may also multiply by $-\sqrt{8}$ - 2; try it later on.)

$$\frac{1}{\sqrt{8} - 2} = \frac{1}{(\sqrt{8} - 2)} \frac{(\sqrt{8} + 2)}{(\sqrt{8} + 2)}$$

$$= \frac{\sqrt{8} + 2}{(\sqrt{8})(\sqrt{8}) + 2\sqrt{8} - 2\sqrt{8} - 4} \qquad \text{<----After some practice, you may skip writing this step.}$$

$$= \frac{\sqrt{8} + 2}{\sqrt{64} + 0 - 4} \qquad (2\sqrt{8} - 2\sqrt{8} = 0)$$

$$= \frac{\sqrt{8} + 2}{8 - 4}$$

$$= \frac{2\sqrt{2} + 2}{4}$$

$$= \frac{2(\sqrt{2} + 1)}{4}$$

$$= \frac{\sqrt{2} + 1}{2} \quad \text{or} \quad \frac{1}{2} + \frac{\sqrt{2}}{2}$$

Note above that $\dfrac{1}{\sqrt{8} - 2} = \dfrac{1}{-2 + \sqrt{8}}$

* Perhaps, it would be better to say "a" conjugate instead of "the" conjugate since it does not matter which terms differ in sign, so far as the rationalization of denominators is concerned.

Lesson 45 Exercises

A 1. Simplify: $\dfrac{8 + \sqrt{48}}{4}$ **2.** $\dfrac{12 - \sqrt{18}}{3}$

Answers: **1.** $2 + \sqrt{3}$;   **2.** $4 - \sqrt{2}$

B Rationalize the denominators:

1. $\dfrac{1}{\sqrt{3}}$;   **2.** $\dfrac{2}{\sqrt{5}}$;   **3.** $\dfrac{5}{\sqrt{2}}$;   **4.** $\dfrac{6}{\sqrt{5} - \sqrt{3}}$;   **5.** $\dfrac{\sqrt{3}}{4 + \sqrt{3}}$;

6. $\sqrt[3]{\dfrac{4}{9}}$;   **7.** $\sqrt[3]{\dfrac{8}{9}}$.   **8.** Is $\dfrac{\sqrt{3}}{\sqrt{2}} = \dfrac{3}{\sqrt{6}}$?

Answers: **1.** $\dfrac{\sqrt{3}}{3}$;   **2.** $\dfrac{2\sqrt{5}}{5}$;   **3.** $\dfrac{5\sqrt{2}}{2}$;   **4.** $3\sqrt{5} + 3\sqrt{3}$;   **5.** $\dfrac{-3 + 4\sqrt{3}}{13}$;   **6.** $\dfrac{\sqrt[3]{12}}{3}$;   **7.** $\dfrac{2\sqrt[3]{3}}{3}$;   **8.** Yes

Lesson 46
Fractional Exponents and Reduction of Indices

Fractional Exponents

Algebraic operations involving radicals can sometimes be made easy by changing the radicals involved to their equivalent exponential forms, and then applying the laws of exponents. The advice is that when a radical is easy to simplify as is, simplify accordingly; but if it is too complicated and not easy to simplify, consider changing to exponential form and then simplifying, followed by conversion back to radical form.

Interconversion between radicals and exponential expressions

Examples

1. $25^{1/2} = \sqrt[2]{25} = \sqrt{25} = 5$

2. $9^{1/2} = \sqrt{9} = 3$

3. $8^{1/3} = \sqrt[3]{8} = 2$

4. $8^{2/3} = (8^{1/3})^2 = (\sqrt[3]{8})^2 = 2^2 = 4$

5. $16^{3/4} = (16^{1/4})^3 = (\sqrt[4]{16})^3 = 2^3 = 8$

Rules: **1**. $\sqrt[n]{a} = a^{1/n}$

2. $\sqrt[n]{a^m} = a^{m/n}$

3. $\sqrt[n]{a^m} = (\sqrt[n]{a})^m$

Example 1   Simplify the following:

1. $a^{1/3} \cdot a^{2/3}$; **2.** $b^{2/3} \cdot b^{1/2}$; **3.** $4^{-1/2}$; **4.** $2^{2/5} \cdot 2^{1/4}$ **5.** $\sqrt[5]{x^2} \cdot \sqrt[5]{x^3}$

6. $\sqrt[5]{3^2} \cdot \sqrt[4]{3}$; **7.** $\sqrt{x} \cdot \sqrt[3]{x^2}$

Solutions

1. $a^{1/3} \cdot a^{2/3} = a^{1/3 + 2/3} = a^1 = a$

2. $b^{2/3} \cdot b^{1/2} = b^{2/3 + 1/2} = b^{7/6}$ $(2/3 + 1/2 = 7/6)$

3. $4^{-1/2} = \dfrac{1}{4^{1/2}} = \dfrac{1}{\sqrt{4}} = \dfrac{1}{2}$

4. $2^{2/5} \cdot 2^{1/4} = 2^{2/5 + 1/4} = 2^{13/20}$

5. $\sqrt[5]{x^2} \cdot \sqrt[5]{x^3} = x^{2/5} x^{3/5} = x^{5/5} = x^1 = x$ or $\sqrt[5]{x^2} \cdot \sqrt[5]{x^3} = \sqrt[5]{x^2 \cdot x^3} = \sqrt[5]{x^5} = x$

6. $\sqrt[5]{3^2} \cdot \sqrt[4]{3} = 3^{2/5} \cdot 3^{1/4} = 3^{13/20} = \sqrt[20]{3^{13}}$ <---(Multiplying radicals with different indices)

7. $\sqrt{x} \cdot \sqrt[3]{x^2} = x^{1/2} \cdot x^{2/3} = x^{7/6} = \sqrt[6]{x^7} = x\sqrt[6]{x}$ <---(Multiplying radicals with different indices)

Example 2   Evaluate. $1000^{-1/3}$

Step 1: Express with positive exponents.

$$1000^{-1/3} = \frac{1}{1000^{1/3}}$$

Step 2:   Change the exponential form to radical form.

$$\text{then, } \frac{1}{1000^{1/3}} = \frac{1}{\sqrt[3]{1000}} = \frac{1}{10}$$

Example 3   Evaluate. $81^{-1/2}$

$$81^{-1/2} = \frac{1}{81^{1/2}}$$
$$= \frac{1}{\sqrt{81}}$$
$$= \frac{1}{9}$$

Example 4   Evaluate $\left(-\frac{8}{27}\right)^{-5/3}$

We will cover two methods.
Method 1

$$\left(-\frac{8}{27}\right)^{-5/3} = \left(\frac{-8}{27}\right)^{-5/3} \quad \text{<--- (We can give the minus sign to either the "8" or the "27")}$$
$$= \frac{1}{\left(\frac{-8}{27}\right)^{5/3}}$$
$$= \frac{1}{\left(\sqrt[3]{\frac{-8}{27}}\right)^5}$$
$$= \frac{1}{\frac{(\sqrt[3]{-8})^5}{(\sqrt[3]{27})^5}}$$
$$= \frac{1}{\frac{-32}{243}} \quad \text{or} \quad 1 \div \frac{-32}{243}$$
$$= -\frac{243}{32}$$

Scrapwork: 1. $\sqrt[3]{-8} = -2$

2. $(-2)^5 = \textbf{-32}$

3. $\sqrt[3]{27} = 3$

4 . $3^5 = \textbf{243}$

Method 2   (Much faster than Method 1)

$$\left(-\frac{8}{27}\right)^{-5/3} = \left(\frac{-8}{27}\right)^{-5/3}$$

$$= \left(\frac{27}{-8}\right)^{5/3} \quad \text{(Interchange the numerator and the denominator and change the sign of the exponent)}$$

$$= \frac{(\sqrt[3]{27})^5}{(\sqrt[3]{-8})^5} = \frac{(3)^5}{(-2)^5} = \frac{243}{-32}$$

$$= -\frac{243}{32}$$

Reduction of Indices

Example 1   Reduce the index: $\sqrt[4]{36}$

Step 1:   Change the radical to exponential form (power form).

$$\text{Then, } \sqrt[4]{36} = \sqrt[4]{6^2} = 6^{2/4}$$

Step 2: Reduce the exponent $\frac{2}{4}$ to lowest terms.

$$\text{then } \frac{2}{4} = \frac{1}{2}$$

Step 3: $6^{2/4} = 6^{1/2}$

Step 4:   Change back to the radical form.

$$\text{then } 6^{1/2} = \sqrt{6} \qquad \textbf{Note: } 6^{1/2} = \sqrt[2]{6} = \sqrt{6}$$

Perfect powers

Examples: **1.** 9 is a perfect square.
  2. 8 is a perfect cube .
  3. $16y^2$ is a perfect square .
  4. $\frac{4}{9}$ is a perfect square .

What is meant by to simplify a radical?

1. It may mean remove all perfect powers from the radicand (page 206); or

2. It may mean rationalize the denominator (page 212)

3. It may mean reduce the index of the radical (page 218)

4. It may mean perform a combination of the above.

Lesson 46 Exercises

A   Simplify the following:

1. $x^{3/4} \cdot x^{1/4}$ **2.** $x^{2/3} \cdot x^{3/4}$ **3.** $3^{1/2} \cdot 3^{2/3}$

Answers: **1.** x ;   **2.** $x^{17/12}$;   **3.** $3^{7/6}$

B   Evaluate the following:

1. $100^{-1/2}$;   **2.** $25^{-1/2}$;   **3.** $\left(\dfrac{16}{81}\right)^{-1/4}$

Answers: **1.** $\dfrac{1}{10}$;   **2.** $\dfrac{1}{5}$;   **3.** $\dfrac{3}{2}$

C Reduce the index:

1. $\sqrt[4]{x^2}$;   **2.** $\sqrt[6]{x^8}$;   **3.** $\sqrt[4]{25}$;   **4.** $\sqrt[4]{4}$

Answers: **1.** $\sqrt{x}$;   **2.** $\sqrt[3]{x^4}$;   **3.** $\sqrt{5}$; **4.** $\sqrt{2}$.

CHAPTER 14

Lesson 47

The Right Triangle, the Pythagorean Theorem and Applications

A **right triangle** is a triangle which has one right angle. In a right triangle, the side opposite the right angle is known as the hypotenuse. Also, the hypotenuse is the longest side of a right triangle.

The Pythagorean theorem

The Pythagorean theorem states that in any right triangle, the square of the length of the hypotenuse is equal to the sum of the squares of the lengths of the other two sides.

Example 1   Find x:

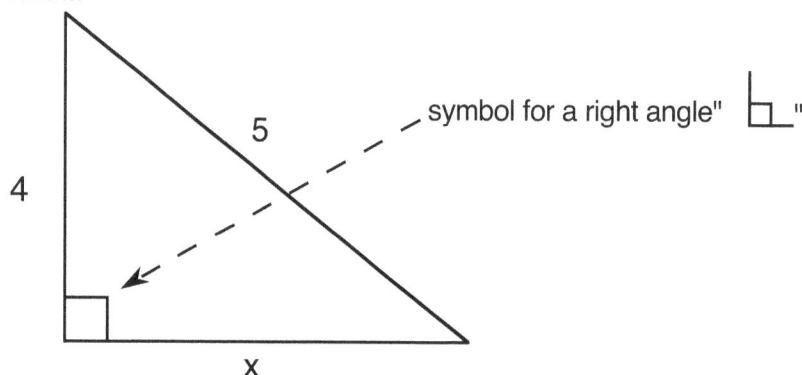

Step 1:   The given triangle is a right triangle (by the right angle symbol shown.)
Pick the length of the hypotenuse first. We pick 5 first.   (The length of the hypotenuse is 5)

Step 2:   Apply the Pythagorean theorem.

$$5^2 = 4^2 + x^2 \longleftarrow$$ ---------- **Note:** Here, the unknown is not on one side of the equation by itself. Always, it is the hypotenuse term which is on one side of the equation by itself.

square of (length) the hypotenuse

The squares of the (lengths) other two sides

Continuing,
$$25 = 16 + x^2$$
$$\underline{-16 \quad -16}$$
$$9 = x^2$$
$$\sqrt{9} = x$$
$$3 = x \ \text{ or } \ x = 3$$

Example 2   Find h

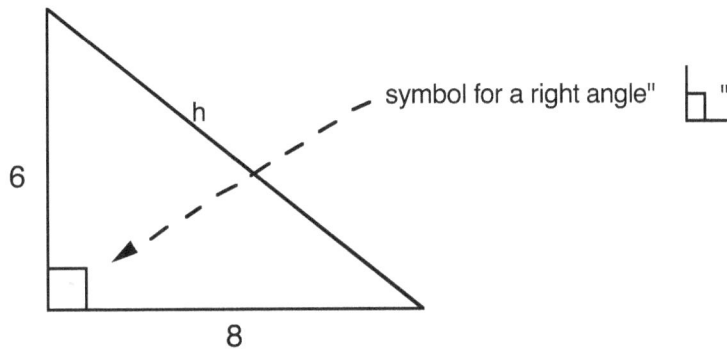

Solution: We will apply the Pythagorean theorem since the triangle is a right triangle.

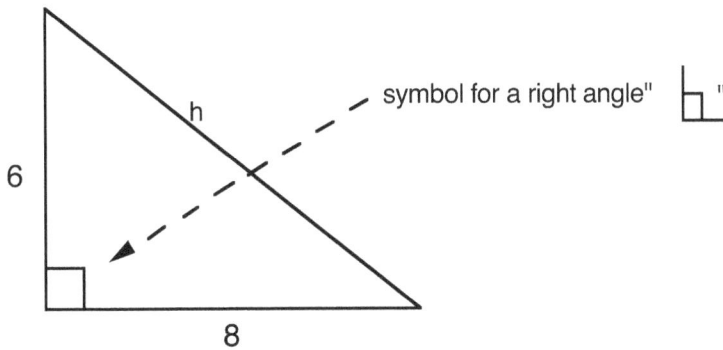

By the Pythagorean theorem,

$h^2 = 6^2 + 8^2$ (The length of the hypotenuse is h)

$h^2 = 36 + 64$

$h^2 = 100$

$h = \sqrt{100}$

$h = 10$

Example 3 Find x

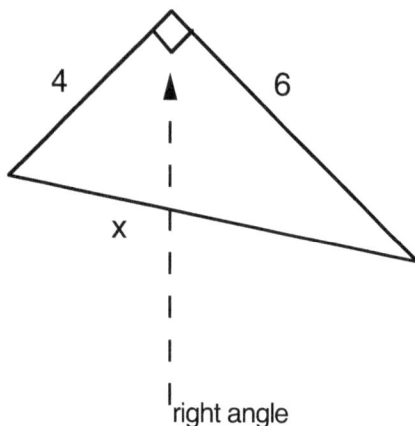

right angle

Step 1: Since the given triangle is a right triangle, we can apply the Pythagorean theorem. The length of the hypotenuse is x.

Step 2: Applying the Pythagorean theorem,

$$x^2 = 4^2 + 6^2$$
$$x^2 = 16 + 36$$
$$x^2 = 52$$

$$x = \sqrt{52}$$

Note: 52 is not a perfect square but we can simplify the radical $\sqrt{52}$.

$$x = \sqrt{(4)(13)}$$ (Remove any perfect square in 52)
$$x = \sqrt{4}\sqrt{13}$$
$$x = 2\sqrt{13}$$

Note: It is a good practice to always pick the hypotenuse first, whether or not it is known. Sometimes, some students pick the unknown first and this may lead to the misapplication of the Pythagorean theorem.
   Therefore, remember to always pick the hypotenuse (the side opposite the right angle) first. The Pythagorean theorem is **valid only for a right triangle** and therefore, we should **not** apply it to any triangle.

Question: How do we know if the triangle is a right triangle?

Answer: The given information may state so, or the given information may be used to deduce that the triangle is a right triangle.

The following symbolic form of the **Pythagorean theorem** will be useful in the future.
By convention, c is the length of the hypotenuse, a and b are the lengths of the other two sides.

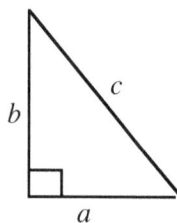

The Pythagorean theorem: $$c^2 = a^2 + b^2$$

Lesson 47 Exercises

A

1. Find c

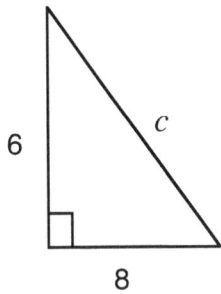

6

c

8

2. Find b

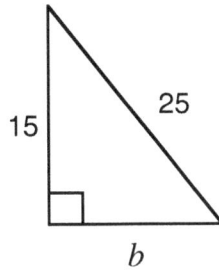

15

25

b

Answers: **1.** $c = 10$ **2.** $b = 20$

B In each of the following right triangles, find x:

1.

12

13

x

2.

8

x

15

3.

x

$\sqrt{6}$

5

4.

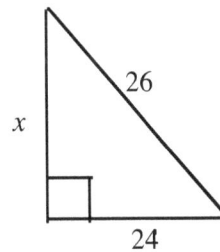

4

x

26

24

Answers: **1.** $x = 5$; **2.** $x = 17$;   **3.** $x = \sqrt{31}$;   **4.** $x = 10$

CHAPTER 15
Areas and Perimeters

Lesson 48: **Area and Perimeter of Triangles, Rectangles, and Trapezoids**

Lesson 49: **Area and Perimeter of a Circle; Arc Length and Sector of a Circle; Area and Perimeter of Composite Figures**

Lesson 48
Area and Perimeter of Triangles, Rectangles, and Trapezoids

Example 1 Find the area of figure (triangle) below.

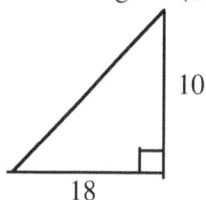

Solution : The figure is a right triangle with base 18 units and altitude 10 units.

$$\text{The area of a triangle} = \frac{1}{2}\,\text{base} \times \text{altitude}$$

$$= \frac{1}{2}(18)(10)$$

$$= 90 \text{ square units}$$

Note: The altitude and the base always meet at right angles. In the above problem, it does not matter which of 18 or 10 we call the the base. We could have taken the base to be 10 and then, the altitude would have been 18. The altitude of the triangle is also known as the height of the triangle.

Example 2 In the figure below:
(a) Find the area of the figure.
(b) Find the perimeter of figure.

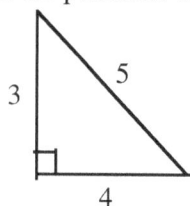

(a) $b = 4, h = 3$

$$\text{Area of the triangle} = \frac{1}{2}\,bh$$

$$= \frac{1}{2}(4)(3)$$

$$= 6 \text{ sq. units}$$

(b) Perimeter of the triangle = 3 + 4 + 5

$$= 12 \text{ units}$$

Example 3   In the figure below:
(a) Find the area of the figure.
(b) Find the perimeter of figure.

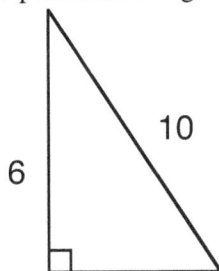

Solution

(a)

Let the base be b, and the altitude h.

Area of the triangle $= \frac{1}{2} bh$.

$h = 6$, but we are **not** given b; however

Step 1: We can find b by using the Pythagorean theorem since the triangle is a right triangle.

$$10^2 = 6^2 + b^2$$
$$100 = 36 + b^2$$
$$64 = b^2$$
$$8 = b$$
$$b = 8$$

Step 2: Now, $b = 8, h = 6$

Area of the triangle $= \frac{1}{2} bh$

$$= \frac{1}{2}(8)(6)$$
$$= 24 \text{ sq. units}$$

The area of the triangle is 24 sq. units

(**b**)   From figure below,
Perimeter $= 6 + 8 + 10$
$= 24$ units

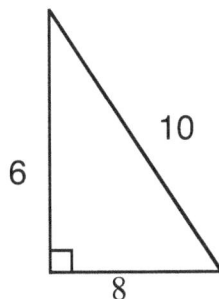

Note: Obtaining the same numerical answers in (**a**) and (**b**) is a coincidence; and therefore, the perimeter is **not** always numerically equal to the area.

Area and Perimeter of a Trapezoid

Example 4   Find the area of the figure shown below.

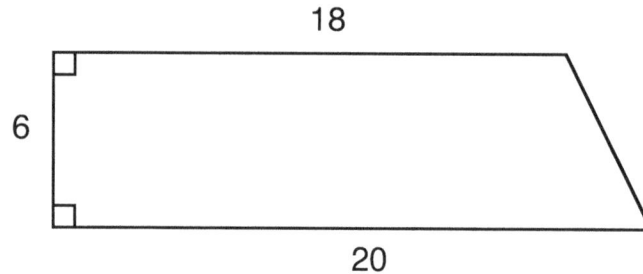

18

6

20

Method 1: The figure is a trapezoid.
The formula for the area, A, of a trapezoid is given
by $A = \frac{1}{2}(b_1 + b_2)h$, where b_1 and b_2 are the bases
(the bases are the parallel sides) and h is the altitude.

$b_1 = 20, b_2 = 18, h = 6$

$A = \frac{1}{2}(20 + 18)(6)$

$= \frac{1}{2}(38)(6)$

$= 19(6)$

$= 114$ square units

The area of the figure is 114 square units.

Method 2: We break up the figure into a rectangle and a triangle.

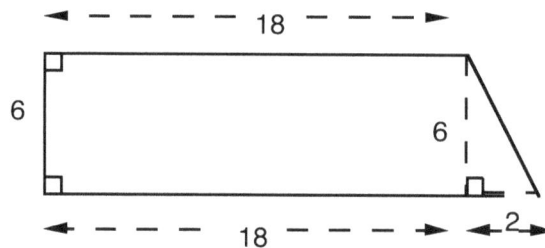

18

6

6

18

2

The area of the rectangle $= 6 \times 18$
$= 108$ sq. units

The area of the triangle $= \frac{1}{2}(2)(6)$
$= 6$ sq. units

Area of the figure = Area of the rectangle + Area of the triangle
$= (108 + 6)$ sq. units
$= 114$ sq. units

Again, we obtain the same answer as in Method 1.

Example 5 Find the area of the figure below.

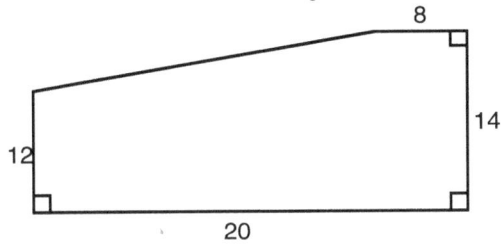

Solution   We break up the figure into a rectangle and a trapezoid.

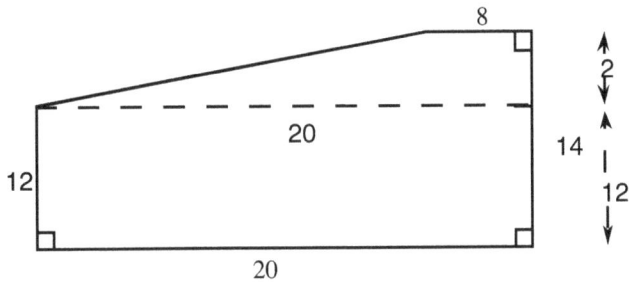

The area of the rectangle = 12 (20)

$\qquad$ =240 sq. units

The area of the trapezoid = $\frac{1}{2} (b_1 + b_2)h$

$\qquad = \frac{1}{2}(20 + 8)(2)$ $\qquad\qquad b_1 = 20 , b_2 = 8, \ h = 2$

$\qquad = \frac{1}{2}(28)(2)$

$\qquad = 28$ sq. units

Total area = area of the rectangle + area of the trapezoid

$\qquad$ =(240 + 28) sq. units

$\qquad$ = 268 sq. units

Note in the above problem that the original figure given is **not** a trapezoid, since it has five sides. By definition, a trapezoid is a quadrilateral (a four-sided figure) which has exactly a pair of parallel sides.

Note also that in the above problems, given a complex figure for which we do not know any formulas, we can break up the figure into simpler figures (such as a rectangle, a triangle, or a trapezoid), whose formulas we know, and then apply the appropriate formulas accordingly to find the areas, which we can then add together. Sometimes, we can extend the sides of the figures to form "known" figures and then, in this case, subtract the extensions of the areas.

Lesson 48 Exercises

A Find (a) the area and (b) the perimeter of the figure below.

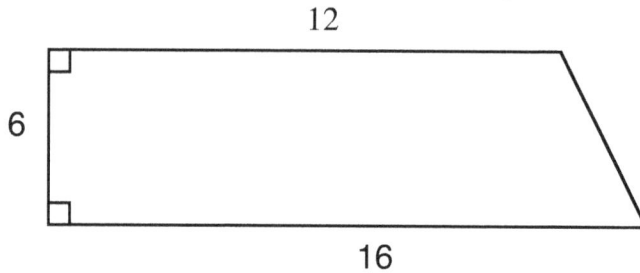

Answers: (a) 84 sq. units; (b) $(34 + 2\sqrt{13})$ units

B

1. Find (a) Area; (b) Perimeter of right triangle ABC.
2. Find (a) Area; (b) Perimeter of right triangle *DEF*.

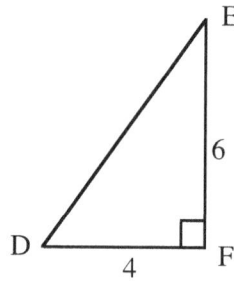

Answers: Triangle ABC:(a). 96 sq. units; (b). 48 units;   Triangle *DEF*: (a) 12 sq. units; (b) $10 + 2\sqrt{13}$

Lesson 49

Area and Perimeter of a Circle; Arc Length and Sector of a Circle; Area and Perimeter of Composite Figures

Area and Perimeter of a Circle

Area of a circle

The area, A, of a circle of radius , r, is given by $A = \pi r^2$.
π is approximately equal to 3.1416; in calculations, we will leave our answers in terms of π, unless instructed otherwise.

Note also that the radius of a circle $= \frac{1}{2}$ of the diameter of the circle.

Area of a semicircle (half-circle) $= \frac{1}{2}$ of the area of the given circle.

Area of a quarter-circle $= \frac{1}{4}$ of the area of the given circle.

Perimeter (or Circumference) of a Circle

The perimeter of a circle is the distance around the circle (i.e., the length of the circle).
The perimeter, P, of a circle of radius, r, is given by $P = 2\pi r$.

The perimeter of a semicircle $= \frac{1}{2}$ of the perimeter of the given circle.

The perimeter of a quarter circle is $\frac{1}{4}$ of the perimeter of the given circle.

Example Find the area and the perimeter of a circle of radius 6 units.

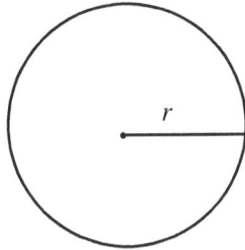

Finding the area
$r = 6$
Area of a circle $= \pi r^2$.
Area of the given circle $= \pi(6)^2$
$\qquad\qquad = \pi(36)$
$\qquad\qquad = 36\pi$ sq. units

Finding the perimeter
$r = 6$ units
Perimeter of a circle $= 2\pi r$
Perimeter of the given circle $= 2\pi(6)$
$\qquad\qquad\qquad = 12\pi$ units

Note: In the above problem, the area of a **semicircle** of radius, $r = 6$ is $\frac{1}{2}(36\pi) = 18\pi$ sq. units;

and the perimeter of the semicircle of radius, $r = 6$, is $\frac{1}{2}(12\pi)$
$\qquad\qquad\qquad\qquad = 6\pi$ units.

Arc Length and Sector of a Circle

Arc length: Arc length is the length of a part of a circle between any two points on the circle. In a circle of radius r, the arc length s cut off by a central angle θ (θ in degrees) is given by

$s = \dfrac{\theta}{360}$ of the circumference of the circle.

$$\boxed{s = \dfrac{\theta}{360}(2\pi r) \text{ or } \dfrac{\theta\pi r}{180}}.$$

However, if the central angle is in radians, ($360° = 2\pi$ radians, or $180^o = \pi\ radians$)

$$\boxed{s = r\theta}$$

Sector of a circle: A sector of a circle is the part of the interior of the circle bounded by two radii and the intercepted arc.

In a circle of radius r, the area A of a sector with central angle θ (θ in degrees) is given by

$A = \dfrac{\theta}{360}$ of the area of the circle.

$$\boxed{A = \dfrac{\theta}{360}(\pi r^2)}$$

However, if θ is in radians,

$$\boxed{A = \tfrac{1}{2}r^2\theta}$$

Example In the circle shown below, if the diameter is 6 and the central angle is 30°,
(a) Find the circumference of the circle.
(b) Find the arc length of CB
(c) Find the area of sector COB.

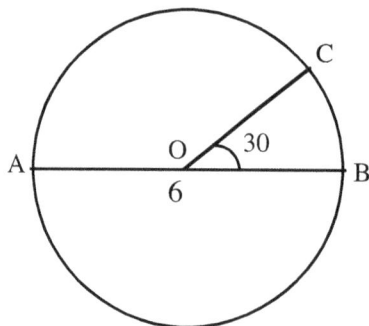

Figure 1

Solution
(a) Circumference of a circle of radius r is given by $2\pi r$.

Radius of a circle $= \dfrac{1}{2}$ of the diameter of the circle.

Therefore, $r = \dfrac{1}{2}(6) = 3$

Circumference of the given circle is $2\pi r. = 2\pi(3) = 6\pi$ units.

(b) arc length, s, of CB $= \dfrac{\theta}{360}$ of the circumference of the circle

$$s = \dfrac{30}{360} \cdot 2\pi(3) = \dfrac{\pi}{2} \text{ units}$$

OR if θ is in radians, $s = r\theta = 3 \cdot \dfrac{\pi}{6}$ (**Note**: $\theta = 30° = \dfrac{\pi}{6}$; $r = 3$.)

$$s = \dfrac{\pi}{2} \text{ units.}$$

(c) Area, A , of the shaded sector $= \dfrac{30°}{360°}$ of the area of the circle.

$$A = \dfrac{30°}{360°} (\pi r^2)$$

$$= \dfrac{30°}{360°} (\pi \cdot 3^2)$$

$$= \dfrac{30}{360} (9\pi)$$

$$= \dfrac{3\pi}{4} \text{ sq. units}$$

OR if θ is in radians, $A = \dfrac{1}{2} r^2 \theta$ (**Note**: $\theta = 30° = \dfrac{\pi}{6}$; $r = 3$.)

$$= \dfrac{1}{2} \cdot 3^2 \cdot \dfrac{\pi}{6}$$

$$= \dfrac{1}{2} \cdot \dfrac{9}{1} \cdot \dfrac{\pi}{6}$$

$$= \dfrac{3\pi}{4} \text{ sq. units.}$$

Areas and Perimeters of Composite Figures

Example 1   In the figure below:
(a) Find the area of the figure.
(b) Find the perimeter of figure.

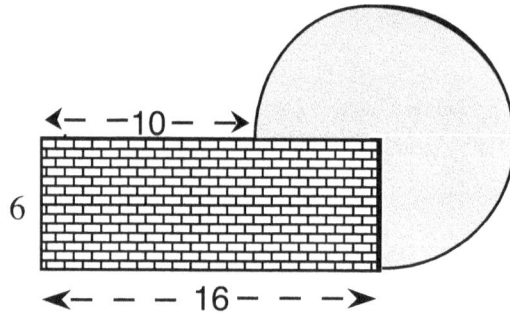

Solution

(a) Finding the area

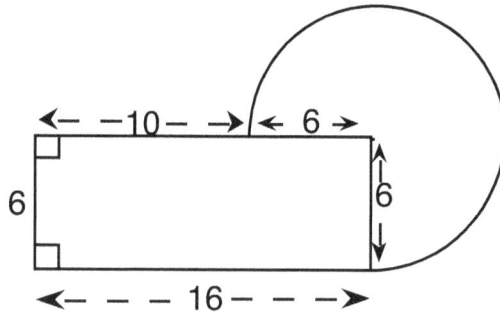

Radius, r, of the three-quarter circle = 6

Area of the three-quarter circle $= \frac{3}{4}\pi r^2$ ($\frac{3}{4}$ of the area of a circle of radius 6 units)

$$= \frac{3}{4}\pi(6)^2$$

$$= \frac{3}{4}\pi(36)^{9}$$

$$= 27\pi \text{ sq. units}$$

Area of the 16 by 6 rectangle $= 16 \times 6$ sq. units
$$= 96 \text{ sq. units}$$
Area of the **whole figure** = area of rectangle + area of three-quarter circle
$$= (96 + 27\pi) \text{ sq. units}$$

(b) Finding the perimeter

Perimeter of the three-quarter circle $= \dfrac{3}{4} 2\pi r$

$$= \dfrac{3}{4}(2\pi)(6)$$

$$= 9\pi \text{ units}$$

Perimeter of the whole figure $= 6 + 16 + 9\pi + 10$

$$= (32 + 9\pi) \text{ units}$$

Begin here and go around counterclockwise

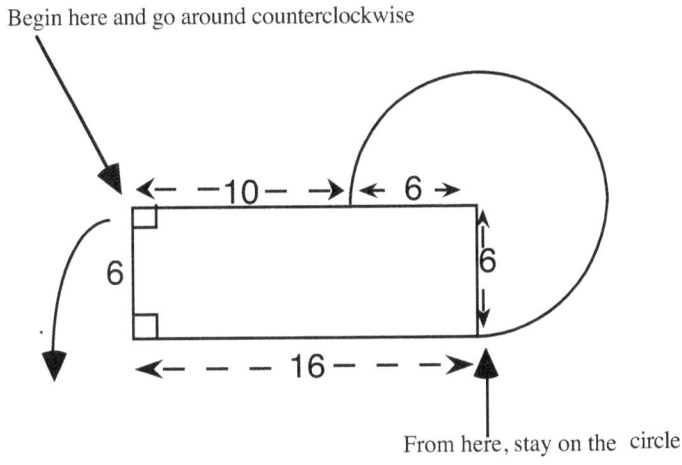

From here, stay on the circle

Note that in finding the perimeter we did **not** add the parts (6 and 6) of the rectangle which are also radii of the three-quarter circle, because the perimeter is the distance around the figure.

Lesson 49 Exercises

A

In the circle below, if the diameter of the circle is 18 units and the central angle has a measure 60°,
(a) Find the circumference of the circle.
(b) Find the area of sector COB.
(c) Find the arc length of CB.

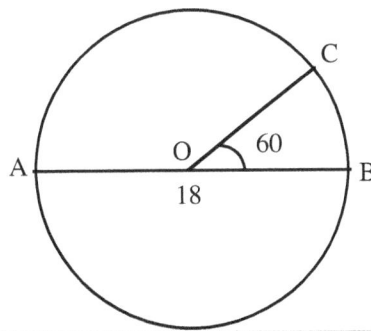

Answers (a) 18π units; (b) $\dfrac{27\pi}{2}$ sq. units; (c) 3π units

Lesson 49: Area, Perimeter, and Sector of a Circle; Arc Length; Composites Figures

B

(a) Find the area of the figure.
(b) Find the perimeter of figure.

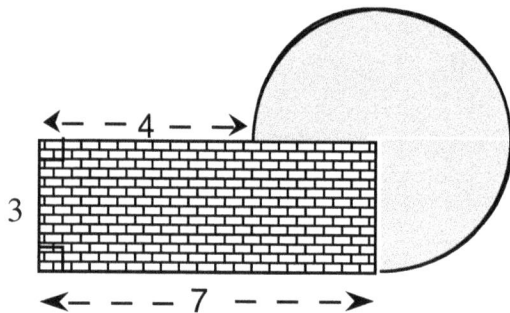

Answers: **(a)** $(21 + \frac{27\pi}{4})$ sq. units; **(b)**. $(14 + \frac{9\pi}{2})$ units

Test # 12 -Student's Self-Test (**Always**, **Test yourself before you are tested**)

Attempt all questions on clean sheets of paper, **Do not write in the book** Show all necessary work.

---

1. Simplify:

(a) $\sqrt{12}$;   (b) $\sqrt{48}$;   (c) $\sqrt{75}$.

Simplify:

2. (a) $\sqrt{162}$;   (b) $\sqrt{128}$;

(c) $\sqrt{96}$;   (d) $\sqrt{\dfrac{16}{81}}$.

---

3. Simplify:

(a) $\sqrt{36x^8}$;   (b) $\sqrt{4x^6y^4}$;   (c) $\sqrt{8x^7y^2}$

4. Simplify the following:

(a) $\sqrt{8x^5y^7}$;   (b) $\sqrt{18x^{10}y}$;   (c) $\sqrt{32x^6y^4z^5}$

---

5. Add or subtract:

(a) $5\sqrt{3} + 3\sqrt{3}$

(b) $2\sqrt{6} + \sqrt{18} + \sqrt{24} - 2\sqrt{72}$

(c) $\sqrt{12} + \sqrt{48} + \sqrt{75}$.

6. Add or subtract:

(a) $4\sqrt{50} - \sqrt{24} + 5\sqrt{2}$

(b) $3x\sqrt{25y} - 2x\sqrt{16y}$

---

7. Multiply and simplify:

(a) $4\sqrt{5}$ and $2\sqrt{6}$;   (b) $2\sqrt{5}$ and $3\sqrt{30}$;

(c) $3\sqrt{2}$ and $6\sqrt{2}$

8. Multiply and simplify:

(a) $5\sqrt{7}$ and $6\sqrt{7}$;   (b) $3x\sqrt{2y}$ and $\sqrt{3y}$.

---

9. Find c

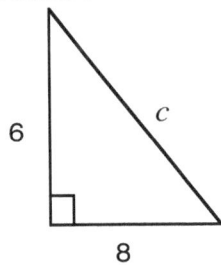

6

c

8

10. Find b

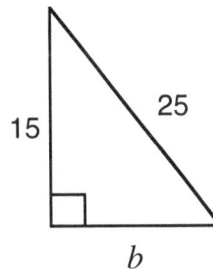

15   25

b

---

11. Fund x:

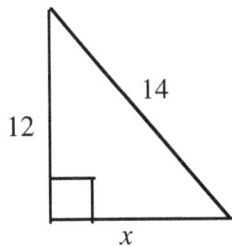

14

12

x

12. Find x:

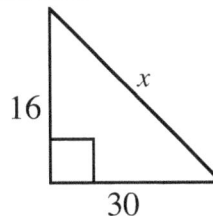

x

16

30

13. Find x:

14. Find x:

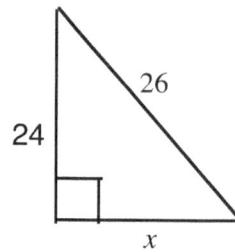

15. Find (a) Area; (b) Perimeter of right triangle ABC

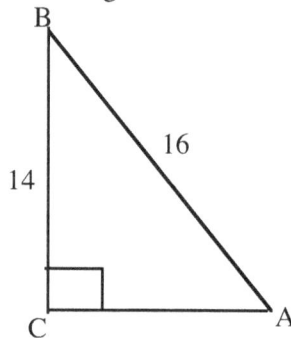

16. Find (a) Area; (b) Perimeter of right triangle DEF.

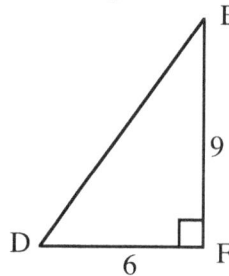

17. Find (a) the area and (b) the perimeter of the figure below.

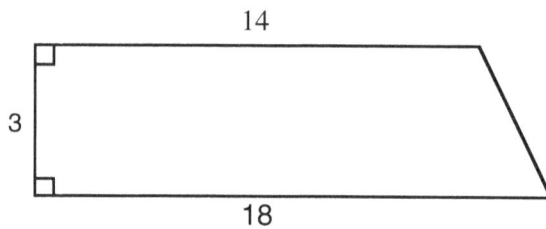

18. a) Find the area of the figure.
(b) Find the perimeter of figure.

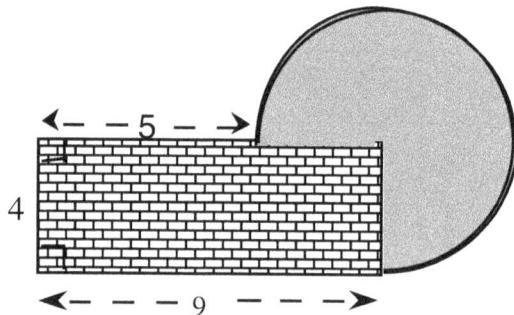

Answers: **1**. (a) $2\sqrt{3}$; (b) $4\sqrt{3}$; (c) $5\sqrt{3}$; **2**. (a) $9\sqrt{2}$; (b) $8\sqrt{2}$; (c) $4\sqrt{6}$; (d) $\frac{4}{9}$;

3. (a) $6x^4$; (b) $2x^3y^2$; **(c)** $2x^3y\sqrt{2x}$; **4**. (a) $2x^2y^3\sqrt{2xy}$; (b) $3x^5\sqrt{2y}$; (c) $4x^3y^2z^2\sqrt{2z}$;

5. (a) $8\sqrt{3}$; (b) $4\sqrt{6} - 9\sqrt{2}$; (c) $11\sqrt{3}$; **6**. (a) $25\sqrt{2} - 2\sqrt{6}$; (b) $7x\sqrt{y}$;

7. (a) $8\sqrt{30}$; (b) $30\sqrt{6}$; (c) 36; **8**. (a) 210 ; (b) $3xy\sqrt{6}$; **9**. $c = 10$; **10** $b = 20$;

11. $x = 2\sqrt{13}$; **12**. $x = 34$; **13**. $\sqrt{69}$; **14**. $x = 10$; **15**. (a) $14\sqrt{15}$; (b) $(30 + 2\sqrt{15})$ units;

16. (a) 27 sq. units; (b) $(15 + 3\sqrt{13})$ units; **17**. (a) 48 sq. units; (b) 40 units ;

18. (a) $(36 + 12\pi)$ sq. units; (b) $(18 + 6\pi)$ units.

CHAPTER 16
ALGEBRAIC FRACTIONS

Lesson 50: **Reduction, Multiplication, Division of Algebraic Fractions**
Lesson 51: **Addition of Fractions; Simplification of Complex Fractions**

Lesson 50
Reduction, Multiplication, and Division of Algebraic Fractions

Reduction of Fraction to Lowest Terms

The most important (and perhaps the only) prerequisite in being able to reduce algebraic fractions to lowest terms is **knowing how to factor polynomials.** Therefore, go back and review the factoring of polynomials.

Example 1   Reduce to lowest terms: $\dfrac{x^2-5x}{x^2-4x-5}$

Step 1   Factor both the numerator and the denominator

$$\frac{x^2-5x}{x^2-4x-5}$$
$$=\frac{x(x-5)}{(x+1)(x-5)}$$

Step 2:   Cancel (divide out) the common factor in the numerator and the denominator.

$$\frac{x\cancel{(x-5)}}{(x+1)(\cancel{x-5})}$$

$$=\frac{x}{x+1}$$

Example 2   Reduce to lowest terms

$$\frac{x^2+10x+24}{3x+12}$$

Step 1: Factor both the numerator and the denominator.

$$\frac{x^2+10x+24}{3x+12}=\frac{(x+4)(x+6)}{3(x+4)}$$

Step 2: Cancel the common factors in the numerator and the denominator

$$\frac{x^2+10x+24}{3x+12}=\frac{\cancel{(x+4)}(x+6)}{3\cancel{(x+4)}}$$
$$=\frac{x+6}{3}$$

Example 3 Reduce to lowest terms $\dfrac{x^2 + 2x - 24}{12 - 3x}$

Solution

Step 1: Factor the numerator, factor the denominator:

Then, $\dfrac{x^2 + 2x - 24}{12 - 3x} = \dfrac{(x-4)(x+6)}{3(4-x)}$

Step 2: $= \dfrac{(x-4)(x+6)}{-3(x-4)}$ Note that $4 - x = -(-4+x) = -(x-4)$

Step 3: $= \dfrac{\cancel{(x-4)}(x+6)}{-3\cancel{(x-4)}}$

Step 4: $= \dfrac{x+6}{-3}$

Step 5: $= -\dfrac{x+6}{3}$

A note about cancellation (dividing out):   When may we cancel?
We may cancel when the following two conditions have been satisfied simultaneously.

1. All quantities (factors) in the numerator are multiplying each other, and all the quantities (factors) in the denominator are multiplying each other.

2. There are common quantities (factors) in both the numerator and the denominator.

Also, we consider each expression (such as a binomial) within the parentheses as one quantity.

Example in which we **can** cancel: $\dfrac{2x(x+2)(x-1)(x+3)}{x(x-1)(x+3)} = \dfrac{2\cancel{x}(x+2)\cancel{(x-1)}\cancel{(x+3)}}{\cancel{x}\cancel{(x-1)}\cancel{(x+3)}} = 2(x+2)$

Example in which we **cannot** cancel as is: $\dfrac{2x(x+2) + (x-1)(x+3)}{x(x-1)(x-3)}$

because of the "+" sign between the parentheses in the numerator.

Multiplication and Division of Fractions

As it was in the case of reduction of fractions to lowest terms, the most important (and perhaps the only) prerequisite in being able to multiply and divide algebraic fractions is **knowing how to factor polynomials.** Therefore, go back and review the factoring of polynomials.

Perform the indicated operations and simplify.

Example 1 $\dfrac{x}{x-2} \cdot \dfrac{3x-6}{6x}$

Solution: Factor and divide out (cancel) the common factors in the numerators and the denominators

$\dfrac{x}{x-2} \cdot \dfrac{3x-6}{6x} = \dfrac{x}{x-2} \cdot \dfrac{3(x-2)}{6x}$ Note: $3x - 6 = 3(x-2)$

Consider $(x-2)$ as a single quantity

$= \dfrac{\overset{1}{\cancel{x}}}{\underset{1}{\cancel{x-2}}} \cdot \dfrac{\overset{1}{\cancel{3(x-2)}}}{\underset{2}{\cancel{6x}}{}^{1}}$

$= \dfrac{1}{2}$

Example 2   Divide: $\dfrac{3y^2}{y+2} \div \dfrac{6y}{y^2-4}$

$\dfrac{3y^2}{y+2} \div \dfrac{6y}{y^2-4} \quad = \dfrac{3y^2}{y+2} \cdot \dfrac{y^2-4}{6y}$ (Changing to multiplication and inverting the divisor)

$= \dfrac{\cancel{3}y^2\,y}{\cancel{y+2}} \cdot \dfrac{(y+2)(y-2)}{\cancel{6y}_2}$ Scrapwork: $y^2-4=(y+2)(y-2)$

$= \dfrac{y(y-2)}{2}$ or $\dfrac{y^2-2y}{2}$

Example 3   Simplify: $\dfrac{x^2-49}{x^4} \cdot \dfrac{x^2}{3x+21}$

Solution: Factor and cancel the common factors.

$\dfrac{x^2-49}{x^4} \cdot \dfrac{x^2}{3x+21} = \dfrac{(x+7)(x-7)}{x^4} \cdot \dfrac{x^2}{3(x+7)}$

$= \dfrac{\cancel{(x+7)}(x-7)}{\cancel{x^4}\,x^2} \cdot \dfrac{\cancel{x^2}^{\;1}}{3\cancel{(x+7)}_1}$

$= \dfrac{x-7}{3x^2}$

Lesson 50 Exercises

A Reduce to lowest terms:

1. $\dfrac{x+4}{x^2+7x+12}$;   **2.** $\dfrac{x^2-8x+15}{x^2-x-20}$ **3.** $\dfrac{6-x}{x^2-9x+18}$;   **4.** $\dfrac{2x^2-32}{x^2-11x+28}$

Answers: **1.** $\dfrac{1}{x+3}$;   **2** $\dfrac{x-3}{x+4}$ **3.** $-\dfrac{1}{x-3}$ **4.** $\dfrac{2(x+4)}{x-7}$

B Reduce the following to lowest terms:

1. $\dfrac{2x-6}{x-3}$;   **2.** $\dfrac{x^2+x-6}{x-2}$;   **3.** $\dfrac{4-x}{x^2+x-20}$;   **4.** $\dfrac{x^2+x-30}{x^2-x-20}$

5. $\dfrac{x^2+4x-21}{x^2-49}$;   **6.** $\dfrac{x^2-81}{x^2-11x+18}$;   **7.** $\dfrac{x^2-8x+12}{4x^2-32x+48}$ **8.** $\dfrac{2x^2-32}{2x^2+2x-40}$

Answers: **1.** 2;   **2.** $x+3$;   **3.** $-\dfrac{1}{x+5}$;   **4.** $\dfrac{x+6}{x+4}$;   **5.** $\dfrac{x-3}{x-7}$;   **6.** $\dfrac{x+9}{x-2}$;   **7.** $\dfrac{1}{4}$;   **8.** $\dfrac{x+4}{x+5}$

C Perform the indicated operations:

1. $\dfrac{x}{x-3} \cdot \dfrac{4x-12}{x+3}$;

2. $\dfrac{10x^2}{x-5} \div \dfrac{5x}{x^2-25}$;

3. $\dfrac{x^2-81}{x^6} \cdot \dfrac{x^4}{x^2+6x-27}$

Answers: **1.** $\dfrac{4x}{x+3}$;   **2.** $2x^2 + 10x$;   **3.** $\dfrac{x-9}{x^2(x-3)}$

D Simplify the following:

1. $\dfrac{6x^4}{7y^2} \cdot \dfrac{21y}{3x}$;

2. $\dfrac{x^2}{x+3} \div \dfrac{x}{2x+6}$;

3. $\dfrac{8-x}{x^2-64} \cdot \dfrac{3x+24}{x+3}$

4. $\dfrac{x^2-9}{4x^2-36} \cdot \dfrac{2x^2-50}{3x+15}$;

5. $\dfrac{8-x}{x^8} \div \dfrac{x^2-4x-32}{x^6+4x^5}$

Answers: **1.** $\dfrac{6x^3}{y}$;   **2.** $2x$;   **3.** $-\dfrac{3}{x+3}$;   **4.** $\dfrac{x-5}{6}$;   **5.** $-\dfrac{1}{x^3}$

Lesson 51

Addition of Fractions; Simplification of Complex Fractions

Addition and Subtraction of Fractions

Like Fractions (Fractions with the same denominator)

Example   Add: $\dfrac{2x}{x-3} + \dfrac{4}{x-3}$

Solution Since the above fractions already have a common denominator, we add the numerators and keep the common denominator.

$$\dfrac{2x}{x-3} + \dfrac{4}{x-3}$$

$$= \dfrac{2x+4}{x-3}$$

Unlike Fractions (Fractions with different denominators)

In this case, we have to change the fractions to (equivalent) like fractions, add the numerators and then reduce the fraction to lowest terms if possible.

Example 1   Add: $\dfrac{5}{x^2 + 7x + 12} + \dfrac{x+2}{x+3}$

Step 1:   Factor the denominators if factorable.

$$\dfrac{5}{(x+3)(x+4)} + \dfrac{x+2}{x+3}$$

Step 2: We will make the denominators the same by looking for missing factors. The factors of the denominator of the first fraction are $(x+3)$ and $(x+4)$, but the denominator of the second fraction has the factor $x+3$. To make the denominators of the two fractions identical, we multiply both the denominator and the numerator of the second fraction by $x+4$, since the denominator of the second fraction is "missing" $x+4$, compared to the denominator of the first fraction.

Note that any time we multiply the denominator by any quantity, we must also multiply the numerator by the same quantity (in order **not** to change the value of the original fraction.).

$$= \dfrac{5}{(x+3)(x+4)} + \dfrac{(x+2)(x+4)}{(x+3)(x+4)}$$

Step 3: Since the denominators are now the same, add the numerators. (We have also produced the LCD)

$$= \dfrac{5 + (x+2)(x+4)}{(x+3)(x+4)}$$ The LCD we produced (in Step 1) is $(x+3)(x+4)$

$$= \dfrac{5 + x^2 + 4x + 2x + 8}{(x+3)(x+4)}$$

$$= \dfrac{x^2 + 6x + 13}{(x+3)(x+4)}$$ (Simplifying the numerator)

Note that if the numerator were factorable, we would have factored it to determine if there are common factors in the numerator and the denominator, and in which case, we would have canceled the common factors. In this example, the numerator is not factorable. It is preferable to leave the denominator in the factored form.

Example 2 Add: $\dfrac{3}{x+2} + \dfrac{2}{x^2-4}$

Step 1: Factor the factorable denominator(s).

Then, $\dfrac{3}{x+2} + \dfrac{2}{x^2-4}$

$= \dfrac{3}{x+2} + \dfrac{2}{(x+2)(x-2)}$

Step 2: The denominator of the first fraction is "missing" $(x-2)$ compared to the denominator of the second fraction. Therefore, multiply both denominator and numerator of the first fraction by $(x-2)$; add the numerators of the resulting like fractions and reduce if possible.

$= \dfrac{3(x-2)}{(x+2)(x-2)} + \dfrac{2}{(x+2)(x-2)}$ \qquad (We have produced the LCD which is $(x+2)(x-2)$)

$= \dfrac{3(x-2)+2}{(x+2)(x-2)}$

$= \dfrac{3x-6+2}{(x+2)(x-2)}$

$= \dfrac{3x-4}{(x+2)(x-2)}$

Example 3 Combine into a single fraction: $\dfrac{3x}{5y} + \dfrac{x}{2y}$

Solution

Step 1: Make the denominators identical by " looking for missing factors" and multiplying accordingly.
(Multiply the first fraction by 2 and the second fraction by 5.)

$\dfrac{3x}{5y} + \dfrac{x}{2y}$

$= \dfrac{3x(2)}{5y(2)} + \dfrac{x(5)}{2y(5)}$

$= \dfrac{6x}{10y} + \dfrac{5x}{10y}$ <------Like fractions   (The LCD produced is $10y$)

Step 2: Add the numerators since we have like fractions.

Then, $\dfrac{6x}{10y} + \dfrac{5x}{10y}$

$= \dfrac{11x}{10y}$

Example 4   Add: $\dfrac{7}{5} + \dfrac{3}{4n}$

Solution

Step 1:   Make the denominators the same by looking for missing factors and multiplying accordingly.
(Multiply the first fraction by 4n, and multiply the second fraction by 5)

$$\dfrac{7}{5} + \dfrac{3}{4n}$$

$$= \dfrac{7(4n)}{5(4n)} + \dfrac{3(5)}{4n(5)}$$

$$= \dfrac{28n}{20n} + \dfrac{15}{20n} \quad \longleftarrow \text{--------Like fractions (The LCD produced is } 20n)$$

Step 2:   Add the numerators.

Then, $\dfrac{28n}{20n} + \dfrac{15}{20n}$

$$= \dfrac{28n + 15}{20n}$$

Example 5   Combine into a single fraction: $\dfrac{4}{x} + \dfrac{1}{2x} + \dfrac{2}{5}$

$$\dfrac{4}{x} + \dfrac{1}{2x} + \dfrac{2}{5}$$

Step 1:   Multiply accordingly by 2, 5, and x.

$$= \dfrac{4(2)(5)}{x(2)(5)} + \dfrac{1(5)}{2x(5)} + \dfrac{2(x)(2)}{5(x)(2)} \qquad \text{(Making the denominators the same)}$$

$$= \dfrac{40}{10x} + \dfrac{5}{10x} + \dfrac{4x}{10x} \quad \longleftarrow \text{----Like fractions}$$

Step 2:   Add the numerators, since we have like fractions.

Then, $\dfrac{40}{10x} + \dfrac{5}{10x} + \dfrac{4x}{10x}$

$$= \dfrac{4x + 45}{10x}$$

Example 6. Add: $\dfrac{5}{x+4} - \dfrac{3x}{x-4} + \dfrac{2}{x^2-16}$

$$= \dfrac{5}{x+4} - \dfrac{3x}{x-4} + \dfrac{2}{(x+4)(x-4)}$$

$$= \dfrac{5(x-4)}{(x+4)(x-4)} - \dfrac{3x(x+4)}{(x-4)(x+4)} + \dfrac{2}{(x+4)(x-4)}$$

Now, since we have like fractions (all the denominators are the same), we add the numerators.

$$= \dfrac{5(x-4) - 3x(x+4) + 2}{(x+4)(x-4)}$$

$$= \dfrac{5x - 20 - 3x^2 - 12x + 2}{(x+4)(x-4)}$$

$$= \dfrac{-3x^2 - 7x - 18}{(x+4)(x-4)}$$

$$= -\dfrac{3x^2 + 7x + 18}{(x+4)(x-4)}$$

Example 7 Simplify: $\dfrac{3x+4}{x-3} - \dfrac{x-1}{x+1}$

Solution

$$= \dfrac{3x+4}{x-3} - \dfrac{x-1}{x+1}$$

$$= \dfrac{(x+1)(3x+4)}{(x+1)(x-3)} - \dfrac{(x-1)(x-3)}{(x+1)(x-3)} \quad \text{(Making the denominators the same and producing the LCD)}$$

$$= \dfrac{3x^2 + 4x + 3x + 4 - (x^2 - 3x - x + 3)}{(x+1)(x-3)} \quad \text{(The minus sign before the parentheses is important)}$$

$$= \dfrac{3x^2 + 7x + 4 - (x^2 - 4x + 3)}{(x+1)(x-3)} \quad \text{(The minus sign before the parentheses is important)}$$

$$= \dfrac{3x^2 + 7x + 4 - x^2 + 4x - 3}{(x+1)(x-3)}$$

$$= \dfrac{2x^2 + 11x + 1}{(x+1)(x-3)}$$

Example 8 Simplify : $3x + \dfrac{2x}{y}$

$$3x + \dfrac{2x}{y}$$

$$= \dfrac{3x(y)}{1y} + \dfrac{2x}{y}$$

$$= \dfrac{3xy}{y} + \dfrac{2x}{y}$$

$$= \dfrac{3xy + 2x}{y}$$

Example 9 Simplify: $x - 2 + \dfrac{3}{x - 4}$

$$= \dfrac{x-2}{1} + \dfrac{3}{x-4} \qquad \text{or} \qquad \dfrac{x}{1} - \dfrac{2}{1} + \dfrac{3}{x-4}$$

$$= \dfrac{(x-4)(x-2)}{(x-4)1} + \dfrac{3}{x-4} \qquad \text{or} \qquad \dfrac{x(x-4)}{1(x-4)} - \dfrac{2(x-4)}{1(x-4)} + \dfrac{3}{x-4}$$

$$= \dfrac{(x-4)(x-2)+3}{x-4} \qquad \text{or} \qquad \dfrac{x(x-4)-2(x-4)+3}{x-4}$$

$$= \dfrac{x^2-2x-4x+8+3}{x-4} \qquad \text{or} \qquad \dfrac{x^2-4x-2x+8+3}{x-4}$$

$$= \dfrac{x^2-6x+11}{x-4} \qquad \text{or} \qquad \dfrac{x^2-6x+11}{x-4}$$

The above "or" means that you may follow the steps on the left or the steps on the right.

Example 10 Add: $x + 4 + \dfrac{3}{x-2}$

$$= \dfrac{x}{1} + \dfrac{4}{1} + \dfrac{3}{x-2}$$

$$= \dfrac{x(x-2)}{1(x-2)} + \dfrac{4(x-2)}{1(x-2)} + \dfrac{3}{x-2}$$

$$= \dfrac{x(x-2)}{(x-2)} + \dfrac{4(x-2)}{(x-2)} + \dfrac{3}{x-2}$$

(Now, we can add the numerators since we have like fractions, and write the denominator only once.)

$$= \dfrac{x(x-2)+4(x-2)+3}{x-2}$$

$$= \dfrac{x^2-2x+4x-8+3}{x-2}$$

$$= \dfrac{x^2+2x-5}{x-2}$$

Example 11 Add : $\dfrac{x^2 + 7x + 12}{x^2 + 8x + 15} + \dfrac{3}{x - 5}$

Solution

Step 1: Factor and reduce the first expression before proceeding to add. (Failure to reduce to lowest terms first, in this problem, will result in having to deal with higher powers of x.)

$$\dfrac{x^2 + 7x + 12}{x^2 + 8x + 15} + \dfrac{3}{x - 5}$$

$$= \dfrac{(x + 3)(x + 4)}{(x + 3)(x + 5)} + \dfrac{3}{x - 5}$$

$$= \dfrac{x + 4}{x + 5} + \dfrac{3}{x - 5}$$

Step 2: Now, proceed to add as before.

$$= \dfrac{(x + 4)(x - 5)}{(x + 5)(x - 5)} + \dfrac{3 \cdot (x + 5)}{(x - 5)(x + 5)}$$

$$= \dfrac{x^2 - 5x + 4x - 20 + 3x + 15}{(x + 5)(x - 5)}$$

$$= \dfrac{x^2 + 2x - 5}{(x + 5)(x - 5)}$$

Note above that sometimes a given fraction may **not** be in its lowest terms; and in this case, it is good practice to reduce to lowest terms before proceeding to add.

Simplification of Complex Fractions

Example 1   Simplify: $\dfrac{\dfrac{3x}{ax}}{\dfrac{5x}{bx}}$

Solution $\dfrac{\dfrac{3x}{ax}}{\dfrac{5x}{bx}} = \dfrac{3x}{ax} \div \dfrac{5x}{bx}$

$$= \dfrac{3x}{ax} \cdot \dfrac{bx}{5x} \qquad \text{(inverting the divisor, } \dfrac{5x}{bx}\text{).}$$

$$= \dfrac{3b}{5a} \qquad \text{(canceling the common factors).}$$

Example 2 Simplify:
$$\dfrac{\dfrac{y}{y^2 + 4\,y + 3}}{\dfrac{1}{y^2 - 1} + 1}$$

Solution   **Method 1** (Adding the denominator first)

$$\dfrac{\dfrac{y}{y^2 + 4\,y + 3}}{\dfrac{1}{y^2 - 1} + 1} \quad = \quad \dfrac{y}{y^2 + 4\,y + 3} \div \dfrac{1}{y^2 - 1} + 1$$

Scrapwor k

$$\dfrac{1}{(y + 1)(\ y - 1)} + \dfrac{1}{1}$$

$$= \dfrac{1 + (\ y + 1)(\ y - 1)}{(y + 1)(\ y - 1)}$$

$$= \dfrac{y}{y^2 + 4\,y + 3} \div \dfrac{y^2}{(y + 1)(\ y - 1\ )}$$

$$= \dfrac{1 + \ y^2 - 1}{(y + 1)(\ y - 1)}$$

$$= \dfrac{y}{y^2 + 4\,y + 3} \cdot \dfrac{(y + 1)(\ y - 1)}{y^2}$$

$$= \dfrac{y^2}{(y + 1)(\ y - 1)}$$

$$= \dfrac{\overset{1}{\cancel{y}}}{\underset{1}{\cancel{(y + 1)}}(\ y + 3)} \cdot \dfrac{\overset{1}{\cancel{(y + 1)}}(\ y - 1)}{\underset{y}{\cancel{y^2}}}$$

$$= \dfrac{y - 1}{y(y + 3)}$$

Method 2: Multiply both the numerator and denominator of the complex fraction by the LCD of all the fractions in the numerator and the denominator. The LCD = $(y + 1)(y - 1)(y + 3)$

$$\dfrac{\dfrac{y}{y^2 + 4y + 3}}{\dfrac{1}{y^2 - 1} + 1}$$

$$= \dfrac{\dfrac{y(y + 1)(\ y - 1)(\ y + 3)}{(y + 1)(\ y + 3)}}{\dfrac{(y + 1)(\ y - 1)(\ y + 3)}{(y + 1)(\ y - 1)} + \dfrac{1(y + 1)(y - 1)(\ y + 3)}{1}}$$

$$= \dfrac{\dfrac{y\cancel{(y + 1)}(\ y - 1)\cancel{(y + 3)}}{\cancel{(y + 1)}\cancel{(y + 3)}}}{\dfrac{\cancel{(y + 1)}\cancel{(y - 1)}(y + 3)}{\cancel{(y + 1)}\cancel{(y - 1)}} + \dfrac{(y + 1)(\ y - 1)(\ y + 3)}{1}}$$

$$= \dfrac{y(y - 1)}{(y + 3) + (y^2 - 1\ )(y + 3)}$$ Note : $y^2 - 1 = (y + 1)(y - 1)$

$$= \dfrac{y(y - 1)}{(y + 3)[1 + \ y^2 - 1]}$$

$$= \dfrac{y(y - 1)}{(y + 3)\ y^2}$$

$$= \dfrac{\overset{1}{\cancel{y}}(y - 1)}{(y + 3)\ \underset{y}{\cancel{y^2}}}$$

$$= \dfrac{(y - 1)}{y(y + 3)}$$

Lesson 51 Exercises

A Carry out the indicated operation and simplify:

1. $\dfrac{3}{x-4} + \dfrac{x+1}{x^2-16}$;

2. $\dfrac{2}{xy} - \dfrac{4}{y}$;

3. $\dfrac{3x}{x-4} + \dfrac{2}{x-4}$;

4. $x-2+\dfrac{2}{x-3}$

5. $\dfrac{5}{x^2-4} + \dfrac{2x}{x+2} - \dfrac{4}{x-3}$.

Answers: **1.** $\dfrac{4x+13}{(x-4)(x+4)}$; **2.** $\dfrac{-4x+2}{xy}$; **3.** $\dfrac{3x+2}{x-4}$; **4.** $\dfrac{x^2-5x+8}{x-3}$; **5.** $\dfrac{2x^3-14x^2+17x+1}{(x-3)(x-2)(x+2)}$;

B Carry out the indicated operation and simplify:

1. $\dfrac{x}{3} + \dfrac{4}{x+2}$

2. $\dfrac{2}{x-3} - \dfrac{x}{4}$

3. $\dfrac{x}{y} + \dfrac{2}{n}$

4. $\dfrac{2}{x} + \dfrac{3}{y}$;

5. $\dfrac{2}{x-5} + \dfrac{3}{x-5}$;

6. $\dfrac{x+1}{x+2} + \dfrac{x-3}{x-2}$;

7. $\dfrac{2x-1}{x+4} - \dfrac{x-5}{x-3}$

Answers: **1.** $\dfrac{x^2+2x+12}{3(x+2)}$; **2.** $\dfrac{-x^2+3x+8}{4(x-3)}$; **3.** $\dfrac{nx+2y}{ny}$; **4.** $\dfrac{3x+2y}{xy}$;

5. $\dfrac{5}{x-5}$; **6.** $\dfrac{2x^2-2x-8}{(x+2)(x-2)}$; **7.** $\dfrac{2x^2-6x+23}{(x-3)(x+4)}$

C Simplify the following:

1. $\dfrac{\dfrac{x}{x^2+7x+12}}{\dfrac{16}{x^2-16}+1}$;

2. $\dfrac{\dfrac{2ab}{cd}}{\dfrac{cd}{ab}}$;

3. $\dfrac{3-\dfrac{2}{x}}{9x}$;

4. $\dfrac{\dfrac{x}{y}+3}{\dfrac{y}{x}-3}$;

5. $\dfrac{\dfrac{5c^2y}{3bx}}{\dfrac{25cy}{6bx^2}}$

Answers: **1.** $\dfrac{x-4}{x(x+3)}$; **2.** $\dfrac{2a^2b^2}{c^2d^2}$; **3.** $\dfrac{3x-2}{9x^2}$; **4.** $\dfrac{x^2+3xy}{y(y-3x)}$; **5.** $\dfrac{2cx}{5}$

CHAPTER 17
Inequalities

Lesson 52: **Set Notation, Set Operations, Interval Notation**
Lesson 53: **Linear Inequalities**; **Compound Inequalities**; **Word Problems**

Lesson 52
Set Notation, Set Operations, Interval Notation

Sets, Set Notation

Set   A set is a well-defined collection of objects or things. The objects or things are called the elements or members of the set. A set may contain many elements; it may contain only one element; or it may contain no element. We call the set which contains no element the empty set or the null set.

Representation of sets: We will cover the roster method and the set-builder notation.

Roster method: In the roster method, we list the elements of the set and enclose them by braces.
 Example: We can denote the set of numbers $2, 3, 4$, as $\{ 2, 3, 4\}$.
 We separate the elements by commas. Note that the order in which the elements are listed does not matter. We may **name** a set by a capital letter . For example, if we denote the set of numbers $2, 3, 4$ by A, then we write $A = \{ 2, 3, 4\}$.
We read this as "A is the set whose elements are $2, 3, 4$.

Set-builder notation: In the set builder notation, we state the conditions which the elements of the set must satisfy.
Example: Let E be the set of all odd integers between 4 and 12. Then if we use x to represent an arbitrary element of this set, we write $E = \{x \mid x \text{ is an odd integer between 4 and 12}\}$.
This is read " E is the set of numbers (elements) x such that x is an odd integer between 4 and 12".
We read the vertical line " |" as "such that".
(Note that by the roster method, the set E in this example would be represented by
$E = \{5, 7, 9, 11\}$

We denote the empty set by $\{ \}$ or $\varnothing$.

We will use the **set-builder notation** in stating the solutions of inequalities.

Definition

An **inequality** is a mathematical statement that two expressions are not equal.

An inequality, like an equation, has three parts, namely the left-hand side, the inequality symbol which may be " $>$, $<$, $\geq$, $\leq$ or $\neq$", and the right-hand side.

Examples 1. $2x - 5 > 6x + 3$ read " $2x$ minus 5 **is greater than** $6x$ plus 3".

2. $2x + 3 \geq 6x + 7$ read " $2x$ plus 3 **is greater than or equal to** $6x$ plus 7".

3. $4x + 5 < 8x - 2$ read "$4x$ plus **five is less than** $8x$ minus 2".

4. $3x - 1 \leq 5x + 7$ read "$3x$ minus 1 **is less than or equal to** $5x$ plus 7 ".

5. $x > 6$ read " x is greater than 6"

Sense of an inequality (direction or order of an inequality)
 The sense of an inequality refers to whether the inequality symbol is the greater than symbol " $>$" or the less than symbol "$<$".

Set Notation, Interval Notation and Graphs

Set notation Interval notation

$\downarrow$ $\downarrow$

$\{x \mid x > a\} = (a, +\infty)$

$\{x \mid x \geq a\} = [a, +\infty)$

$\{x \mid x < b\} = (-\infty, b)$

$\{x \mid x \leq b\} = (-\infty, b]$

$\{x \mid a < x < b\} = (a, b)$

$\{x \mid a \leq x \leq b\} = [a, b]$

$\{x \mid a < x \leq b\} = (a, b]$

$\{x \mid a \leq x < b\} = [a, b)$

$\{x \mid x \text{ is a real number}\}$

$\quad = (-\infty, +\infty)$

Graphs

Example 1 Draw the graph for $\{x \mid x > 2\} = (2, +\infty)$

Solution

Graph for $\{x \mid x > 2\} = (2, +\infty)$

Note: The hollow circle at 2 indicates that 2 is **not** part of the solution set.

Example 2 Draw the graph for $\{x \mid x \geq 2\} = [2, +\infty)$

Solution

Graph for $\{x \mid x \geq 2\} = [2, +\infty)$

Note: The solid circle at 2 indicates that 2 is part of the solution set.

Example 3

$\{x \mid x \leq -8\} = (-\infty, -8]$

Graph for $\{x \mid x \leq -8\} = (-\infty, -8]$

Set Operations Involving Inequalities

Union of two sets

The union of two sets A and B, written $A \cup B$, is the set containing all the elements that belong to either set A or set B (or belong to both sets).

Symbolically, $A \cup B = \{x \mid x \in A, \text{ or } x \in B\}$

Example 1   If $A = \{x \mid x < -3\}$ and $B = \{x \mid x > 4\}$ then

$\qquad A \cup B = \{x \mid x < -3\} \cup \{x \mid x > 4\}$

$\qquad\qquad = \{x \mid x < -3 \text{ or } x > 4\}$

Graph for $\{x \mid x < -3 \text{ or } x > 4\}$

Intersection of two sets

The intersection of two sets A and B, written $A \cap B$ is the set containing the elements that are common to A and B. Thus the elements of the intersection are the elements that belong to both sets simultaneously.

Symbolically, $A \cap B = \{x \mid x \in A \text{ and } x \in B\}$

Example 2 If $A = \{x \mid x > -3\}$ and $B = \{x \mid x < 2\}$ then

$$A \cap B = \{x \mid x > -3\} \cap \{x \mid x < 2\}$$

$$A \cap B = \{x \mid -3 < x < 2\}$$

Example 3 If $A = \{x \mid x < -3\}$ and $B = \{x \mid x > 4\}$ then

$$A \cap B = \{x \mid x < -3\} \cap \{x \mid x > 4\}$$

$A \cap B = \{\ \}$, the empty set.

(Since a number cannot be less than -3 and greater than 4 at the same time.)

Compare the solutions to Examples 1 and 3, above, and note that even though they have identical graphs, Example 1 has a solution but Example 3 has no solution.

Lesson 52 Exercises

1. What is a set?
2. What is an inequality?
3.What is meant by the sense or direction or the order of an equality?
4. Draw graphs for the following:
a. $\{x \mid x > 3\}$; **b.** $\{x \mid x \geq 3\}$; **c.** $\{x \mid x < 3\}$; **d.** $\{x \mid x \leq 3\}$; **e.** $\{x \mid x < -5\}$

Answers:**4.**

(a)

$\{x \mid x > 3\}$

(b)

$\{x \mid x \geq 3\}$

(c)

$\{x \mid x < 3\}$ $\{x \mid x \leq 3\}$

(e)

$\{x \mid x < -5\}$

Lesson 53

Solving Linear Inequalities; Compound Inequalities; Word Problems

Solving Linear Inequalities

The solution set of a linear inequality is the set of real numbers each of which when substituted for the variable makes the inequality true.

The techniques for solving linear inequalities are similar to the techniques for solving linear equations, except that when an inequality is divided or multiplied by a negative number, the sense of the inequality must be reversed as follows: The symbol " > " when reversed becomes " < "; the symbol " < " when reversed becomes " > "; the symbol " $\geq$ " when reversed becomes " $\leq$ "; and the symbol " $\leq$ " when reversed becomes " $\geq$ ".

Example 1 Solve and graph the solution set of the inequality,
$$3x - 4 > 0$$
Step 1: To undo the -4, add + 4 to both sides of the inequality.

$$\begin{array}{r} 3x - 4 > 0 \\ \underline{+ 4 \ +4} \\ 3x > 4 \end{array}$$

Step 2: Divide both sides of the inequality by 3.

$$\frac{3x}{3} > \frac{4}{3} \qquad x > \frac{4}{3}$$

The solution set is $\{x \mid x > \frac{4}{3}\}$.   (The solution set is all real numbers greater than $\frac{4}{3}$)

Graph for $x > \frac{4}{3}$

Note: The hollow circle at $\frac{4}{3}$ indicates that $\frac{4}{3}$ is **not** part of the solution set.

Example 2 Solve and graph the solution set of the inequality
$$3x - 4 \geq 0$$
Step 1: To undo the -4, add + 4 to both sides of the inequality.

$$\begin{array}{r} 3x - 4 \geq 0 \\ \underline{+4 \ +4} \\ 3x \geq 4 \end{array}$$

Step 2: Divide both sides of the inequality by 3.

$$\frac{3x}{3} \geq \frac{4}{3}$$
$$x \geq \frac{4}{3}$$

The solution set is $\{x \mid x \geq \frac{4}{3}\}$. (The solution set is all real numbers greater than or equal to $\frac{4}{3}$)

Graph for $x \geq \frac{4}{3}$ **Note**: The solid circle at $\frac{4}{3}$ indicates that $\frac{4}{3}$ is part of the solution set.

Example 3 $5x + 5 > 9x - 6$

Solution

Step 1: We undo the $9x$ by adding $-9x$ to both sides of the inequality.

$$5x + 5 > 9x - 6$$
$$\underline{-9x \qquad - 9x}$$
$$-4x + 5 > - 6$$

Step 2: To undo the 5, add -5 to both sides of the inequality.

$$-4x + 5 > - 6$$
$$\underline{\quad -5 \quad - 5}$$
$$-4x > -11$$

Step 2: Divide both sides of the inequality by -4 and **reverse the sense** of the inequality.

$$\frac{-4x}{-4} < \frac{-11}{-4} \quad \text{(change this } " > \text{' to that } " < \text{ ")}$$

$$x < \frac{11}{4}$$

The solution set is $\{x \mid x < \frac{11}{4}\}$.   (The solution set is all real numbers less than $\frac{11}{4}$.)

$$\text{Graph for } x < \frac{11}{4}$$

Example 4 Solve for x: $2(2x + 1) \geq 5(x + 2)$

Solution

$$2(2x + 1) \geq 5(x + 2)$$
$$4x + 2 \geq 5x + 10$$
$$\underline{-5x \qquad -5x} \cdot$$
$$-x + 2 \geq 10$$
$$\underline{\quad -2 \qquad -2}$$
$$-x \geq 8$$
$$x \leq -8$$

(Multiplying or dividing both the left-hand side and the right-hand side by -1 and reversing the sense of the inequality or changing the signs of both sides and reversing the sense of the inequality)

The solution set is $\{x \mid x \leq -8\}$.

Graph for $x \leq -8$

Compound Inequalities

A compound inequality is formed by connecting two inequalities with the words "**and**" or " **or**".

Example We will call the following type an "**AND**" problem. (More formally, a **conjunction** problem)
$$3x - 2 < 4 \text{ and } 2x + 10 > 4$$
Because of the connective "and", we want a solution set (a set of numbers, if any) which is common to both solution sets of the two inequalities (i.e., the intersection of the solution sets). Any solution set must satisfy both inequalities simultaneously.

Example Solve for x: $3x - 2 < 4$ and $2x + 10 > 4$

Procedure: Solve each inequality separately and find the intersection (common set of numbers) of the solutions.

$$
\begin{array}{ll}
3x - 2 < 4 \quad \text{and} & 2x + 10 > 4 \\
\underline{+2\ +2} & \underline{\ -10\ -10} \\
3x < 6 & 2x > -6 \\
x < 2 \quad \text{and} & x > -3
\end{array}
$$

The solution is the intersection of $\{x \mid x < 2\}$ and $\{x \mid x > -3\}$ i.e., $\{x \mid x < 2\} \cap \{x \mid x > -3\}$

Graph the solutions to determine if the solutions intersect.

The solution is $\{x \mid -3 < x < 2\}$

Graph for $-3 < x < 2$

Another "**AND**" problem

Example 2 Solve for x: $-5 < 2x + 1 \le 7$

Solution: The above inequality is equivalent to the compound inequality
$$-5 < 2x+1 \text{ and } 2x + 1 \le 7.$$

We can solve this compound inequality using the method used in Example 1 above, however, we will use a faster method called the "**Condensed**" or "**continued inequality**" method as follows:

$$
\begin{array}{l}
-5 < 2x + 1 \le 7 \\
\underline{-1 \qquad\ -1\ \ -1} \qquad \text{(Adding -1 to all three sides of the inequality)} \\
-6 < \ 2x \qquad \le 6
\end{array}
$$

$$\frac{-6}{2} < \frac{2x}{2} \le \frac{6}{2} \qquad \text{(Dividing each of all the three sides of the inequality by 2)}$$
$$-3 < x \le 3$$

Graph for $-3 < x \le 3$ (Note the hollow circle at -3 and the solid circle at 3)

Example 3 Solve for x: $-6x > 12$ and $x + 1 > 5$

Solution: For the first inequality, divide by -6 and reverse the sense of the inequality.

For the second inequality, add -1 to both sides of the inequality.
$$-6x > 12 \text{ and } x + 1 > 5$$
$$x < -2 \quad \text{and} \quad x > 4$$

Graph the solution set for $x < -2$ and for $x > 4$ and determine if the two graphs intersect. If they intersect, there is a solution, but if they do not intersect, then there is no solution.

$x < -2$ $\qquad\qquad$ $x > 4$

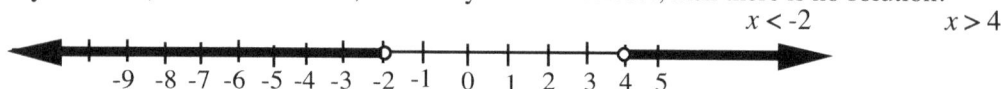

From the graph, the two solution sets do not intersect. (i.e., there are no numbers which belong to both solutions). Therefore, there is no solution or the solution set is the empty set.

An "OR " Compound Inequality Problem (More formally, a **disjunction** problem)
Example 4 Solve for x: $2x - 3 > 5$ or $3x + 1 < -8$

Because of the connective "or" we want the solution set (if any) to be the union of the solution sets of the two inequalities. The solution is the set of real numbers which satisfies either inequality or both.

$$2x - 3 > 5 \text{ or } 3x + 1 < -8$$
$$\underline{+ 3 + 3} \qquad \underline{-1 \ -1}$$
$$2x > 8 \qquad 3x < -9$$
$$x > 4 \quad \text{or} \quad x < -3$$

Same as $x < -3$ or $x > 4$

Graph

Solution set is $\{x \mid x > 4\} \cup \{x \mid x < -3\} = \{x \mid x < -3 \text{ or } x > 4\}$.

Comparison of Example 3 and Example 4

Note that even though the two problems have similar graphs, Example 3 has no solution because it is an " **and**" problem but Example 4 has a solution because it is an " **or** " problem".

More on Intervals   (Finite Intervals)

Open interval The **open** interval (Figure) from a to b , written (a, b) or $a < x < b$, (where x is real number between a and b) is the set of all numbers between a and b but excluding the end-points a and b. (" open" means the endpoints are **excluded**)

Closed interval The **closed** interval from a to b , written $[a, b]$ or $a \leq x \leq b$ (where x is real number between a and b) is the set of all numbers between a and b and including the end-points a and b. (" closed " means the endpoints are **included**)

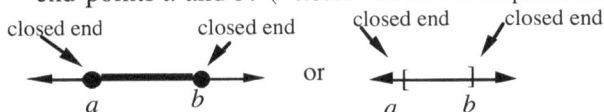

Half-closed and half-open Interval The half-closed and half-open interval from a to b is the set of all numbers between a and b, including either the endpoint a or the endpoint b but not both.

In Figure 1   the interval half-open on the left (or half-closed on the right) is the set of all numbers between a to b , including the endpoint b but excluding the endpoint a.. This interval is symbolized $(a, b]$ or $a < x \leq b$.

Figure 1

In Figure **2**, the interval half-open on the right (or half-closed on the left) is the set of all numbers between a to b , including the endpoint a but excluding the endpoint b.

This interval is symbolized $[a, b)$ or $a \leq x < b$.

Figure 2

Note: Within any interval, there are infinitely many rational numbers and infinitely many irrational numbers.

Applications (Word Problems Using Inequalities)

Example During the semester James scored 85, 70, 75 on the first three tests. He wants his class average for the term to be at least 80. There is one more test to take. What is the score he needs on the last test ?

Solution Let x be the last score.

Then, the total score for the four tests will be $85 + 70 + 75 + x$.

$$\text{Average score} = \frac{\text{Total score}}{\text{Number of tests}} = \frac{85 + 70 + 75 + x}{4}$$

Now, we want this average score to be at least 80 (i.e., 80 or more).

$$\frac{85 + 70 + 75 + x}{4} \geq 80 \qquad\qquad \text{Scrapwork}$$

$$\frac{230 + x}{4} \geq 80 \qquad\qquad 85+70+75=230$$

We solve for x.

$$\frac{4(230 + x)}{4} \geq 80(4)$$

$$230 + x \geq 320$$
$$\underline{-230 \qquad\quad -230}$$
$$x \geq 90$$

$\therefore$ James needs at least a score of 90 on the last test. (i.e., 90 or more).

Some words to help translate inequality word problems

1. " is at least " translates to " $\geq$ ".

2. " is no less than" translates to " $\geq$ ".

3 " is greater than or equal to " translates to " $\geq$"

4 " is more than" translates to " $>$ ".

5. " is greater than" translates to " $>$ ".

6. " is over " translates to " $>$ "

7. " is no more than" translates to " $\leq$ ".

8. " is at most " translates to " $\leq$ ".

9. " is less than or equal to " translates to " $\leq$ ".

10 " is less than " translates to " $<$ ".

11. " is under " translates to " $<$ "

Also: x is between a and b translates to " $a < x < b$ " .

x is greater than a but less than b translates to " $a < x < b$ " .

Note the difference between "**more than**" and "**is more than** " as in the following:

Example: 3 **more than** twice a number **is more than** 5 less than the number

If the number is x, the translation is $2x + 3 > x - 5$.   ("**Is more than**" translates to " $>$")

Lesson 53 Exercises

A Solve and graph the solution set:
1. $3x - 6 > 0$;   2. $-4x + 1 > 13$;   3. $2x - 1 \leq 7x - 11$;   4. $2(x + 4) \geq 3(x - 2)$

5. $x + 4 > -5$;   6. $x + 4 < 6$;   7. $x + 4 \leq 6$;   8. $x + 4 \geq 6$;   9. $-4x > 12$

10. $-5x \leq -30$;   11. $\frac{-x}{4} \leq 9$;   12. $\frac{x}{4} - 6 > 14$;   13. $2(x - 4) + 3 \leq 17$

Answers; **1.** $\{x \mid x > 2\}$;   **2.** $\{x \mid x < -3\}$;   **3.** $\{x \mid x \geq 2\}$;   **4.** $\{x \mid x \leq 14\}$;   **5.** $\{x \mid x > -9\}$;   **6.** $\{x \mid x < 2\}$;
7. $\{x \mid x \leq 2\}$; **8.** $\{x \mid x \geq 2\}$; **9.** $\{x \mid x < -3\}$; **10.** $\{x \mid x \geq 6\}$; **11.** $\{x \mid x \geq -36\}$; **12.** $\{x \mid x > 80\}$; **13.** $\{x \mid x \leq 11\}$;

B Solve for: **1.** $2x - 3 < 1$ and $3x + 5 > 2$;   **2.** $-4 < 3x - 1 < 5$;   **3.** $2x - 3 < 1$ and $3x - 5 > 7$
4. $-1 \leq 2x + 5 < 3$;   **5.** $3x + 7 < 1$ and $2x - 2 < 5x + 7$

Answers: **1.** $\{x \mid -1 < x < 2\}$;   **2.** $\{x \mid -1 < x < 2\}$ **3.** No solution: $x < 2$ and $x > 4$ do not intersect.
4. $\{x \mid -3 \leq x < -1\}$;   **5.** $\{x \mid -3 < x < -2\}$

C Solve for x:
1. $3x - 2 > 4$ or $3x - 1 > -7$;   **2.** $3x + 4 > 1$ or $2x + 1 < -9$;   **3.** $2x - 3 < 1$ or $3x - 5 > 7$
4. $2x + 3 < 5$ or $x + 1 > 4$;   **5.** $3x - 2 > 4$ or $x - 2 < 6$;   **6.** $x - 1 < \frac{1}{2}$ or $x + 3 > 5$

Answers: **1.** $\{x \mid x > -2\}$;   **2.** $\{x \mid x < -5$ or $x > -1\}$;   **3.** $\{x \mid x < 2$ or $x > 4\}$; **4.** $\{x \mid x < 1$ or $x > 3\}$;
5. All real numbers since $x > 2$ or $x < 8$; **6.** $\{x \mid x < \frac{3}{2}$ or $x > 2\}$;

D During the semester, Betty scored 75 and 78 on the first two tests. She wants her class average for the term to be at least 80. There is one more test to take. What is the score she needs on the last test ?

Answer: Betty needs at least a score of 87 on the last test

E Solve for x:
1. $\frac{x}{2} + 1 < x - 14$;   **2.** $3x - 2 < \frac{10 - x}{-2}$;   **3.** $\frac{x + 3}{2} \leq \frac{2 + x}{-2}$

Answers **1.** $x > 30$;   **2.** $x < -\frac{6}{5}$;   **3.** $x \leq -\frac{5}{2}$

Extra Chapters

CHAPTER 18: Quadratic Equations p.259

CHAPTER 19: Functions, Relations, Functional Notation p.271

CHAPTER 20: Sketching Parabolas p.278

CHAPTER 21: Solving Fractional (Rational Equations) p.285

CHAPTER 22: Solving Radical Equations p.289

CHAPTER 23: Basic Review For Geometry p.292

CHAPTER 18
Quadratic Equations

Lesson 54: **Solving by Factoring**
Lesson 55: **Solving by the Square Root Method**
Lesson 56: **Solving by Completing the Square**
Lesson 57: **Solving by Quadratic Formula**
Lesson 58: **Applications of the Quadratic Equation**

Standard form of the quadratic equation : $ax^2 + bx + c = 0$, where a, b, and c are constants and $a \neq 0$

Examples **1.** $x^2 - 3x - 28 = 0$ <--------in standard form
2. $x^2 - 2x = 0$ <--------- in standard form
3. $x^2 = 5x + 14$ <---------**not** in standard form
4. $5x^2 = 8x$ <---------**not** in standard form

We shall consider four methods:

Method 1: **By factoring** (For easily recognizable factors and for cases in which the constant term is missing, i.e., $c = 0$)

Method 2: **By the square root method** (For cases in which the x-term is missing i.e., $b = 0$)

Method 3: **By the quadratic formula** (This always works)

Method 4: **By completing the square** (If asked to use this method, otherwise choose from Methods 1, 2, and 3 above).

Note that Method 4 also always works. **Note also** that the quadratic formula can be derived from the quadratic equation **by completing the square** and solving for x.

Lesson 54
Solving Quadratic Equations by Factoring

Principle of zero products: If $ab = 0$ then either $a = 0$, or $b = 0$ (or both $= 0$).

Example 1   Solve by factoring
$$x^2 - 3x - 28 = 0$$

Step 1:   Factor the quadratic trinomial.

$$(x + 4)(x - 7) = 0$$

Step 2:   Set each factor equal to zero and solve each equation for x.

$$x + 4 = 0 \qquad \text{or} \qquad x - 7 = 0$$
$$\underline{-4 \ -4} \qquad\qquad \underline{+7 \ +7}$$
$$x = -4 \qquad\qquad\qquad x = 7$$

$$\therefore \ x = -4, \text{ or } x = 7$$

The solutions are -4 and 7.

Lesson 54: **Solving by Factoring**

Example 2   Solve by factoring.   260

$$3x^2 - 4x = 0$$

Solution

Step 1: Factor. $3x^2 - 4x = 0$

$x(3x - 4) = 0$ (performing common monomial factoring).

Step 2: Set each factor equal to zero and solve for x. (Only each factor containing a variable is set to zero)

$$x = 0 \quad \text{or} \quad 3x - 4 = 0$$

$$\underline{+ 4 \quad +4}$$

$$\frac{3}{3}x = \frac{4}{3}$$

$$x = \frac{4}{3}$$

$\therefore x = 0, \text{ or } x = \frac{4}{3}$ **Note** that 0 is also a solution .

The solutions are 0 and $\frac{4}{3}$

Note: A common **wrong** approach in the above problem:

$3x^2 - 4x = 0$

$3x^2 = 4x$

$3x = 4$ (Canceling, that is dividing out an x on both sides of the equation ; such a cancellation excludes $x = 0$ as a solution.)

$x = \frac{4}{3}$ (Of course, you still obtain one of the solutions but lose the other solution)

Example 3   Solve for x by factoring.

$$\frac{x^2}{2} + \frac{10x}{3} + 2 = 0$$

Step 1: Undo the denominators by multiplying the equation by the LCM of 2 and 3 which is 6.

$$(6)\frac{x^2}{2} + (6)\frac{10x}{3} + 2(6) = 6(0)$$

$$3\cancel{(6)}\frac{x^2}{\cancel{2}}_1 + \frac{2\cancel{(6)}10x}{\cancel{3}}_1 + 2(6) = 0$$

$$3x^2 + 20x + 12 = 0 \text{ ....(A)}$$

Step 2:   Factor by the substitution method or otherwise

$$3(3x^2) + 20x(3) + 12(3) = 0$$

$$9x^2 + 20(3x) + 36 = 0$$

$$(3x)^2 + 20(3x) + 36 = 0 \text{ ....(B)}$$

Step 3:   Let $3x = s$ in equation (B), and factor.

$$s^2 + 20s + 36 = 0$$

$$(s + 2)(s + 18) = 0\text{.....................................(C)}$$

Step 4:   Replace s by $3x$ in equation (C)

$$(3x + 2)(3x + 18) = 0 \text{ ....(D)}$$

Step 5:   Set each factor equal to zero and solve for x.

$$3x + 2 = 0 \qquad\qquad\qquad 3x + 18 = 0$$
$$\underline{\;-2\quad -2\;} \qquad\qquad\qquad \underline{\;- 18\quad -18\;}$$
$$\frac{3}{3}x = \frac{-2}{3} \qquad\qquad\qquad \frac{3}{3}x = \frac{-18}{3}$$
$$x = -\frac{2}{3} \qquad\qquad\qquad x = -6$$

The solutions are $-\frac{2}{3}$ and -6.   or solution set = $\{-\frac{2}{3}, -6\}$

Note that in the above problem , since it is an equation, it was not necessary to factor completely the left-hand side of equation (D) of Step 4.

Factorability of a quadratic trinomial

A quadratic trinomial is factorable if the discriminant $b^2 - 4ac$ is a perfect square. (see also p.138.)

Lesson 54 Exercises

A   Solve by factoring: **1.** $x^2 - 11x + 18 = 0$;   **2.** $x^2 - 5x - 36 = 0$;
    3. $x^2 - 18x = 0$;   **4.** $3x^2 = 24x$

   Solutions 1. {2, 9};   **2.** {- 4, 9};   **3.** {0,18};   **4.** {0,8}

B   Solve by factoring: **1.** $x^2 - 4x - 21 = 0$;   **2.** $x^2 + 4x - 45 = 0$;   **3.** $x^2 - 25 = 0$

   4. $9x^2 - 6x = 0$ **5.** $14t = 7t^2$ **6.** $x^2 = 11x - 18$

   7. $2x^2 + 2x - 144 = 0$ **8.** $9x^2 - 36 = 0$ **9.** $ax^2 = -bx$

   10. $x^2 - mx + nx - mn = 0$

   Answers: **1.** { -3, 7} ;   **2.** {-9, 5} ;   **3.** {-5, 5} ; **4.** {0, $\frac{2}{3}$} ;   **5.** {0, 2};   **6.** {2, 9} ;   **7.** {-9, 8} ;   **8.** {-2, 2} ;

   9. {0, -$\frac{b}{a}$ } ;   **10.** {$m, -n$}

Lesson 55

Solving Quadratic Equations by the Square Root Method
(For cases in which $b = 0$, i.e., the x-term is missing)

Principle: If $x^2 = k$, then $x = \pm\sqrt{k}$ (i.e., $x = +\sqrt{k}$ or $x = -\sqrt{k}$)

Example 1 Solve by the square root method.
$$x^2 - 9 = 0$$

Solution
$$x^2 - 9 = 0$$
$$\underline{+9 \quad +9}$$
$$x^2 = 9$$
$$x = \pm\sqrt{9}$$
$$x = \pm 3 \quad \text{(i.e. } x = 3 \text{ or } x = -3)$$
The solutions are -3 and 3.

Example 2 Solve by the square root method.
$$9x^2 - 36 = 0$$

Solution
$$9x^2 - 36 = 0$$
$$\underline{+36 \quad +36}$$
$$9x^2 = 36$$
$$\frac{9}{9}x^2 = \frac{36}{9}$$
$$x^2 = 4$$
$$x = \pm\sqrt{4}$$
$$x = \pm 2$$
The solutions are -2 and 2.

Lesson 55 Exercises

A Solve by the square root method:

1. $x^2 - 16 = 0$;   2. $4x^2 - 20 = 0$;   3. $x^2 - 49 = 0$;

Read page 427-428 before attempting Problem 4:   **4.** $x^2 + 49 = 0$

Answers: **1.** $\{-4, 4\}$;   **2.** $\{-\sqrt{5}, \sqrt{5}\}$;   **3.** $\{-7, 7\}$; 4. $\{-7i, 7i\}$

B Solve the following for x:

1. $x^2 - 9 = 0$ 2. $-8 + x^2 = 0$ 3. $ax^2 - c = 0$

4. $x^2 + 9 = 0$ 5. $s = ax^2$ 6. $x^2 - \frac{1}{4} = 0$

Answers: **1.** $\{-3, 3\}$;   **2.** $\{-2\sqrt{2}, 2\sqrt{2}\}$;   **3.** $\{-\sqrt{\frac{c}{a}}, \sqrt{\frac{c}{a}}\}$;   **4.** $\{-3i, 3i\}$;

5 $\{-\sqrt{\frac{s}{a}}, \sqrt{\frac{s}{a}}\}$;   **6.** $\{-\frac{1}{2}, \frac{1}{2}\}$

Lesson 56
Solving Quadratic Equations by Completing the Square

Example 1 Solve by completing the square.

$$x^2 - 12x + 8 = 0$$

Step 1: Eliminate the constant term (the "8")
from the left-hand side.   (Note that the 8 ends up on the right-hand side of the equation as -8).

$$x^2 - 12x + 8 = 0$$
$$\underline{\quad\quad -8 \quad -8}$$
$$x^2 - 12x = -8$$

Step 2: Add the square of half the coefficient of the x-term to both sides of the equation.
(i.e., add the square of $\frac{b}{2}$ to both sides of the equation)

$$x^2 - 12x + \left(\tfrac{-12}{2}\right)^2 = -8 + (-6)^2 \quad (b = -12, \tfrac{b}{2} = -6; \text{ and the square of } \tfrac{b}{2} = (-6)^2$$
$$x^2 - 12x + (-6)^2 = -8 + (-6)^2$$

Step 3: Complete the square on the left-hand side of the equation
$$(x - 6)^2 = -8 + 36$$
$$(x - 6)^2 = 28$$

Step 4: " Take the square root " of both sides of the equation.

$$x - 6 = \pm\sqrt{28}$$

Step 5: Solve for x and simplify right-hand side.
$$x - 6 = \pm\sqrt{28}$$
$$\underline{+6 \quad +6}$$
$$x = 6 \pm\sqrt{28}$$
$$\mathbf{x = 6 \pm 2\sqrt{7}}$$

Scrapwork:
$$\sqrt{28} = \sqrt{4}\sqrt{7} = 2\sqrt{7}$$

Example 2   Solve by completing the square.   264

$$3x^2 - 9x - 2 = 0$$

Step 1: $3x^2 - 9x = 2$

Step 2:   Divide the equation by the coefficient of the x^2-term (we want this coefficient to be 1, for this method)

$$\frac{3x^2}{3} - \frac{9x}{3} = \frac{2}{3}$$

$$x^2 - 3x = \frac{2}{3}$$

Step 3:   Add the square of the coefficient of the x-term to both sides of the equation, and complete the square.

$$x^2 - 3x + \left(\frac{-3}{2}\right)^2 = \frac{2}{3} + \left(\frac{-3}{2}\right)^2 \qquad \textbf{Note:} \quad \frac{b}{2} = \frac{-3}{2} \; ; \; \left(\frac{b}{2}\right)^2 = \left(\frac{-3}{2}\right)^2$$

$$\left(x - \frac{3}{2}\right)^2 = \frac{2}{3} + \frac{9}{4}$$

Multiply this out

$$\left(x - \frac{3}{2}\right)^2 = \frac{35}{12}$$

Scrapwork:

$$\frac{2}{3} + \frac{9}{4} = \frac{35}{12}$$

$$x - \frac{3}{2} = \pm\sqrt{\frac{35}{12}}$$

Step 4: Solve for x and simplify right-hand side.

$$x - \frac{3}{2} = \pm\sqrt{\frac{35}{12}}$$
$$+\frac{3}{2} \qquad\qquad +\frac{3}{2}$$
_____

$$x = +\frac{3}{2} \pm \sqrt{\frac{35}{12}}$$

$$x = \frac{3}{2} \pm \frac{\sqrt{105}}{6}$$

Scrapwork: $\sqrt{\frac{35}{12}} = \sqrt{\frac{35 \cdot 3}{12 \cdot 3}}$

Solution is $\left\{\frac{3}{2} \pm \frac{\sqrt{105}}{6}\right\}$ or $\left\{\frac{3}{2} + \frac{\sqrt{105}}{6} , \frac{3}{2} - \frac{\sqrt{105}}{6}\right\}$

Lesson 56 Exercises

A Solve by completing the square:
1. $x^2 + 12x + 10 = 0$;   **2.** $x^2 - 4x + 6 = 0$;   **3.** $x^2 - 3x - 8 = 0$

4. $x^2 - 12x - 35 = 0$

Answers: **1.** $\{ -6 + \sqrt{26}, -6 - \sqrt{26} \}$; **2.** $\{2 + i\sqrt{2},\ 2 - i\sqrt{2}\}$;   **3.** $\{ \frac{3}{2} + \frac{\sqrt{41}}{2}, \frac{3}{2} - \frac{\sqrt{41}}{2} \}$;

4. $\{6 + \sqrt{71}, 6 - \sqrt{71} \}$

B Solve by completing the square:
1. $x^2 - 11x + 18 = 0$;   **2.** $2x^2 + 5x - 12 = 0$;   **3.** $x^2 - 4x - 1 = 0$;   **4.** $x^2 - 4x + 8 = 0$.

5. $4x^2 + 48x + 40 = 0$;   **6.** $2x^2 - 7x - 12 = 0$;   **7.** $\frac{1}{2}at^2 + bt + k = 0$;   **8.** $ax^2 + bx + c = 0$

Solutions: **1.**$\{2, 9\}$; **2.** $\{-4, \frac{3}{2}\}$; **3.** $\{ 2 + \sqrt{5}, 2 - \sqrt{5}\}$;   **4.** $\{2 + 2i, 2 - 2i\}$; **5.** $\{-6 + \sqrt{26}, -6 - \sqrt{26} \}$;

6. $\{ \frac{7}{4} + \frac{\sqrt{145}}{4}, \frac{7}{4} - \frac{\sqrt{145}}{4} \}$;   **7.** $\{-\frac{b}{a} \pm \frac{\sqrt{b^2 - 2ak}}{a} \}$; **8.** $x = -\frac{b}{2a} \pm \frac{\sqrt{b^2 - 4ac}}{2a}$

Lesson 57

Solving Quadratic Equations by the Quadratic Formula

Example 1   Solve for x by the quadratic formula

$$3x^2 - 6x - 2 = 0$$

Step 1: $a = 3,\ b = -6,\ c = -2$

Step 2: $x = \dfrac{-b \pm \sqrt{b^2 - 4ac}}{2a}$

Step 3: $x = \dfrac{-(-6) \pm \sqrt{(-6)^2 - 4(3)(-2)}}{2(3)}$ (Substituting for a, b, and c)

$$= \frac{+6 \pm \sqrt{36 + 24}}{6}$$

$$= \frac{6 \pm \sqrt{60}}{6}$$

$$= \frac{6}{6} \pm \frac{\sqrt{60}}{6}$$

$$= 1 \pm \frac{2\sqrt{15}}{6}$$

$$x = 1 \pm \frac{\sqrt{15}}{3}$$

Solution set is $\left\{ 1 - \dfrac{\sqrt{15}}{3},\ 1 + \dfrac{\sqrt{15}}{3} \right\}$

Example 2 Solve by the quadratic formula:

$$x^2 - 16 = 8 - 2x$$

Step 1: Place the equation in standard form (i.e. rewrite the equation so that the only term on the right-hand side is zero)

$$\begin{array}{rcl} x^2 - 16 & = & 8 - 2x \\ +2x & & +2x \end{array}$$

$$\begin{array}{rcl} x^2 + 2x - 16 & = & 8 \\ -8 & & -8 \end{array}$$

$$x^2 + 2x - 24 = 0 \ \text{<------------(Standard Form)}$$

Step 2: $a = 1, \ b = 2, \ c = -24$

Step 3: $$x = \frac{-b \pm \sqrt{b^2 - 4\ ac}}{2a} \quad \text{<----------The quadratic formula}$$

Step 4 : $$x = \frac{-2 \pm \sqrt{(2)^2 - 4(1)(-24)}}{2(1)} \quad \text{(Substituting for } a, b, \text{ and } c)$$

$$x = \frac{-2 \pm \sqrt{4 + 96}}{2}$$

$$x = \frac{-2 \pm \sqrt{100}}{2}$$

$$x = \frac{-2 \pm 10}{2}$$

$$x = \frac{-2 + 10}{2} , \text{ or } x = \frac{-2 - 10}{2}$$

$$x = \frac{8}{2}, \text{ or } x = \frac{-12}{2}$$

$$x = 4 \text{ or } x = -6$$

The solutions are 4 and -6.

Solving by any method

Example Solve by any method: $x^2 = 6x$
$$x^2 = 6x$$
Step 1: Write the equation in standard form.
$$x^2 - 6x = 0$$
Step 2: We solve by factoring (since $c = 0$, the quadratic is easily factorable)
$$x(x - 6) = 0$$
Step 3: Set each factor equal to 0 and solve for x:
$$\begin{array}{rcl} x = 0 \text{ , or} & x - 6 = & 0 \\ & +6 & +6 \\ & x = & +6 \end{array}$$
$x = 0$, or $x = 6$.   The solutions are 0 and 6; or the solution set is $\{0, 6\}$.

Discriminant of the quadratic equation: nature of roots and graphs of the equation

The expression $b^2 - 4ac$ is called the **discriminant** of the quadratic equation. The value of the discriminant determines the nature of the roots (or solutions) of the quadratic equation:

1. If $b^2 - 4ac > 0$, the equation $ax^2 + bx + c = 0$ has two real and unequal roots. Also if $b^2 - 4ac$ is a perfect square, the roots are rational. Graphically, the curve of the corresponding quadratic function, $f(x) = ax^2 + bx + c$, crosses the x-axis at two different points.

2. If $b^2 - 4ac = 0$, the quadratic equation has real equal roots (double root or repeated root). We therefore have only one real solution. Graphically, the curve of the corresponding quadratic function $f(x) = ax^2 + bx + c$, touches the x-axis but does not cross it.

3. If $b^2 - 4ac < 0$, the equation has two non-real (complex) roots. Graphically, the curve of the corresponding quadratic function, $f(x) = ax^2 + bx + c$, does not touch or cross the x-axis. The curve is either entirely above the x-axis or entirely below the x-axis.

Note also that a quadratic equation is factorable if $b^2 - 4ac$ is a perfect square. (see also p.138.)

Lesson 57 Exercises

A Solve by the quadratic formula:
1. $x^2 - 11x + 18 = 0$;   **2.** $2x^2 + 5x - 12 = 0$;   **3.** $x^2 - 4x - 1 = 0$; **4.** $x^2 - 4x + 8 = 0$.
5. $x^2 + 12x + 10 = 0$;   **6.** $x^2 + 6 - 4x = 0$ **7.** $x^2 = 3x + 8$

Solutions: **1.** $\{2, 9\}$;   **2.** $\{-4, \frac{3}{2}\}$;   **3.** $\{2 + \sqrt{5}, 2 - \sqrt{5}\}$;   **4.** $\{2 + 2i, 2 - 2i\}$; **5.** $\{-6 \pm \sqrt{26}\}$

6. $\{2 \pm i\sqrt{2}\}$; **7.** $\{\frac{3 \pm \sqrt{41}}{2}\}$

B Solve by any applicable method:
1. $x^2 - 6x + 4$;   **2.** $6x^2 + 11x - 10 = 0$;   **3.** $2x^2 - 3x + 6 = 0$;   **4.** $2x (x - 4) = 6$

Solutions: **1.** $\{3 + \sqrt{5}, 3 - \sqrt{5}\}$;   **2.** $\{\frac{2}{3}, -\frac{5}{2}\}$;   **3.** $\{\frac{3 + i\sqrt{39}}{4}, \frac{3 - i\sqrt{39}}{4}\}$ **4.** $\{2 + \sqrt{7}, 2 - \sqrt{7}\}$

Lesson 58

Applications of the Quadratic Equation

Example 1   The width of a rectangle is 3 units less than the length. If the area of the rectangle is 28 sq. units, determine the dimensions of this rectangle.

Solution
Let the length of the rectangle $= x$.
Then, the width of this rectangle $= (x - 3)$
Area of a rectangle is given by LW (Where L is the length and W is the width).
The required equation: $x(x - 3) = 28$.
$$x^2 - 3x - 28 = 0$$
$$(x - 7)(x + 4) = 0$$
$$x = 7 \text{ or } x = -4 \qquad \text{(Solving by factoring)}$$
We reject the negative value, -4, since the dimension of a rectangle cannot be negative.
When the length is 7 units, the width is 7 - 3 = 4 units.

(**Check**: If the length is 7 and the width is 4, the area is 7(4) = 28 sq. units; and also the width, 4, is 3 less than the length, 7. Therefore, the conditions in the original word problem have been satisfied.)

The length is 7 units and the width is 4 units.

Example 2   The perimeter of a rectangle is 34 units, and its area is 30 sq. units. Find the dimensions of this rectangle.

Preliminaries:
Perimeter of a rectangle $= 2L + 2W$ (where L is the length, and W is the width)
Area of a rectangle $= LW$

Solution

Let the dimensions be x and y, where the larger of x and y is the length and the smaller of x and y is the width.

$$2x + 2y = 34 \text{ <-----------Perimeter equation} \qquad (1)$$
$$xy = 30 \text{ <-----------Area equation} \qquad (2)$$

We will solve the above system of equations simultaneously by the substitution method.
Step 1 : Solve for y from equation (1).
    Then, $y = (17 - x)$ (3)   (obtained by subtracting $2x$ from both sides of the equation followed by dividing the equation by 2)

Step 2: Substitute for y from (3) in (2).
$$x(17 - x) = 30$$
$$17x - x^2 = 30$$
$$17x - x^2 - 30 = 0$$
$$x^2 - 17x + 30 = 0 \qquad \text{(Multiplying the equation by -1 and rearranging terms)}$$

Step 3: Solve this quadratic equation by any method.
       We solve by factoring since the factors are easily recognizable.
       $(x - 2)(x - 15) = 0$
       $x = 2$ or $x = 15$

Step 4: When $x = 2$,
       $2y = 30$; and from which $y = 15$.
       when $x = 15$,
       $15y = 30$; and from which $y = 2$.
From Step 4, we obtain the dimensions 15 and 2.

The dimensions are 15 units and 2 units. (The length is 15 units and the width is 2 units)

Lesson 58 Exercises

A The perimeter of a rectangle is 36 units, and its area is 77 sq. units.
  Find the dimensions of this rectangle.

Answers: Dimensions are 11 and 7.

B 1. A number is 4 more than another number. If their product is 165, find them.

2. The sum of two numbers is 20, and their product is 96. Find these numbers.

3. The length of a rectangle is 5 units more than the width. If the area of this rectangle is 126 sq, units, find the length and width of this rectangle.

Ans: **1.** 15 and 11; and also -11 and -15;   **2.** 8 and 12;   **3.** Length = 14, width = 9.

CHAPTER 19
FUNCTIONS

Lesson 59: **Sets, Relations, Functions, Comparison of Relations and Functions**

Lesson 60: **Functional Notation; Defined Functions; Excluded Values, Domain and Range**

Lesson 59

Sets, Relations, Functions, Comparison of Relations and Functions

Ordered Pair

An **ordered pair** of numbers is an arrangement of two numbers in a specified order. In an x-y rectangular coordinate system of axes, the first element (or component) is the x-value and the second element is the y-value.

Example 1   (*a*) $(1, 2)$ <--- $(x = 1, y = 2)$
           (*b*) $(2, 3)$ <---- $(x = 2, y = 3)$
           (*c*) $(5, -1)$ <---- $(x = 5, y = -1)$

Note that each ordered pair represents a point in an x-y coordinate system of axes.

Set of numbers

A **set of numbers** is a well-defined collection of numbers. The numbers are called the elements or members of the set.

Example 2: If we denote the set of the numbers 2, 5 and 6 by A, then we may write $A = \{2, 5, 6\}$

Example 3: The set B of the ordered pairs $(1, 2), (2, 3),$ and $(5, -1)$ is given by
         $B = \{(1, 2), (2, 3), (5, -1)\}$.

Relation

A **relation** is a set of ordered pairs. (A collection of ordered pairs of numbers)

Example 4: The set $E = \{(2, 3), (2, 5), (4, 6)\}$ is a relation, <---There are three ordered pairs.

Example 5: The set $C = \{(6, 2), (7, 4), (11, 5)\}$ is a relation.

Definition of a Function

A function may be defined in a number of ways, namely,

(a) in terms of ordered pairs; (b) in terms of a rule involving two variables;
(c) in terms of a rule for inputs and outputs; (d) in terms of correspondence of two sets; (e) as a graph

Definition 1: In terms of ordered pairs
A **function** is a relation in which no two distinct ordered pairs have the same first component; or a function is a set of ordered pairs in which for any two different ordered pairs, the first elements are different. The set in Example 5, above, is a function but the set in Example 4 is not a function, because the first two ordered pairs have the same first element, namely 2.

The set of all the first elements of the ordered pairs is called the **domain** of the function; and the set of all the second elements is called the **range** of the function.

Example 6: In the function $C = \{(6, 2), (7, 4), (11, 5)\}$. The domain, D = $\{6, 7, 11\}$ (first elements)
                                     The range, R = $\{2, 4, 5\}$. (second elements)

Other definitions of a function

Definition 2: If x and y are two variables. then we say that y is a function of x if there is a rule which gives just one corresponding value of y for **each** value of x. The variable x is called the independent variable, and a variable y is called the dependent variable. The rule may be specified in the form of a set, in the form of a graph, in the form of a table, or in the form of an equation or formula.

The **domain** of a function is the set of numbers that can be assigned to x (the independent variable).
The **range** of a function is the set of all the corresponding numbers y (the dependent variable) associated by the function (rule) with the numbers, x, in the domain .

We symbolize that f is a function of x by $f(x)$, where x is called the independent variable, y is called the dependent variable. **Note**: $f(x)$s is read as f of x.

The following are examples of how the rules for functions may be specified:

(a) In the form of an equation or a formula: $y = 2x$.
(b) In the form of a set: $\{(2, 3), (1, 4), (7, 5)\}$.
(c) In the form of a table for x and y: See Table 1.
(d) In the form of a graph. See Figure

Table 1:: $y = 2x$

$x =$	0	1	2	3	4
$y =$	0	2	4	6	8

Figure: Graph of $y = 2x$

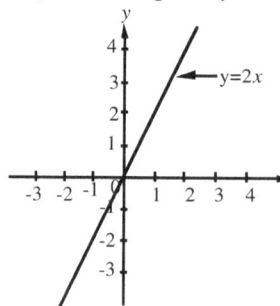

Definition 3 In terms of the correspondence of two sets (Fig. 1)

A function is a correspondence between a first set, say set A and a second set, say Set B such that **each** element of set A corresponds to exactly one element of set B. The set of all the elements of set A is a called the **domain** of the function, and the set of all the corresponding elements of set B is called the **range** of the function.

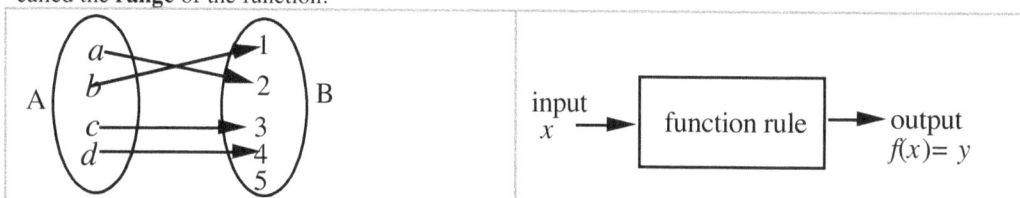

Fig 1   **Fig 2**

Definition 4   In terms of a rule for inputs and outputs (Fig. 2)

A function is a rule which assigns to each input number exactly one output number. The set of all input numbers that the rule is applicable to is called the **domain** of the function; and the set of all the corresponding output numbers is called the **range** of the function.

A variable representing an input number is called the independent variable, and a variable representing an output number is called the dependent variable.

Given a graph   (Vertical line test)

A given graph is that of a function if every possible vertical line drawn to intersect the graph intersects (cuts) the graph exactly once (i.e., at one point only).

Comparison of a Function and a Relation

Similarities: Each is a set of ordered pairs.
Differences: In a relation, two or more ordered pairs may have the same first component; but in a function, no two distinct ordered pairs may have the same first component.
Example: The set $D = \{(1, 6), (3, 4), (3, 5), (4, 6)\}$ is only a relation and **not** a function. because the second and third ordered pairs have the same first component, which is 3.

Example: The set $E = \{(1, 2), (2, 3), (4, 5), (7, 5)\}$ is a function (even though the second components of the third and fourth ordered pairs are the same).

A function is a relation, but a relation is not necessarily a function.

Determining if a given graph is a relation or a function

We will use the so-called **vertical line test**.
Procedure
Step 1: Draw as many vertical lines as possible (This can be done visually.) to intersect the graph.

Step 2: If any of the possible lines intersects (cuts) the graph at more than one point, then the given graph is not a function but a relation. However, if each of the possible vertical lines intersects the graph only once (at one point only), then the graph represents a function. Figure.. is a graph which is a relation but not a function.

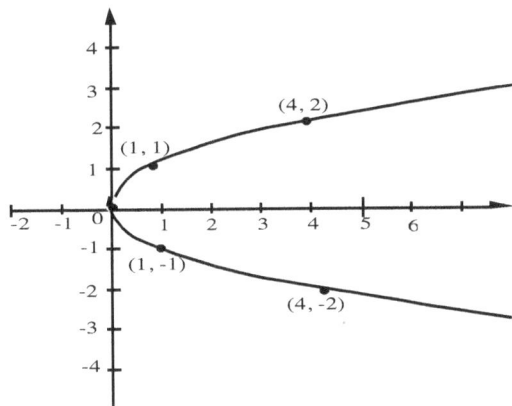

Figure: Graph of $y = \pm\sqrt{x}$ or $x = y^2$
This graph is a relation but not a function

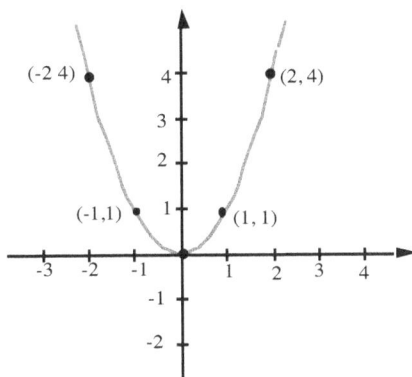

Figure: Graph of $y = x^2$.
This graph is that of a function.

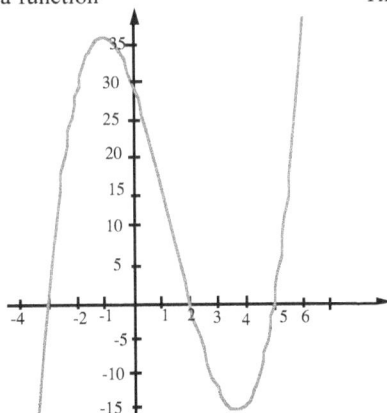

Figure: Graph of $y = (x - 5)(x - 2)(x + 3)$. This graph is that of a function.

Lesson 59 Exercises

Determine which of the following are graphs of functions.

Figure (a)

Figure (b)

Figure (c)

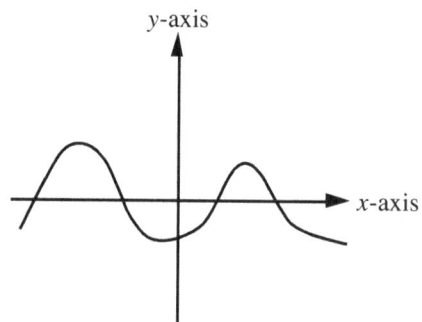

Figure (d)

Answers: (a) A function; (b) Not a function; (c) Not a function; (d) A function

Lesson 60
Functional Notation

Functional Notation

Let a function $f(x)$ be specified by the rule $f(x) = x^2 + 3$.   (1)

In equation (1), $f(x)$ is read "f of x" or f is a function of x.

To evaluate a function for a particular value of x, we substitute that value of x in the rule that defines $f(x)$. Note that $f(x)$ does not mean f times x but that $f(x)$ is written as a symbol.

Example 1   Given that $f(x) = x^2 + 3$, find (a) $f(-1)$.; (b) $f(x_0 + h) - f(x_0)$

Solution: (a) $f(x) = x^2 + 3$

$\qquad\qquad f(-1) = (-1)^2 + 3$ $\qquad\qquad\qquad$ (replacing x in the given equation by -1)
$\qquad\qquad\qquad = 1 + 3$
$\qquad\qquad f(-1) = 4$

(b) (Replace x in the given equation by $(x_0 + h)$ and x_0, accordingly in $f(x) = x^2 + 3$

$$f(x_0 + h) \; - \; f(x_0) = [(x_0 + h)^2 + 3] - (x_0{}^2 + 3)$$
$$= [x_0{}^2 + 2hx_0 + h^2 + 3] - x_0{}^2 - 3$$
$$= x_0{}^2 + 2hx_0 + h^2 + 3 - x_0{}^2 - 3$$
$$= 2hx_0 + h^2$$

Example 2 Find $f(2)$, given that $f(x) = x + 7$

Solution

$\qquad\qquad f(x) = x + 7$
$\qquad\qquad f(2) = 2 + 7$
$\qquad\qquad\quad\; = 9$

Example 3   If $f(x) = 2 - \dfrac{1}{x-4}$, find $(a) f(-3)$; (b) $f(x_0 + h)$; (b) $f(-x)$.

Solution (a) $f(x) = 2 - \dfrac{1}{x-4}$

$\qquad\qquad f(-3) = 2 - \dfrac{1}{(-3)-4}$ $\qquad\qquad$ (replacing x in the given equation by -3)

$\qquad\qquad\qquad = 2 - \dfrac{1}{-7}$

$\qquad\qquad\qquad = 2 + \dfrac{1}{7}$

$\qquad\qquad\qquad = 2\frac{1}{7}$

(c)   (Replace x in the given equation by $x_0 + h$

$$f(x) = 2 - \frac{1}{x - 4} \quad <---- \text{given equation}$$

$$f(x_0 + h) = 2 - \frac{1}{(x_0 + h) - 4} \quad <--- \text{replacing } x \text{ by } x_0 + h$$

$$= 2 - \frac{1}{x_0 + h - 4}$$

$$= \frac{2(x_0 + h - 4) - 1}{x_0 + h - 4}$$

$$= \frac{2x_0 + 2h - 8 - 1}{x_0 + h - 4}$$

$$= \frac{2x_0 + 2h - 9}{x_0 + h - 4}$$

(c)   (Replace x in the given equation by $-x$)

$$f(x) = 2 - \frac{1}{x - 4} \quad <---- \text{given equation}$$

$$f(-x) = 2 - \frac{1}{(-x) - 4} \quad <----- (replacing\ x\ by - x)$$

$$= 2 - \frac{1}{-x - 4}$$

$$= \frac{2(-x - 4) - 1}{-x - 4}$$

$$= \frac{-2x - 8 - 1}{-x - 4}$$

$$= \frac{-2x - 9}{-x - 4}$$

$$= \frac{-(2x + 9)}{-(x + 4)} \qquad (factoring\ out - 1)$$

$$= \frac{2x + 9}{x + 4}$$

Note above that in (b) the final result contains x. This is so, because we replaced x by $-x$.
In the case of (a), we replaced x by the integer -3 , and the final result was purely a numerical
value. Furthermore, in Example 3, $f(-a) = \dfrac{2a + 9}{a + 4}$

Lesson 60 Exercises

A

1. If $f(x) = x^2 - 5x + 2$, find (a) $f(-2)$; (b) $f(-1)$, (c) $f(-x)$, (d) $f(a+h)$, (e) $\dfrac{f(a+h) - f(a)}{h}$;

2. If $f(x) = 4 - \dfrac{1}{x-2}$, find (a) $f(-3)$; (b) $f(-x)$; (c) $f(a)$.

3. If $f(x) = x^3 - x^2 - x - 1$, find $f(-1)$.

Answers: **1.** (a) 16; (b) 8; (c) $x^2 + 5x + 2$; (d) $a^2 - 5a + 2ah - 5h + h^2 + 2$; (e) $2a + h - 5$

2. (a) $4\frac{1}{5}$; (b) $\dfrac{4x+9}{x+2}$; (c) $\dfrac{4a-9}{a-2}$; **(3)** -2

CHAPTER 20

Sketching Parabolas

Lesson 61: **By General method**

Lesson 62: **By location of vertex, axis of symmetry and intercepts.**

The quadratic functions $y = f(x) = ax^2 + bx + c$ $(a \neq 0)$ and their graphs

We will cover two methods: a general method and the method of the location of the vertex, axis of symmetry and the x- and y-intercepts.

Lesson 61

General method of sketching the graph of the quadratic function

Example 1   Sketch the graph of $y = x^2 - 5$ for $x = 0, \pm 1, \pm 2, \pm 3$.

ordered pairs

Step 1: When $x = 0$, $y = (0)^2 - 5$
$\quad\quad\quad\quad y = -5$ $\quad\}\longrightarrow (0,-5)$

When $x = 1$, $y = (1)^2 - 5$
$\quad\quad\quad\quad y = 1 - 5$
$\quad\quad\quad\quad y = -4$ $\quad\}\longrightarrow (1,-4)$

When $x = 2$, $y = (2)^2 - 5$
$\quad\quad\quad\quad y = 4 - 5$
$\quad\quad\quad\quad y = -1$ $\quad\}\longrightarrow (2,-1)$

When $x = 3$, $y = (3)^2 - 5$
$\quad\quad\quad\quad y = 9 - 5$
$\quad\quad\quad\quad y = 4$ $\quad\}\longrightarrow (3,4)$

When $x = -1$, $y = (-1)^2 - 5$
$\quad\quad\quad\quad y = 1 - 5$
$\quad\quad\quad\quad y = -4$ $\quad\}\longrightarrow (-1,-4)$

When $x = -2$, $y = (-2)^2 - 5$
$\quad\quad\quad\quad y = 4 - 5$
$\quad\quad\quad\quad y = -1$ $\quad\}\longrightarrow (-2,-1)$

When $x = -3$, $y = (-3)^2 - 5$
$\quad\quad\quad\quad y = 9 - 5$
$\quad\quad\quad\quad y = 4$ $\quad\}\longrightarrow (-3,4)$

Step 2: Plot the points $(0,-5)$, $(1,-4)$, $(2,-1)$, $(3,4)$, $(-1,-4)$, $(-2,-1)$ and $(-3,4)$ on a rectangular coordinate system of axes (on graph paper) and connect the points by a smooth curve noting that the parabola is U-shaped and that since the coefficient of the x^2-term is positive, the parabola opens upwards (**Figure 1**).

Note above that the y-values are the same for $x = 1$ and $x = -1$; the same for $x = 2$ and $x = -2$ and the same for $x = 3$ and $x = -3$.

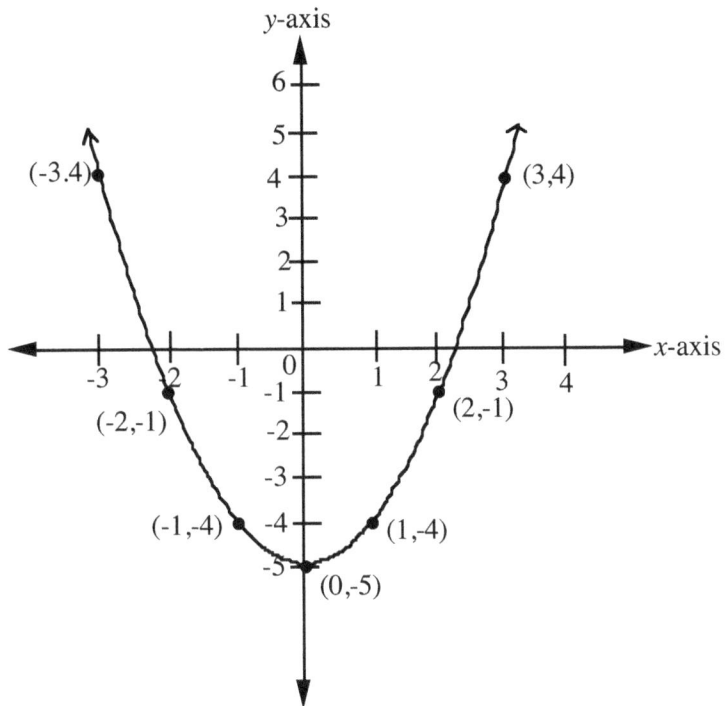

Figure 1:   Graph of $y = x^2 - 5$

Lesson 61: By General method

Example 2 Sketch the graph of $y = -x^2 + 5$ for $x = 0, \pm 1, \pm 2, \pm 3$.

Step 1: See Example 1 above and perform similar calculations to obtain
(0,5), (1,4), (2,1), (3,-4), (-1,4), (-2,1), (-3,-4).

Step 2: Plot the points obtained from Step 1 on a rectangular coordinate system of axes (on graph paper) and connect the points by a smooth curve noting that the parabola is U-shaped and that since the coefficient of the x^2-term is negative (the coefficient of $-x^2$ is -1) , the parabola opens downwards (inverted U-shape). See Figure 2 below.

Figure 2: Graph of $y = -x^2 + 5$

Lesson 61 Exercises

Sketch the graphs of the following:
1. $y = x^2 - 4$ for $x = 0, \pm 1, \pm 2, \pm 3$;   **2.** $y = -x^2 + 1$ for $x = 0, \pm 1, \pm 2, \pm 3$.

1. Parabola opens upwards; **2**. parabola opens downwards

Lesson 62
By location of vertex, axis of symmetry and intercepts

An alternative method of sketching the graphs of quadratic functions is by the location of the vertex, axis of symmetry and the x- and y-intercepts.

Example 1 Given the equation of the parabola $y = f(x) = x^2 - 6x + 8$, perform the following:
1. Find the x- and y-intercepts.
2. Find the axis of symmetry.
3. Find the coordinates of the vertex (or turning point)
4. Find the domain and range.
5. Sketch the graph of $y = f(x) = x^2 - 6x + 8$ (The graph of a quadratic function is called a parabola.)

Solution

1. To find the x-intercept, let $y = 0$ in $y = x^2 - 6x + 8$.

Then, $0 = x^2 - 6x + 8$ or
$x^2 - 6x + 8 = 0$ <-------This is a quadratic equation. Solve by any method
$(x - 2)(x - 4) = 0$ <------We solve by factoring since the factors are easily recognizable.
Solving, $x = 2$, or $x = 4$
The x-intercepts are 2 and 4 . (The x-intercepts are at the points (2,0), and (4,0))

To find the y-intercept, let $x = 0$ in $y = x^2 - 6x + 8$
Then, $y = (0)^2 - 6(0) + 8$
$y = 0 - 0 + 8$
$y = 8$

The y-intercept $= 8$.   (The y-intercept is at the point (0,8).)

2. $a = 1$ (the coefficient of the x^2-term), $b = -6$ (the coefficient of the x-term).

Axis of symmetry is the line $x = -\dfrac{b}{2a}$ <------formula.

$= -\dfrac{-6}{2(1)}$

$= 3$

The axis of symmetry is the line $x = 3$. (see also p. 168)
The x-coordinate of the vertex $= 3$,

The y-coordinate of the vertex $= y = f(-\dfrac{b}{2a}) = f(3) = 3^2 - 6(3) + 8$

$= 9 - 18 + 8$
$y = -1$.

3. The vertex is at (3,-1). This point is the turning point on this curve. It is also a minimum point, since a is positive. This minimum point is also the lowest point on this curve.

4. The domain is all real x. (The given function is a polynomial.)
The range is specified by using the y-coordinate of the vertex obtained in Step 2.
The range is given by the inequality $y \geq -1$.
The range consists of all real numbers greater than or equal to -1.
(Note: If we did not know the x-coordinate of the vertex, we could obtain the range by

Substituting $a = 1$. $b = -6$, $c = 8$, in $y \geq \dfrac{4ac - b^2}{4a}$ (Note the " $\geq$") to obtain

$y \geq \dfrac{4(1)(8) - (-6)^2}{4(1)} \geq \dfrac{32 - 36}{4} \geq \dfrac{-4}{4} \geq -1$; That is $y \geq -1$)

5. To sketch the graph: On graph paper and using a rectangular coordinate system of axes,

Step 1: Locate the x- and y-intercepts from (**1**) above, by plotting the points (2,0), (4,0) and (0,8).

Step 2: Using a broken line draw the line $x = 3$, the axis of symmetry (from (**2**) above)

Step 3: Locate the vertex by plotting the point (3,-1). (from (**3**) above).

Step 4: Locate the point B using symmetry as follows: From the point A (the y-intercept), count horizontally say h units ($h = 3$ in this problem) to the axis of symmetry (the line $x = 3$). Next, continue and count horizontally h units (h = 3) from the axis of symmetry and stop at the point B (this is another point on the curve). You may also locate the point B by the intersection of the lines $y = 8$ and $x = 6$.

Step 5: Connect the above points by a smooth U-shaped curve (**Figure 1**), noting that the parabola opens upwards (is concave up), since a is positive ($a = 1$), and also noting that the graph is symmetric about the axis of symmetry (the line $x = 3$, in this example).

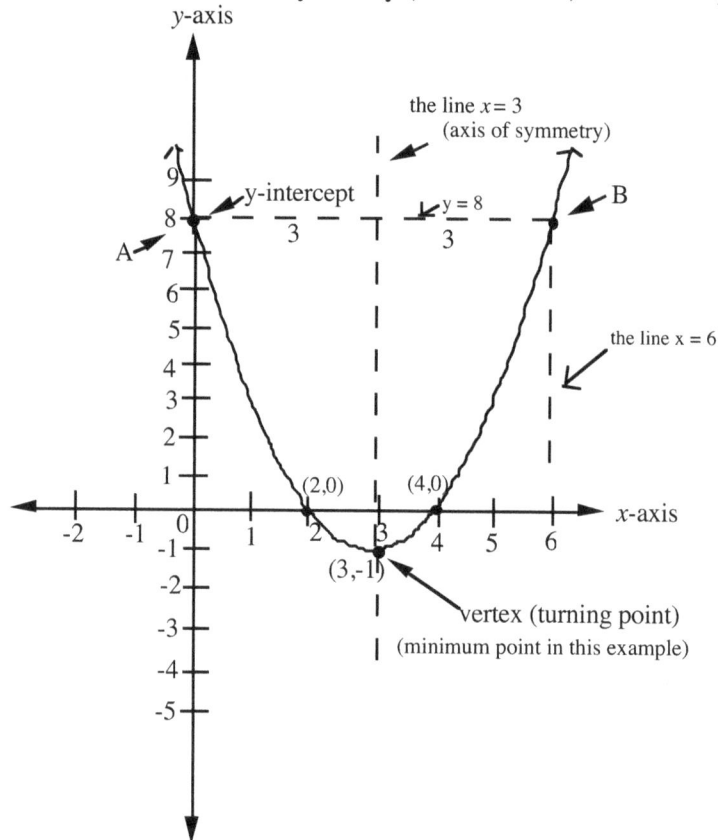

Figure 1: Graph of $y = x^2 - 6x + 8$

Note above that in sketching the graph, you may choose convenient x-values on either side of the axis of symmetry, calculate corresponding y-values to obtain additional points, especially, when there are no (real) x-intercepts.(This occurs when in solving the quadratic equation, we obtain solutions involving imaginary numbers.)
Again: When in doubt as to the location of the curve, pick x-values, calculate y, plot the points and connect them.

Example 2 Given the equation of the parabola $y = f(x) = -x^2 + 6x - 8$, perform the following:

1. Find the x- and y-intercepts.   **2**. Find the axis of symmetry.
3. Find the coordinates of the vertex (or turning point).   **4**. Find the domain and range.
5. Sketch the graph of $y = f(x) = -x^2 + 6x - 8$

Solution

1. Using the procedure in Example 1,

The x-intercepts are 2 and 4.   (The x-intercepts are at the points (2,0), and (4,0))

The y-intercept = -8.   (The y-intercept is at (0,-8))

2. $a = -1$ (the coefficient of the x^2-term), $b = +6$ (the coefficient of the x-term)

Axis of symmetry is the line $x = -\dfrac{b}{2a} = -\dfrac{+6}{2(-1)} = 3.$

The axis of symmetry is the line $x = 3$. The x-coordinate of the vertex = 3,

and the y-coordinate of the vertex = $y = f(-\dfrac{b}{2a}) = f(3) = -3^2 + 6(3) - 8 = 1$

3. The vertex is at the point (3, 1). Since a is negative ($a = -1$), the vertex is a maximum point.

4. The domain is all real x. (The given function is a polynomial)

The range is specified by using the y-coordinate of the vertex obtained in Step 2.

The range is given by the inequality $y \leq 1$. (Note the " $\leq$ ")

The range consists of all real numbers less than or equal to 1.

(Note: If we did not know the x-coordinate of the vertex, we could obtain the range by substituting. Substituting

$a = -1.\ b = +6, c = -8$ in $y \leq \dfrac{4ac - b^2}{4a}$ (Note the " $\leq$ ") to obtain $y \leq \dfrac{4(-1)(-8) - (+6)^2}{4(-1)}$; $y \leq 1$.)

5. To sketch the graph, plot the points from **(1)**, **(2)**, **(3)** above; similarly locate the point B (as was done in Example 1) and connect the points by a smooth U-shaped curve, noting that the parabola opens downwards (**Figure 2**), since a is negative ($a = -1$). Also note the symmetry as in Example 1.

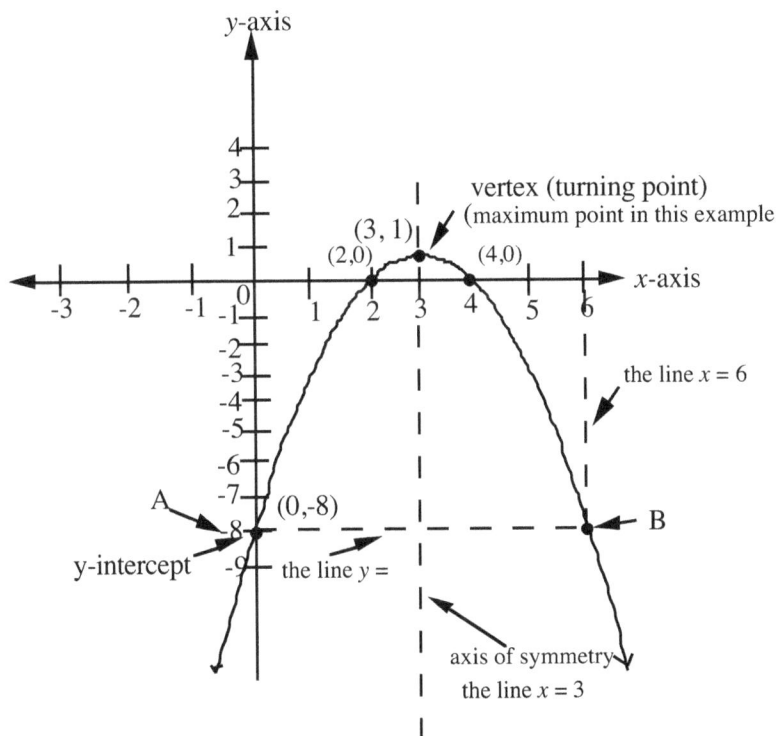

Figure 2: Graph of $y = -x^2 + 6x - 8$

EXTRA: Axis of Symmetry , Vertex and Minimum Value

Example 1

Given $f(x) = x^2 - 12x + 8$ (A)

Let $y = x^2 - 12x + 8$ (B)

$= x^2 - 12x + (-6)^2 - (-6)^2 + 8$

$= x^2 - 12x + (-6)^2 - 36 + 8$

$y = (x - 6)^2 - 28$ <---- knowing this

$y + 28 = (x - 6)^2$ <----- or this

Vertex is at $(6, -28)$

Axis of symmetry is the line $x = 6$
 Minimum value is -28.
 Alternatively

From (B), $a = 1$, $b = -12$

Axis of symmetry is given by $x = -\frac{b}{2a}$

$= -\frac{-12}{2(1)}$

$= \frac{12}{2}$

$x = 6$

$f(6) = 6^2 - 12(6) + 8$ (subst.. in (A)

$= 36 - 72 + 8$

$= -36 + 8$

$y = -28$

when $x = 6$. $y = -28$

Vertex is at $(6, -28)$

Axis of symmetry is the line $x = 6$
Minimum value is -28.

Example 2

Given $f(x) = x^2 - 6x + 8$ (A)

Let $y = x^2 - 6x + 8$ (B)

$= x^2 - 6x + (-3)^2 - (-3)^2 + 8$

$= x^2 - 6x + (-3)^2 - 9 + 8$

$y = (x - 3)^2 - 1$ <---- knowing this

$y + 1 = (x - 3)^2$ <----- or this

Vertex is at $(3, -1)$

Axis of symmetry is the line $x = 3$
 Minimum value is -1.
 Alternatively

From (B), $a = 1$, $b = -6$

Axis of symmetry is given by $x = -\frac{b}{2a}$

$= -\frac{-6}{2(1)}$

$= \frac{6}{2}$

$x = 3$

$f(3) = 3^2 - 6(3) + 8$ (subst.. in (A)

$= 9 - 18 + 8$

$= -9 + 8$

$y = -1$

when $x = 3$. $y = -1$

Vertex is at $(3, -1)$

Axis of symmetry is the line $x = 3$
Minimum value is -1.

Lesson 62 Exercises

Sketch the graphs of the following by the location of the vertex, axis of symmetry and the intercepts.:

1. $y = x^2 - 4$;   **2.** $y = -x^2 + 1$. **3.** $y = x^2 - 6x + 8$

Answers: See above

CHAPTER 21
Lesson 63
Solving Fractional or Rational Equations

Example 1 Solve for x: $\dfrac{4}{x-1} = 5$

Solution

Step 1:   Determine the excluded values. These are the x-values which make the denominators zero. Set the denominator to zero and solve for the variable.

$$x - 1 = 0$$
$$\underline{+1 \quad +1}$$
$$x = 1$$

The excluded value is 1.

Step 2: Multiply both LHS and RHS of the equation by $x - 1$

$$\frac{4}{x-1} = 5$$

$$\frac{\cancel{(x-1)}\,4}{\cancel{(x-1)}} = \frac{5(x-1)}{1}$$

$$4 = 5x - 5$$
$$\underline{+5 \qquad +5}$$
$$9 = 5x$$
$$\frac{9}{5} = x$$

Since $\dfrac{9}{5}$ is not an excluded value, it may be accepted as a solution; however, for good practice, we check it in the original equation.

Step 3: $\dfrac{4}{\frac{9}{5} - 1} \overset{?}{=} 5$

$$\frac{4}{1\frac{4}{5} - 1} \overset{?}{=} 5$$

$$\frac{4}{\frac{4}{5}} \overset{?}{=} 5$$

$$\frac{4}{1} \times \frac{5}{4} \overset{?}{=} 5$$

$$5 = 5 \qquad \text{True} \ \ (\text{RHS} = \text{LHS})$$

$\therefore$ the solution is $\dfrac{9}{5}$.

Example 2 Solve for x: $\dfrac{4}{x-3} = \dfrac{3}{x-4}$

Step 1: Determine the excluded values.

Set $x - 3 = 0$ and solve for x to obtain $x = 3$

Set $x - 4 = 0$ and solve for x to obtain $x = 4$

The excluded values are 3 and 4. Thus in solving the above equation, any value of x obtained may be checked for acceptance as a solution **except** 3 and 4.

Step 2: Multiply both sides of the equation by the LCM of the denominators.

(The LCD = $(x - 3)(x - 4)$

$$\frac{4}{(x-3)} \cdot \frac{(x-3)(x-4)}{1} = \frac{3}{(x-4)} \cdot \frac{(x-3)(x-4)}{1}$$

$$4(x-4) = 3(x-3)$$

$$\begin{array}{r} 4x - 16 = 3x - 9 \\ \underline{-3x \qquad\quad -3x} \\ x - 16 = -9 \\ \underline{+16 \quad +16} \\ x = 7 \end{array}$$

This value, 7, is not one of the excluded values (from Step 1) and so we may accept it as a solution but for good practice, we check it in the original equation.

$$\frac{4}{7-3} \overset{?}{=} \frac{3}{7-4}$$

$$\frac{4}{4} \overset{?}{=} \frac{3}{3} \qquad \text{True} \qquad (\text{Also both the LHS and RHS of the equations are defined.})$$

The solution is 7.

Example 3   Solve for y.

$$y + \frac{2y}{y-2} = \frac{4}{y-2}$$

Step 1: Set $y - 2 = 0$ and solve for y .

Then, $y - 2 = 0$, and $y = 2$

The excluded value is 2.

Step 2: Multiply every term of the equation by $y - 2$ (the LCD of the fractions).

$$y(y-2) + \frac{2y(y-2)}{y-2} = \frac{4(y-2)}{y-2}$$

$$y(y-2) + \frac{2y(y-2)}{y-2} = \frac{4(y-2)}{y-2}$$

$$y^2 - 2y + 2y = 4$$

$$y^2 = 4$$

$$y = \pm\sqrt{4}$$

$$y = \pm 2$$

i.e. $y = +2$, or $y = -2$

Since +2 is an excluded value (from Step 1), it is rejected; and only -2 may be accepted as a solution.

$$\text{Check:} \quad -2 + \frac{2\,(-2)}{-2\,-2} \overset{?}{=} \frac{4}{-2\,-2}$$

$$-2 + \frac{(-4)}{-4} = \frac{4}{-4}$$

$$-2 + 1 = -1$$

$$-\ 1 = -1 \quad \text{True}$$

Therefore, the solution. is -2.

Note above that +2 is an extraneous solution.

Example 4   Solve for x.

$$\frac{3x}{x+5} - \frac{x+1}{x+3} = 2$$

Step 1. Find the excluded values.

Set $x+5=0$ and solve for x to obtain $x = -5$.
-5 is an excluded value.

Similarly, set $x+3=0$, and solve for x to obtain $x = -3$

The excluded values are -5 and -3. Note above that we could obtain the excluded values by inspection (mentally).

Step 2: Multiply each term of the equation by the LCM of the denominators (to undo the denominators). The LCM is $(x+5)(x+3)$.

$$(x+5)(x+3)\,\frac{3x}{x+5} - (x+5)(x+3)\frac{(x+1)}{x+3} = 2(x+5)(x+3)$$

$$\cancel{(x+5)}(x+3)\cdot\frac{3x}{\cancel{x+5}} - (x+5)(\cancel{x+3})\frac{(x+1)}{\cancel{x+3}} = 2(x+5)(x+3)$$

Step 3:   Simplify and solve for x.

$$(x+3)(3x) - (x+5)(x+1) = 2(x+5)(x+3)$$

$$3x^2 + 9x - (x^2+6x+5) = 2(x^2+8x+15)$$

$$3x^2 + 9x - x^2 - 6x - 5 = 2x^2 + 16x + 30 \quad \text{(simplifying the left-hand side)}$$
$$2x^2 + 3x - 5 = 2x^2 + 16x + 30$$
$$2x^2 - 2x^2 + 3x -16x - 5 - 30 = 0 \quad \text{(eliminating the right-hand-side)}$$

$$-13x - 35 = 0$$
$$\frac{-35x}{-13} = \frac{+35}{-13}$$
$$x = -\frac{35}{13}$$

After checking (not shown), the solution is $-\frac{35}{13}$.

Note in Example **4** above, that in Step 3, the equation contained x^2-terms; but when we proceeded to solve the equation, the x^2-terms dropped out. Sometimes, this happens; in some cases these terms do not drop out, as in Example 3 above.

Note:

The excluded values for $\dfrac{8}{x^2-4}=1$ are -2 and +2 (obtained by setting $x^2-4=0$ and solving for x);

however, note that for $\dfrac{8}{x^2+4}=1$, x^2+4 is positive for all real values of x and never zero, since the

square of any nonzero real number is always positive. Therefore, there are no real excluded values

for $\dfrac{8}{x^2+4}=1$. In fact the solution to $\dfrac{8}{x^2+4}=1$ is the set $\{-2,2\}$.

Lesson 63 Exercises

A Solve for x:

1. $\dfrac{9}{x+4}=1$;   **2.** $\dfrac{2}{x-3}=\dfrac{5}{x-4}$;   **3.** $\dfrac{x^2}{x+2}=\dfrac{4}{x+2}$ **4.** $\dfrac{7}{x-5}+x=\dfrac{3}{x-5}$

Answers: **1.** $\{5\}$;   **2.** $\{\dfrac{7}{3}\}$ **3.** $\{2\}$ Note: $x\neq -2$;   **4.** $\{1,4\}$

B Solve for x and state the excluded values:

1. $\dfrac{4}{x}+\dfrac{1}{2}=\dfrac{2}{5}$;   **2.** $\dfrac{2}{3}-\dfrac{5}{x}=\dfrac{3}{4}$; **3.** $\dfrac{2x}{2x-5}=2$;   **4.** $\dfrac{1}{x-3}+\dfrac{1}{x+3}=\dfrac{9}{x^2-9}$;   **5.** $\dfrac{1}{x-2}=\dfrac{x^2}{x-2}+1$

Answers: **1.** -40 $(x\neq 0)$; **2.** -60 $(x\neq 0)$; **3.** 5 $(x\neq \dfrac{5}{2})$; **4.** $\dfrac{9}{2}$ $(x\neq -3$ or $3)$; **5.** $-\dfrac{1}{2}\pm\dfrac{\sqrt{13}}{2}$ $(x\neq 2)$

CHAPTER 22
Lesson 64
Solving Radical Equations

We define **a radical equation** as an equation in which one of the variables occurs under one or more radical signs.

Examples   **1.** $\sqrt{x} = 6$

2. $\sqrt{x-3} = 4$

3. $\sqrt{2x+4} = \sqrt{x+2}$

Principle of powers: If $a = b$

Then, $a^n = b^n$. However, this principle does not always produce equivalent equations, and therefore, if we use this principle to solve an equation, we must check the solutions in the original equation.

Procedure   The technique here involves the elimination of the radical signs by raising the radical expressions to integral (integer) powers: both sides of an equation are to be raised to the same integral power. We will then solve the resulting non-radical equation for the variable. If there are more than one radical, we may sometimes have to "undo" the radical signs in a number of steps. It will also be necessary that we test any solutions in the original equation for **extraneous solutions**. Before raising each side of an equation to an integral power, the radical whose radical sign we want to eliminate must be on one side of the equation by itself.

Example 1   Solve for x: $\sqrt{x} = 6$.

Solution

Step 1:   Square both sides of the equation to undo the square root symbol (i.e., raise both sides of the equation to the second power).
Then, $(\sqrt{x})^2 = (6)^2$
$x = 36$.
Step 2: Check the solution in the original equation.
$\sqrt{36} \overset{?}{=} 6$
$6 = 6$ True (left-hand side equals right-hand side, i.e., LHS = RHS).
∴ the solution is 36.

The above problem was so simple that we could have asked: If the square root of a number is 6, what is the number? Of course, the number is 36.

Example 2 Solve for x: $\quad \sqrt{x-3} = 4$

Solution

Step 1: Square both sides of the equation to undo the square root symbol.

Then, $(\sqrt{x-3})^2 = (4)^2$

$$x - 3 = 16$$

Step 2: Solve for x.

$$x - 3 = 16$$
$$\underline{\quad +3 \quad +3}$$
$$x = 19$$

Step 3: Check the solution 19 in the original equation.

$$\sqrt{19-3} \overset{?}{=} 4$$
$$\sqrt{16} \overset{?}{=} 4$$
$$4 = 4 \text{ True} \quad \text{(LHS = RHS)}$$

$\therefore$ the solution is 19.

Note above that squaring the square root of a number eliminates the square root symbol.

Example 3 Solve for x:

$$\sqrt{2x+4} = \sqrt{x+2}$$

Solution

Step 1: Square both sides of the equation.

$$(\sqrt{2x+4})^2 = (\sqrt{x+2})^2$$

$$2x + 4 = x + 2$$

Step 2: Solve for x.

$$2x + 4 = x + 2$$
$$\underline{-x \qquad -x \qquad}$$
$$x + 4 = 2$$
$$\underline{\quad -4 \quad -4}$$
$$x = -2$$

Step 3: Check for $x = -2$ in the original equation

$$\sqrt{2(-2)+4} \overset{?}{=} \sqrt{(-2)+2}$$
$$\sqrt{-4+4} \overset{?}{=} \sqrt{0}$$
$$\sqrt{0} \overset{?}{=} \sqrt{0}$$
$$0 = 0 \qquad \text{True} \quad \text{(RHS = LHS)}$$

$\therefore$ the solution is -2.

Example 4 Solve for x: $\sqrt{2x-5} = x - 4$

Solution

Step 1:   Square both sides of the equation.
$$(\sqrt{2x-5})^2 = (x-4)^2$$
$$2x - 5 = x^2 - 8x + 16$$
$$0 = x^2 - 10x + 21$$
or $x^2 - 10x + 21 = 0$

Step 2:   Solve the quadratic equation by any method.
We will solve by factoring, since the factors are easily recognizable.
$$x^2 - 10x + 21 = 0$$
$$(x - 3)(x - 7) = 0$$
$$x - 3 = 0 \text{ or } x - 7 = 0$$
$$x = 3 \text{ or } x = 7 \text{ (Setting each factor equal to zero and solving for } x)$$

Step 3: Checking for $x = 3$:
$$\sqrt{2(3)-5} \overset{?}{=} 3 - 4$$
$$\sqrt{6-5} \overset{?}{=} -1$$
$$\sqrt{1} \overset{?}{=} -1$$
$$1 = -1 \text{ False } \text{(LHS } \textbf{not} \text{ equal to RHS)}$$
Therefore 3 is not a solution. It is an extraneous root.

Checking for $x = 7$:
$$\sqrt{2(7)-5} \overset{?}{=} 7 - 4$$
$$\sqrt{14-5} \overset{?}{=} 3$$
$$\sqrt{9} \overset{?}{=} 3$$
$$3 = 3 \quad \text{True} \quad \text{(RHS = LHS)}$$
Therefore, 7 is a solution.
Since 3 is an extraneous root , the only solution is 7.

Lesson 64 Exercises

Solve for x:
1. $\sqrt{x} = 7$;   **2.** $\sqrt{x-4} = 5$ **3.** $\sqrt{3x+5} = \sqrt{x+9}$

Answers: 1. $x = 49$;   **2.** $x = 29$;   **3.** $x = 2$;

Solve and check: 1 $\sqrt{x-5} + 6 = 2$;   2. $2\sqrt{x} = 6$;   3. $\sqrt{x^2 - 8x} = 3$;

4. $3\sqrt{x+2} = 4\sqrt{x-5}$;   **5.** $\sqrt{x+5} - \sqrt{x} = 3$;   **6.** $\sqrt[3]{x+1} = 2$

Answers: **1**. No solution (21 is extraneous) ;   **2.** { 9 } ;   **3.** { -1 , 9 } ;   **4.** { 14 } ;

5. No solution. ($\frac{4}{9}$ is extraneous) ;   **6.** {7}

CHAPTER 23
Basic Review For Geometry

Lesson 65: **Points, Lines, Angles, Congruency, Complimentary and Supplementary Angles; Parallel and Perpendicular Lines**

Lesson 66: **Triangles, Classification of triangles, Angles of a triangle**

Lesson 65

Points, Lines, Angles, Congruency, Complimentary and Supplementary Angles; Perpendicular Lines

The undefined terms in geometry are a point, a line, and a plane.
By using these terms we can define other geometric terms.

Point: We use a point to indicate a position. On a piece of paper, we represent a point by a dot. A point has no size. (The "thinner" the dot, the better it represents a point; the tip of a well-sharpened pencil will be an acceptable size). We name a point by a capital letter written near the point (See the points P, A and B below).

Figure 1

Line: We may consider a line as an infinite set of points. A line may be a straight line or a curved line. A straight line may be extended indefinitely in both directions.
We may name a line by using a single lower case letter or two capital letters representing two points on the line, with a double-headed arrow over the capital letters (see Figure 2 below)

The line passing through the points A and B may be named as $\overleftrightarrow{AB}$ or l

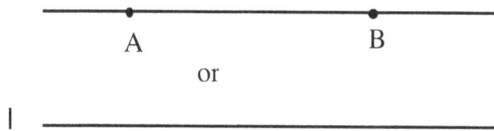

Figure 2

Line segment: A line segment or segment is a part of a line between two points on the line, including the two points (end points). We denote a line segment by placing a bar over the two capital letters which represent the two end points (see Figure 3 below).

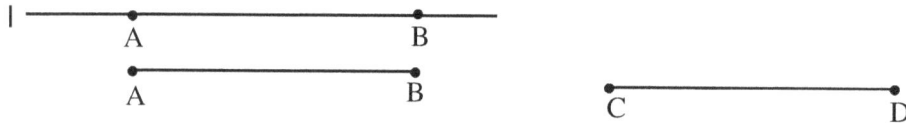

Line segment AB is represented by $\overline{AB}$ Line segment CD is represented by $\overline{CD}$

Figure 3

Ray: A ray is a part of a line to one side of a point on the line. This point is the endpoint of the ray. The ray begins at the end point and extends indefinitely in one direction. The arrow drawn begins at the end point . We name a ray by using two capital letters: the first letter represents the end point and the other letter represents any other point on the ray (Fig.4).

ray PQ is represented by $\overrightarrow{PQ}$ ray AB is represented by $\overrightarrow{AB}$

Figure 4

Measurement of a line segment : The **length** of a line segment is the distance between its end points.

The length of line segment $\overline{AB}$ is denoted by AB (i.e. without a bar over AB)

Figure 5

Angle: An angle (symbol: $\angle$) is the union of two rays having the same end point, the endpoint being called the vertex of the angle. An angle is formed when two line segments (or lines) meet at a common point (Figure 6).

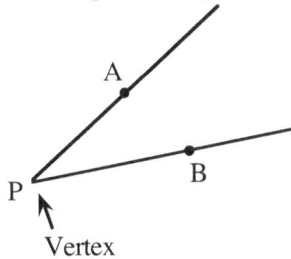

Vertex

Figure 6

Naming of angles   We may name an angle by using capital letters or lowercase letters as follows:

   1. If the angle is the only angle at the vertex, then we can use either a single capital letter or three capital letters. However, if there are two or more angles at the vertex, then we must use three capital letters. In using three capital letters, the middle letter is the letter at the vertex.

   2. We may use a lower case letter , a Greek letter, or a number .The letter or the number is placed inside the angle (as shown in Figure 7).

Examples: $\angle$ 2 and $\angle$ DAB   denote the same angle
         $\angle$ 1 and $\angle$ CAD   denote the same angle.

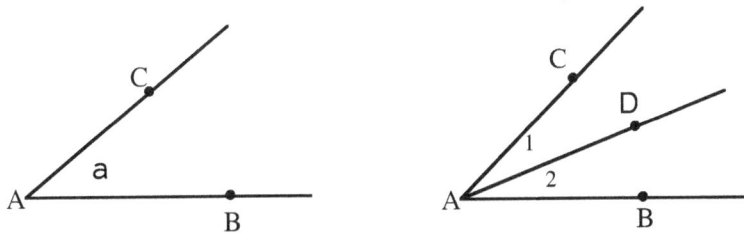

Figure 7

Congruent Figures: Congruent figures have the same shape and size.
Congruent figures can be made to coincide.
The symbol for "congruent" or "congruent to" is ≅ .

Congruency of Line Segments: As shown in Figure 8, if we can place $\overline{AB}$ and $\overline{CD}$ so that
they coincide, then, we say that $\overline{AB}$ and $\overline{CD}$ are congruent,
and we write $\overline{AB}$ ≅ $\overline{CD}$ (read: line segment $\overline{AB}$ is
congruent to line segment $\overline{CD}$)

Congruent line segments have the same length.

$$\overline{AB} ≅ \overline{CD} \text{ if and only if } AB = CD$$

(that is, $\overline{AB}$ is congruent to $\overline{CD}$ if the length of $\overline{AB}$ is equal to the length of $\overline{CD}$, and conversely,
the length of $\overline{AB}$ is equal to the length of $\overline{CD}$, if $\overline{AB}$ is congruent to $\overline{CD}$).

Indication of congruency of line segments: We may use the same number of slashes or strokes
(as shown in Figure 8) to indicate that two line segments are congruent.

Example (a) $\overline{AB} ≅ \overline{CD}$ (b) $\overline{EF} ≅ \overline{GH}$

Figure 8

Congruency of angles: As shown in Figure 9, if ∠ P can be placed on ∠ Q so that the angles
coincide, then we say that ∠ P is congruent to ∠ Q and we write
∠ P ≅ ∠ Q

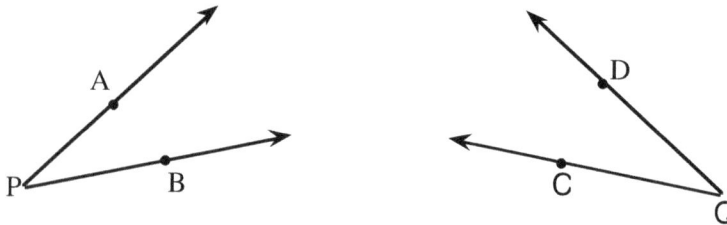

Figure 9

Note that congruent angles have equal measures.

Indication of congruency of angles: We can use arcs or a combination of arcs and slashes to indicate congruency (as shown in Figure 10)

Examples $\angle 1 \cong \angle 2$ (the angle marked 1 is congruent to the angle marked 2)
$\angle 3 \cong \angle 4$ (the angle marked 3 is congruent to the angle marked 4)
$\angle 5 \cong \angle 6$ (the angle marked 5 is congruent to angle marked 6)

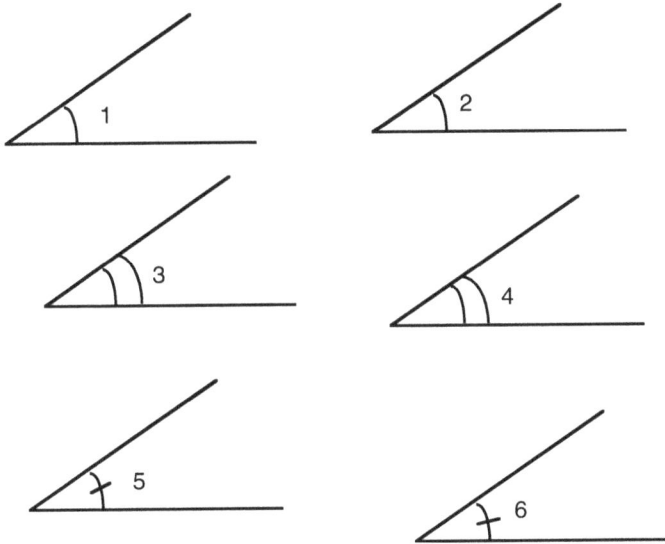

Figure 10

Units of Measuring angles : The main unit of measuring angles in geometry is the degree.
There are smaller units of the degree, namely the minute, and the second.

The degree and the smaller units are as follows: 1 degree = 60 minutes
1 minute = 60 seconds

Classification of angles by the measures of the angles

Straight angle: An angle whose measure is 180°

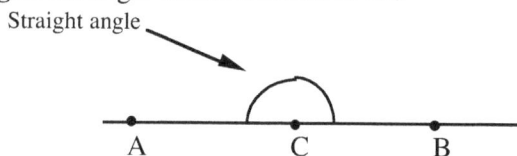

Straight angle

A   C   B

Right angle: A right angle is an angle whose measure is 90°.
In Figure 11, below, $\angle 5$ is a right angle.

5

Figure 11

Acute angle: An angle whose measure is between 0° and 90°.
In Figure 12, below, $\angle 6$ is an acute angle.

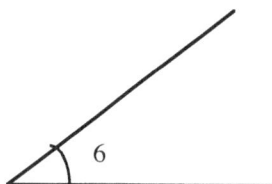

6

Figure 12

Obtuse angle: An angle whose measure is between 90° and 180°.
In Figure 13, below, $\angle 8$ is an obtuse angle.

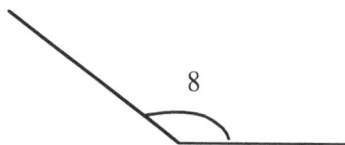

8

Figure 13

Reflex angle: An angle whose measure is between 180° and 360°.
In Figure 14, below, $\angle 7$ is reflex angle.

7

Figure 14

Pairs of Angles

Complementary angles : If the sum of the measures of two angles is 90°, then the two angles are complementary.
The complement of a 40° angle is a 50° angle
In Figure 15, below, $\angle 1$ and $\angle 2$ are complementary angles.(i.e., m $\angle 1$ + m $\angle 2$ = 90°)

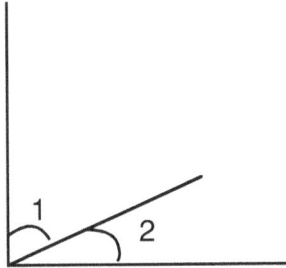

Figure 15

Supplementary angles: If the sum of the measures of two angles is 180°, then the two angles are supplementary.
The supplement of a 70° angle is a 110° angle.

In Figure 16, $\angle 3$ and $\angle 4$ are supplementary angles. (i.e., m $\angle 3$ + m $\angle 4$ = 180°)

Figure 16

Example 1   In the figure below , find x

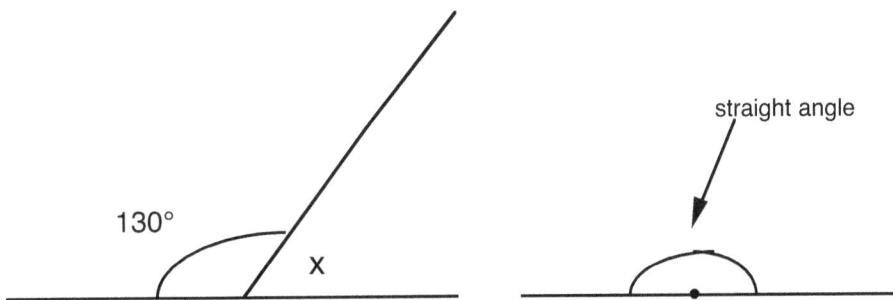

straight angle

Solution: $x + 130° = 180°$ Reason : A straight angle has a measure of 180°
$x = 50°$

Vertical angles: As shown in Fig.17, below, when two lines intersect (cross each other) a pair of opposite angles such as $\angle$ 1 and $\angle$ 3 are called vertical angles. Similarly, $\angle$ 2 and $\angle$ 4 are vertical angles.

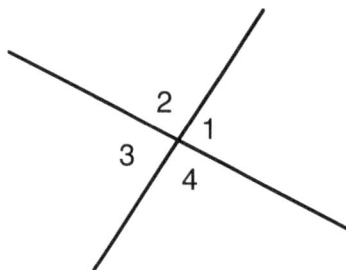

Figure 17

Theorem: When two lines intersect, the vertical angles are congruent.
In Figure 17, $\angle$ 1 and $\angle$ 3 are congruent.
$\angle$ 2 and $\angle$ 4 are congruent

Adjacent angles: Two angles which are in the same plane and which have a common vertex and a common side . In Figure 17, $\angle$ 1 and $\angle$ 2 are adjacent angles .
$\angle$ 2 and $\angle$ 3 are adjacent angles; $\angle$ 3 and $\angle$ 4 are adjacent angles.

Perpendicular lines (symbol for " perpendicular to": $\perp$): Perpendicular lines are two lines which meet at right angles.
In Figure 18,. line ℓ_1 is perpendicular to line ℓ_2 and we write
$$\ell_1 \perp \ell_2$$

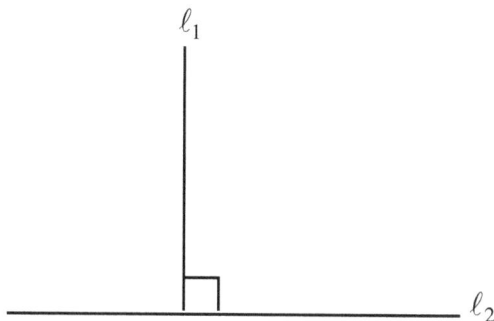

Figure 18

Example 2   In the figure below, find x

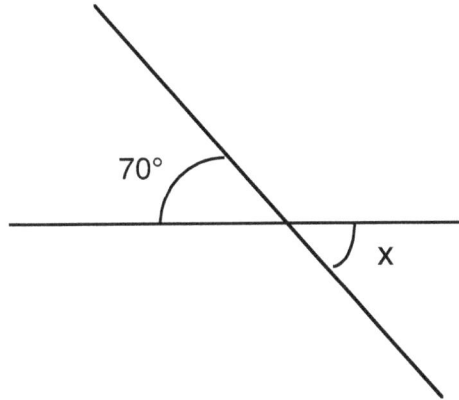

Solution: $x = 70°$ **Reason:** Vertically opposite angles are congruent, and therefore have equal measures

Lesson 65 Exercises

1. In figure (a) below, find x

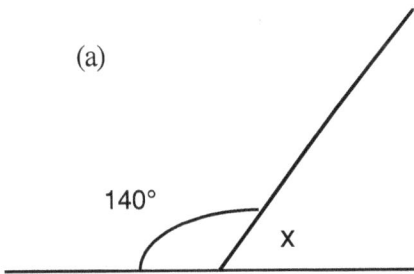

2. In figure (b) below, find x

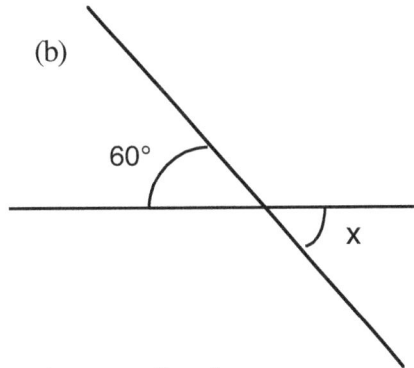

(a)

(b)

Answers: See above

Lesson 66
Triangles

A triangle is a closed figure bounded by three straight line segments.
Triangles may be classified by angles: By this classification, a triangle may be acute, obtuse, or right.

Acute triangle: A triangle in which all the angles are acute.

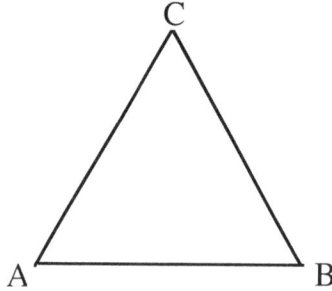

Obtuse triangle: A triangle in which there is an obtuse angle. **Note** that a triangle cannot have more than one obtuse angle.

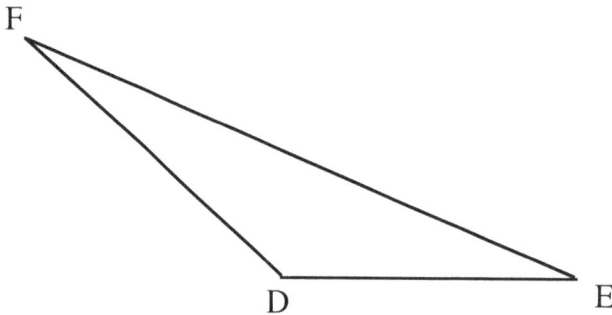

Right triangle (or right-angled triangle): A triangle which has one right angle (90°). **Note** that a triangle cannot have more than one right angle.

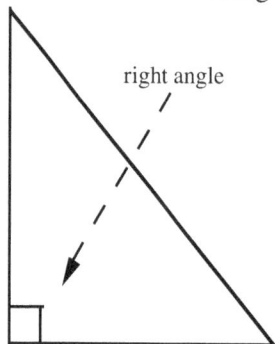

right angle

We can also classify triangles by sides. By this classification, a triangle may be equilateral, isosceles, or scalene.

Equilateral triangle: A triangle in which all three sides are congruent.
(The lengths of all three sides are equal)

All the interior angles of an equilateral triangle are congruent. (Each angle has a measure of 60°).

Isosceles triangle: A triangle in which two sides are congruent. In an isosceles triangle, the angles opposite to the congruent sides are congruent.

Useful relationships concerning the relative measures of the sides and angles of a triangle

1. In any triangle, the **sum** of the **lengths** of any **two sides** is larger than the length of the **third** side.

2. In any triangle, the **longest side** is opposite to the **largest angle**, and the **shortest side** is opposite to the **smallest angle**.

3. In an **acute triangle**, the square of the length of **longest side is smaller than** the sum of the squares of the lengths of the other two sides.

4. In an **obtuse triangle**, the **square** of the length of the **longest side is larger than** the sum of the **squares** of the lengths of the other two sides.

5. In a **right triangle**, the **square** of the length of the **longest side** (the hypotenuse) **is equal** to the sum of the **squares** of the lengths of the other two sides.   (**The Pythagorean theorem**)

Theorem: The sum of the measures of the interior angles of a triangle is 180°.
In the figure below, m $\angle$ A + m $\angle$ B + m $\angle$C = 180°

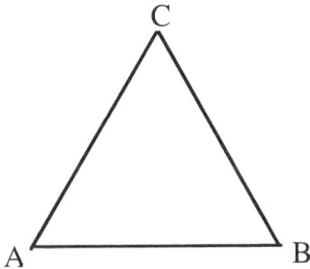

Theorem: If one side of a triangle is extended, the measure of the exterior angle so formed equals the sum of the measures of its remote interior angles.
In the figure below, m $\angle A$ + m $\angle C$ = m $\angle CBE$
($\angle A$ and $\angle C$ are the remote interior angles. $\angle CBE$ is the exterior angle)

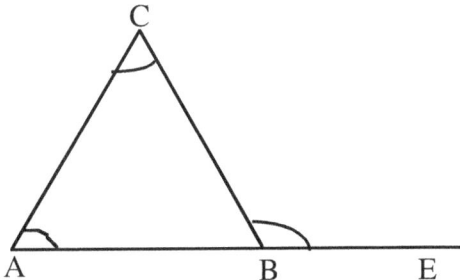

Example 1   In the figure below, find x.

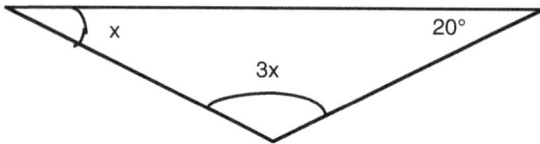

Solution: $x + 3x + 20° = 180°$ (The sum of the measures of the angles of a
                 $4x + 20° = 180°$ triangle is 180º)
                     $4x = 160°$
                      $x = 40º$

Example 2   In the right triangle shown below , find x

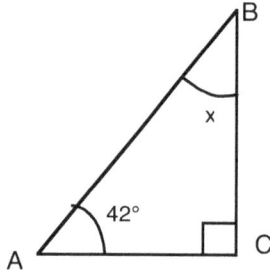

Solution: $m \angle C = 90º$ (A right angle has a measure of 90º)
        $m \angle A + m \angle C + m \angle B = 180º$ (The sum of the measures of the angles of a triangle is 180º)
                 $42° + 90° + x = 180°$
                  $132º + x = 180°$
                      $x = 180° - 132°$
                      $x = 48º$

Example 3   Find x and y in the figure below.

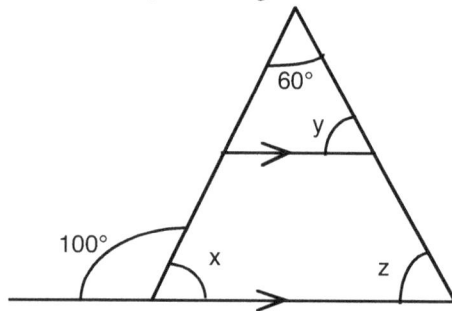

(a) $x + 100° = 180°$ (Supplementary angles)
            $- 100°   - 100°$
               $x = 80°$

  (b) Step 1: $x + 60° + z = 180°$
               $80° + 60° + z = 180°$ (The sum of the measures of the angles of a triangle is 180˚)
               $140° + z = 180°$
              $- 140°   - 140°$
                 $z = 40°$

Step 2: $y = z = 40°$ (Corresponding angles are congruent.)
           $y = 40$

Lesson 66 Exercises

1. Find x;   **2**. Find m $\angle$ A

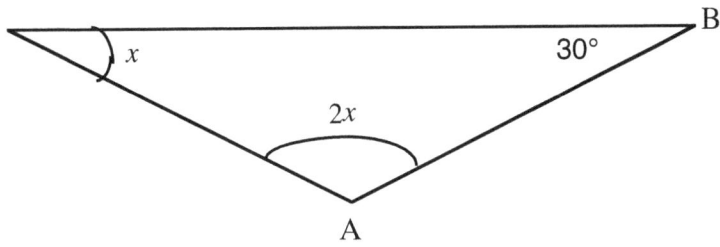

Answers: **1.** $x = 50$;   **2.** m $\angle$ A $= 100$

APPENDIX A

EXTRA: Factoring Perfect Square Trinomials

In factoring, knowing that a given trinomial is a **perfect square** trinomial will speed-up the factoring process. However, note that the methods we have used, so far, in factoring trinomials, can be used to factor perfect square trinomials.

Case 1: The coefficient of the leading term is 1 (assuming the trinomial is properly arranged)

Examples: **1.** $x^2 + 6x + 9 = (x + 3)^2$ same as $(x + 3)(x + 3)$

Observe that (a) the first and third terms are perfect squares, and

(b) $\left(\frac{6}{2}\right)^2 = (3)^2 = 9$ (that is the square of half the coefficient of the x-term is 9)

2. $x^2 - 6x + 9 = (x - 3)^2$ same as $(x - 3)(x - 3)$

Observe that (a) the first and third terms are perfect squares, and

(b)$\left(\frac{-6}{2}\right)^2 = (-3)^2 = 9$ (that is, the square of half the coefficient of the x-term is 9)

3. $x^2 - 4xy + 4y^2 = (x - 2y)^2$ same as $(x - 2y)(x - 2y)$

Observe that (a) the first and third terms are perfect squares, and

(b) $\left(\frac{-4y}{2}\right)^2 = (-2y)^2 = 4y^2$ (that is, the square of half the coefficient of the x-term is $4y^2$)

Case 2: The coefficient of the leading term **is not 1** (assuming the trinomial is properly arranged)

Here, if the first and third terms are perfect squares, the trinomial may or may not be a perfect square trinomial, Obtain the binomial factor using the square roots of the first and third terms and check the factorization by multiplication to see if you obtain the middle term. **Note** that even if the first and third terms are perfect squares, the trinomial may **not** be a perfect square trinomial as in Examples **8** and **9** below.

We can also check for a perfect square trinomial as follows:
Step 1: Find the square root of the first term (the leading term)
Step 2: Divide the middle term by twice the root from step 1:
Step 3: Square the quotient from step 2. If this square equals the third term, then the trinomial is a perfect square trinomial. Factor by taking the square roots of the first and third terms

Simply, if $\left(\frac{\text{middle term}}{2(\text{square root of first term})}\right)^2 =$ Third term, then the trinomial is a perfwct square trinomial

Examples The following are perfect square trinomials
To factor, find the square roots of the first and third terms and use the roots to   form the binomial, with a plus or minus sign between the terms according as the middle term has a plus sign or a minus sign.

1. $4x^2 - 4xy + y^2 = (2x - y)(2x - y) = (2x - y)^2$

2. $4x^2 + 4xy + y^2 = (2x + y)(2x + y) = (2x + y)^2$

3. $25x^2 - 60xy + 36y^2 = (5x - 6y)(5x - 6y) = (5x - 6y)^2$

4. $25x^2 + 60xy + 36y^2 = (5x + 6y)(5x + 6y) = (5x + 6y)^2$

5. $9a^2 - 12ab + 4b^2 = (3a - 2b)^2$

6. $a^2 + 18a + 81 = (a + 9)^2$

7. $4a^2 - 36a + 81 = (2a - 9)^2$

The following are **not** perfect square trinomials, even though the first and third terms are perfect squares. Here, we use the usual methods for factoring.

8. $9x^2 + 30x + 16$ whose factorization is $(3x + 2)(3x + 8)$; but $9x^2 + 24x + 16$ is perfect square. Why?

9. $4x^2 + 15xy + 9y^2$ whose factorization is $(x + 3y)(4x + 3y)$; but $4x^2 + 12xy + 9y^2$ is perfect square.Why?

To show that if $\frac{a}{b} = \frac{c}{d} = \frac{e}{f}$, then $\frac{a}{b} = \frac{c}{d} = \frac{e}{f} = \frac{a+c+e}{b+d+f}$

We proceed in two steps. In Step 1, we consider $\frac{a}{b} = \frac{c}{d}$. In Step 2, We consider the result

from Step 1 and $\frac{e}{f}$.

Step 1: $\frac{a}{b} = \frac{c}{d}$ (Given)

$\frac{a}{b} + \boxed{\frac{c}{b}} = \frac{c}{d} + \boxed{\frac{c}{b}}$ (Adding $\frac{c}{b}$ to both sides of the equation)

$\frac{a+c}{b} = \frac{bc+cd}{bd}$ (Adding on the LHS and on the RHS of the equation)

$\frac{a+c}{b} = \frac{c(b+d)}{bd}$ (Factoring the RHS)

$\frac{a+c}{b+d} = \frac{bc}{bd}$ (Dividing both sides by $b+d$ and multiplying both sides by b)

$\frac{a+c}{b+d} = \frac{c}{d}$ (Dividing out the b on the RHS)

Step 2: $\frac{a+c}{b+d} = \frac{e}{f}$ (Since $\frac{a}{b} = \frac{c}{d} = \frac{e}{f}$, replace $\frac{c}{d}$ on the RHS by $\frac{e}{f}$)

$\frac{a+c}{b+d} + \boxed{\frac{e}{b+d}} = \frac{e}{f} + \boxed{\frac{e}{b+d}}$ (Adding $\frac{e}{b+d}$ to both sides of the equation)

$\frac{a+c+e}{b+d} = \frac{e(b+d)+ef}{f(b+d)}$ (Adding on the LHS and on the RHS of the equation)

$\frac{a+c+e}{b+d} = \frac{be+de+ef}{f(b+d)}$

$\frac{a+c+e}{b+d} = \frac{e(b+d+f)}{f(b+d)}$ (Factoring out the e on the RHS)

$\frac{a+c+e}{b+d+f} = \frac{e(b+d)}{f(b+d)}$ (Dividing both sides by $b+d+f$ and multiplying by $b+d$)

$\frac{a+c+e}{b+d+f} = \frac{e}{f}$ (Dividing out the $b+d$ on the RHS)

Since $\frac{a}{b} = \frac{c}{d} = \frac{e}{f}$,   (given)

$\frac{a}{b} = \frac{c}{d} = \frac{e}{f} = \frac{a+c+e}{b+d+f}$

Application of the above to similar triangles.
Consider two similar triangles ABC and DEF.

If the lengths of the sides of $\triangle$ ABC are a, b, c and the corresponding lengths of the sides of $\triangle$ DEF are d, e, f, then for the ratios of corresponding sides, $\frac{a}{d} = \frac{b}{e} = \frac{c}{f}$ and

$\frac{a}{d} = \frac{b}{e} = \frac{c}{f} = \frac{a+b+c}{d+e+f}$. Thus $\frac{a}{d} = \frac{b}{e} = \frac{c}{f} = \frac{\text{perimeter of } \triangle \text{ ABC}}{\text{perimeter of } \triangle \text{ DEF}}$

Example Let the lengths of sides of $\triangle$ ABC be 2, 3, 4; and let the corresponding lengths of the sides $\triangle$ DEF be 10, 15. 20,. Then $\frac{2}{10} = \frac{3}{15} = \frac{4}{20} = \frac{2+3+4}{10+15+20} = \frac{9}{45} = \frac{1}{5}$

APPENDIX B
About Measurements
Standard Unit, Error, Rounding-off Numbers, Significant Digits, Scientific Notation

To determine the size of a physical quantity, we compare its size with a standard quantity called a unit. Example: To determine the length of the cover of a book in inches or in meters, we can use a ruler with its scale in inches or in meters.
A measurement is the ratio of the magnitude of a physical quantity to that of a standard unit

Standard unit
A standard unit is a measure with which other quantities are compared. A standard unit of measure is defined by a legal authority (such as the US Bureau of Weights and Measures) or by a conference of scientists.

Some universally accepted standards:
1. For mass, the standard (primary standard) is the kilogram (kg).
2. For length, the standard is the meter (m).
3. For time, the standard is the second (s)

Some devices for taking measurements
Examples: Rulers (for length), chemical balances (for mass), stop watches (for time), ammeters, (for electric current) voltmeters (for electric voltage) , thermometers (for temperature) and barometers (for pressure).

Experimental Errors (or Uncertainties)
There are two main types of errors, namely, systematic errors, and random errors.

Systematic Errors (constant errors):
These errors are due to faulty measuring devices. Systematic errors make the measurements either too small or too large:
Examples of faulty devices:
1. Instruments with needles off the zero mark;   2. Faulty clocks (stop clock)
3, Corroded weights;   4. Faulty thermometers. Heat leaking equipment

Random errors (accidental errors or indeterminate errors)
Random errors may be due to chance variations of the physical quantity being measured, or chance variations in the measuring device. Random errors may also be due to failure to take into account variables such as temperature fluctuations; and environmental effects. We can reduce random errors by making a large number of measurements and taking the average of the measurements.

Absolute error (or absolute uncertainty)

Absolute error Experimental value - accepted (true value)
Example: In an experiment to determine the acceleration due to gravity, g, the experimental value of g, was 986 cm/s^2. The accepted value of g is 980 cm/s^2. Find the absolute error.

Solution   Experimental value = 986 cm/s^2

Accepted value = 980 cm/s^2
Absolute error = experimental value - accepted value

= (986 - 980) cm/s^2

= 6 cm/s^2

The positive value indicates that the experimental value is greater than the accepted value.

Note: If the experimental value = 964 cm/s^2

The absolute error = (964 - 980) cm/s^2

$$= -16 \text{ cm}/s^2$$

The negative value means that the experimental value is less than the accepted value.

Relative error (or relative uncertainty)

Relative error = $\dfrac{\text{absolute error}}{\text{accepted value}}$

Example 1: If the absolute error = 6 cm/s^2, and

the accepted value = 980 cm/s^2,

Relative error = $\dfrac{\text{absolute error}}{\text{accepted value}}$

Then relative error = $\dfrac{6 \; cm / s^2}{980 \; cm / s^2}$

$$= 0.00612$$

Example 2: If the absolute error = -16 cm//s^2 and

the accepted value = 980 cm/s^2

Then relative error = $\dfrac{-16 \; cm / s^2}{980 \; cm / s^2}$

$$= - 0.00612$$
$$= - 0.0163$$

Note: If an accepted value is not known, and we have two or more experimental values, then the average of the experimental values would be used as the "accepted value" in calculations.

Rounding-Off Numbers

The rules for rounding-off a number may differ slightly depending upon the field. For instance, in accounting the rule may be slightly different from the rule in chemistry.

1. If the digit or group of digits to be dropped is more than 500...,hen drop that digit or group and add 1 to the last digit retained.

2. If the digit or group of digits to be dropped is less than 500...,then drop that digit or group and leave the last digit retained unchanged.

3. If the digit or group of digits to be dropped is exactly 500..., then drop that portion and add 1 to the last digit retained if this digit is odd but if this digit is even, then this digit remains unchanged.

Rounding off Whole numbers

Procedure:

Step 1: Locate the digit in the round-off place.

(The round-off place is the place to which we want to round-off the number)

Step 2: Drop all digits to the right of the round-off place, and if the digit immediately to the right of the round-off place is more than 5 or is 5 followed by non-zero digits, , add 1 to the round-off place digit (i.e. we round-up); but if the digit immediately to the right of the round-off place is less than 5, the round-off place digit remains unchanged. However, if the digit immediately to the right of the round-off place digit is 5 or 5 followed by zeros, we add 1 to the round-off place digit if it is odd, but if it is even, it remains unchanged.(i.e. we round-down). Also, replace each digit dropped by a zero.

Rounding-off Decimals

The procedure is the same as that for rounding-off whole numbers, except that after the decimal point, we do not replace any digits dropped by zeros.

Procedure:

Step 1: Locate the digit in the round-off place.

(The round-off place is the place to which we want to round-off the number)

Step 2: Drop all digits to the right of the round-off place, and if the digit immediately to the right of the round-off place is more than 5 or is 5 followed by non-zero digits, add 1 to the round-off place digit (i.e., we round-up); but if the digit immediately to the right of the round-off place is less than 5, the round-off place digit remains unchanged. However, if the digit immediately to the right of the round-off place digit is 5 or 5 followed by zeros, we add 1 to the round-off place digit if it is odd, but if it is even, it remains unchanged.(i.e., we round-down).

Rounding-off (Alternatively)

When we round-off a number, we drop some of the digits explicitly or implicitly specified. We must distinguish between rounding-off to a specified number of decimal places (or significant digits) and the implicit rounding-off which we must determine from the numbers involved in the calculation.

The rules for rounding-off a number may differ slightly depending upon the field. For instance, in accounting the rule may be slightly different from the rule in chemistry.

1. If the digit or group of digits to be dropped is more than 500...then drop that digit or group and add 1 to the last digit retained.

2. If the digit or group of digits to be dropped is less than 500...then drop that digit or group and leave the last digit retained unchanged.

3. If the digit or group of digits to be dropped is exactly 500..., then drop that portion and add 1 to the last digit retained if this digit is odd but if this digit is even, then this digit remains unchanged.

Example: The following have been rounded-off to three decimal places.

(1) .4398. ≈ **.440**

(2) .43652 ≈ **.437**

(3) .43637 ≈ **.436**

(4) .43750 ≈ **.438**

(5) .43650 ≈ **.436**

(6) .43946 ≈ **.439**

(7) .43650001 ≈ **.437**

(8) .4365001 ≈ **.437**

Example We round off **85376.7463** to the following places, using the simple "5 or greater or less than 5 rule"

1. 85376.7463 to the nearest **thousandth** becomes **85376.746** (We do **not** replace the 3 dropped by a zero)

2. 85376.7463 to nearest **hundredth** becomes **85376.75** (We added 1 to the digit in the round-off place)

3. 85376.7463 to the nearest **tenth** becomes **85376.7** (The 7 is unchanged since the 4 dropped is less than 5)

4. 85376.7463 the nearest **unit** becomes **85377.** (Adding 1 to the 6)

5. 85376.7463 to the nearest **ten** becomes **85380.** (Replacing the 6 dropped by a zero)

6. 85376.7463 to the nearest **hundred** becomes **85400.** (Replacing the digits (6 and 7) dropped by zeros)

7. 85376.7463 to the nearest **whole number** becomes **85377.** (same as to the nearest unit)

Estimation

In estimation, we round-off the numbers before carrying out the operations of addition, subtraction, multiplication , division etc. For convenience, we will round-off each number to the first non-zero digit, unless specified.

Approximate Numbers, Significant Digits, Scientific Notation,

A measurement consists of a numerical value and a unit of that measurement. Example: 4 kilograms, where the 4 is the numerical value and the kilograms is the unit of the measurement.

Numbers obtained from a measurement are never exact (i.e., are approximate) due to the limitations of the measuring instrument as well as the skill of the person making the measurement. As such, when one records a measurement, one should indicate the reliability of the measurement. All measurements may be assumed to have an uncertainty in at least one unit in the last digit of the measurement, since in making a measurement, we usually estimate the last digit.

Results obtained from calculations using measurements are also as uncertain as the measurements themselves.
In summary, the numbers that we deal with in calculations are obtained from observations. Some of the numbers are exact and some are approximate, The approximate numbers are those numbers obtained from making measurements.

Exact Numbers: An exact number is a number that contains no uncertainties. It is assumed to be infinitely accurate. We can obtain exact numbers from definitions and from direct count. For example, the number of students in a math class by count is 25. In this case, there is no uncertainty, since we know that there are exactly 25 students. Similarly, when one counts 200 dollars, one knows that one has exactly 200 dollars, and there is therefore, no uncertainty. Also by definition, 60 minutes = 1 hour; 2.54 centimeters = 1 inch. Since these numbers are defined, this 60 and 2,54 are exact and contain no uncertainties. We can also add that this 60 has an infinite number of significant digits, (we can rite 60 as 60.000...) and therefore the zero in the 60 is significant. However if you make your own measurement and by coincidence obtain 2.54 centimeters, then this 2.54 would not be exact. We can generalize that all the conversion factors (from tables) are exact.

Significant Digits (Significant Figures), Digits obtained in a measurement

A significant digit (or figure) is one which is known to be reasonably reliable (or correct).
When we make measurements, the digits we read and estimate on a scale are also called significant digits (or significant figures). These digits include digits that we are certain of, and one additional digit that we are uncertain of. This uncertain digit is obtained by the estimation of the fractional part of the smallest subdivision on the scale being used. As such, the rightmost digit is assumed to be uncertain.

Significant figure notation is an approximate method of indicating the uncertainty of a measurement. when recording a measurement.
We agree to the following:

1. The digits 1, 2, 3, 4, 5, 6, 7, 8, 9 are always significant.

2. The digit zero, 0. may or may not be significant according to its position in the number as follows:

(a) Zeros before the first non-zero digits are **not** significant.

For example: (i) .0450 has **three** significant digits: The first zero is not significant; but the last zero is significant since if it were not we would not write it.

(ii) .0012 has **two** significant digits. The first two zeros are not significant

. (b) Zeros between non-zero digits **are** significant.

Example: 3.045 has **four** significant digits. The zero between 3 and 4 is significant.
40.240 has **five** significant digits. The last zero is significant because if it were not we would not write it.

More examples: The numbers referred to below are assumed to have been obtained from measurements.
23.00 has four significant digits: the zeros in this case are significant since we do not have to write the zeros if they were not significant. If the zeros were not significant we would have written 23.
2300. has two significant digits, and 600 has one significant digit; however, in each of these two examples, the number of significant digits is sometimes ambiguous.
It is suggested that when the number of significant digits is in doubt , the maximum number of significant digits is to be assumed. Also, in recording data, if we know the number of significant digits, we will use the scientific notation.
Note above that if 600 and 2300 had been obtained by counting, or by definition, all the zeros would have been significant. As a reminder, significant notation generally pertains to numbers obtained from measurements.
Using **scientific notation** avoids all ambiguities with respect to the number of significant digits..
For instance, if we know that the above number, 600 were measured to two significant digits, then we would write 6.0×10^2. If the measurement were to three significant digits, we would write 6.00×10^2
When we deal with very large or very small numbers we prefer to write the numbers in scientifi c notation form. In this form, the significant digits (digits) including the zeros can be unambiguously indicated.
In scientific notation:

(1) 125000 would be written 1.25×10^5

(2) 1467 would be written 1.467×10^3

(3) .032500 would be written 3.2500×10^{-2}

(4) .0325 would be written 3.25×10^{-2}

(5) If 125000 were known to four significant digits it would be written 1.250×10^5.

Note: Some authors indicate which zeros are significant by underlining the last significant zero.
For example, in 23000 the first two zeros are significant but the last zero is not significant.
Note also that in some books, a decimal point placed after the last zero makes all the zeros significant.
For example, 23000. has five significant digits, but 23000 has two significant digits. However, but there may still be ambiguity if the number is at the end of a sentence.

Accuracy and Precision in Measurements

Two contributions to uncertainty in measurement are limitations of precision, and limitations of accuracy.
Accuracy indicates how close a measured value is to the true value but **precision** indicates how close two measurements of the same quantity are close to each other. Generally, more precision implies more accuracy. However, there are instances in which numbers may be more precise, but may not be more accurate.. For example, if a measuring device is incorrectly calibrated (having incorrect scale).

Accuracy and Precision in Calculations

With respect to significant digits, **accuracy** refers to the number of significant digits but **precision** refers to the number of decimal places. The larger the number of significant digits in a number, the more accurate the number. The larger the number of decimal places, the more precise the number.

Note: In calculations, the approximate numbers determine how the rounding-off is done

Rounding-off to Significant Digits or Figures in Arithmetic Operations
(Implicit Specification)

1. In **multiplication and division** involving significant digits, the product or quotient (answer) should be rounded-off so that the number of significant digits in the answer is equal to the number of significant digits of the number with the least number of significant digits. (In other words, the answer should not be more accurate than the number with the least accuracy)

Example 1: Multiply 2.34 cm by 5.6 cm

Step 1: $2.34 \times 5.6 = 13.104$

Step 2: The number with the fewest number of significant digits is 5.6 and it has two significant digits. Therefore, the product (answer) should contain only two significant digits.
Thus, 13.104 becomes 13.

Answer: $13 \ cm^2$

In the above case the "4" in 2.34 is considered to be reasonably reliable. The " 6" in 5.6 is considered to be reasonably reliable.

Example 2: Maria determined the length of a piece of wood to be 6.47 yards. What is the length of this wood in feet?

Solution

By definition, 3 feet = 1 yard
6.47 yards is used to determine the number of significant digits in the answer,
(The 3 feet is exact and has an infinite number of significant digits)

$$\frac{6.47 \ yards}{1} \times \frac{3 \ feet}{1 \ yard}$$

= 19.41 feet

= 19.4 feet (6.47 has three significant figures)

2. In addition and subtraction, we shall round-off so that the number of decimal place in the answer equals the number of decimal places in the number with the least number of decimal places. (In other words, the answer should not be more precise than the number with the least precision)

Example 1: Add: 143.54, 172.3, and 64.62
Solution
143.54 < - - - - has two decimal places
172.3 < - - - - -has one decimal place (determines the number of decimal places in answer)
  64.62 < - - - - - has two decimal places
380.46
Answer: 380.5 (has one decimal place)

Example 2 Subtract 12.4 from 143.63
Solution
143.63 < - - - - has two decimal places

  12.4 < - - - - -has one decimal place (determines the number of decimal places in answer)

131.23
Answer: 131.2 (has one decimal place)

3. In finding powers and roots, the root or power should be rounded-off so that the number of significant digits in the answer is equal to the number of significant digits in the number.

Example 1: Find $\sqrt{26.9}$
Radicand has three significant digits , and therefore the root should have three significant digits.
From a calculator, $\sqrt{26.9} = 5.1865$
  $\sqrt{26.9} = 5.19$ (has three significant digits. Same as in the radicand)

Note: When two or more different operations are involved, the final operation determines how the final result is rounded-off.
Example Simplify: $38.3 + 12.9(3.58)$
Solution According to order of operations, we multiply 12.9 by 3.58 first and then add 38.3
  $38.3 + 12.9(3.58)$

$= 38.3 + 46.182$

$= 84.485$

$= 84.5$ (has one decimal place as in 38.3)

Since addition was the last step, we use precision (the number of decimal places) to round-off.

Addition and Subtraction Involving Scientific Notation

Before adding or subtracting the numbers must have the **same powers** of 10. We will rewrite the expression so that the power of 10 is that of the highest power in the expression

Example: **1.** $4 \times 10^2 + 3 \times 10^2 = (4 + 3) \times 10^2$
$$= 7 \times 10^2$$

Example: **2.**
$$4 \times 10^2 + 2 \times 10^3 = 0.4 \times 10^3 + 2 \times 10^3$$
$$= (0.4 + 2) \times 10^3$$
$$= 2.4 \times 10^3$$

or
$$4 \times 10^2 + 2 \times 10^3 = 4 \times 10^2 + 20 \times 10^2$$
$$= (4 + 20) \times 10^2$$
$$= 24 \times 10^2$$
$$= 2.4 \times 10^3 \quad \text{(Again, we obtain the same result)}$$

Order of Magnitude (for comparing relative sizes using powers of 10).

The order of magnitude is the power of 10 closest to the given number.

(It is an approximation to the number. Note the sequence, $..., 10^{-2}, 10^{-1}, 10^0, 10^1, 10^2,...$)

If a given quantity is 1000 times another quantity, the given quantity is larger by three orders of magnitude.

Examples:

1. The order of magnitude of 123 is 10^2, since 123 is closer to 100 than to 1000.

2. Find the order of magnitude of 0.00352.

Solution

Step 1: Write the number in scientific notation.
$$0.00352 = 3.52 \times 10^{-3}.$$

Step 2: Since the integer before the decimal point, 3, is less than 5, we replace 3.52 by 1 (since this is closer to 1 or 10^0, than it is to 10 or 10^1)

Step 3: 3.52×10^{-3}
$$= 10^0 \times 10^{-3}$$
$$= 1 \times 10^{-3}$$
$$= 10^{-3}$$

The order of magnitude of 0.00352 is 10^{-3}. (since by definition, the order of magnitude is the power of 10 closest to the given number.).

We can use the order of magnitude in estimation by rounding-off to the orders of magnitude.

More examples: Round-off to the nearest order of magnitude.

1. 1.32×10^2

2. 8.02×10^4

3. 0.0009

4. 0.0302

Solution:

1. Since 1.32 is closer to 1 than to 10,

1.32×10^2

$= 10^0 \times 10^2$

$= 1 \times 10^2$

$= 10^2$

The order of magnitude of 1.32×10^2 is 10^2

2. 8.02 is closer to 10 than to 1

8.02×10^4

$= 10^1 \times 10^4$

$= 10^5$

The order of magnitude is 10^5

3. $\quad 0.0009 = 9 \times 10^{-4}$

$\quad\quad\quad\quad = 10^1 \times 10^{-4}$

$\quad\quad\quad\quad = 10^{-3}$

The order of magnitude is 10^{-3}.

4. $0.0201 = 2.01 \times 10^{-2}$

$\quad\quad\quad = 10^0 \times 10^{-2}$

$\quad\quad\quad = 1 \times 10^{-2}$

$\quad\quad\quad = 10^{-2}$

The order of magnitude is 10^{-2}

Summary for rounding-off a number to the order of magnitude.

Step !: Write the number in scientific notation

Step 2: Ignoring the power of 10, if the integer before the decimal point is 5 or greater, replace the non-power of 10 part by 10 ((i.e. 10^1); but if the integer is less than 5, replace the non-power of 10 part by 1 (10^0)

Step 3: Simplify

International System of Units

The International System of Units (SI) has adopted a set of seven base (or primary) units.

Quantity	Unit	Symbol
Length	meter	m
Mass	kilogram	kg
Time	seconds	s
Electric current	ampere	A
Temperature	Kelvin	K
Amount of substance	mole	mol
Luminous Intensity	candela	cd

Derived units
In addition to the seven base units, there are derived units which are combinations of the base units

Example:
From the SI base unit, m (meter). for length, the unit for area is $m \times m = m^2$,
Since area = length × width, and the unit of length is m and the unit of with is m.

For more practical or convenient units, we use prefixes and multiplication factors (in powers of 10) to express other units.

Example: 1 kilometer = $10^3 m$, where kilo = 10^3

$\qquad$ 1 km = 10^3 m or I km = 1000m.

INDEX

A

Absolute Value 1, 2
Acute angle 296
Acute triangle 300
Addition
of Like Terms 29
of Polynomials 31
of Two Signed Numbers 2
Addition Method 184
Addition of Radicals 208
Adjacent angles 298
Age Problems 80
Algebraic Expression 27
Algebraic expressions
simplifying 34
Algebraic fractions
addition 241
division 238
multiplication 238
reduction to lowest terms 237
Angle 293
Angles
supplementary angles 297
Arc Length 230
Area
of a circle 229
of a composite figure 232
Areas 224
Averages 88
Axioms 45
Axis of symmetry 281, 284

B

Binomial 28

C

Checking Solutions of Equations 49
Circle
area 229
circumference 229
sector 230
Classification
of angles by the measures of the angles 296
Coefficient
numerical 28

of a Term 28
Coin Problems 190
Coin-Like Problems 190, 192
Common Monomial Factoring 129
Complementary angles 297
Completing the square 263
Complex Fractions
simplifying 246
Compound Inequalities 254
Condensed method 254
Congruency
of angles 294
of Line Segments 294
Congruent Figures 294
Conjunction 254
Consecutive Even Integer Problems 78
Consecutive Integer Problems 78
Consecutive Odd Integer Problems 79
Constant 27
continued inequality method 254
coordinate pair of a given point is a solution of an equation of a line. 156
coordinates of a given point satisfy an equation of a line. 156
Cross-multiplication 51
Cube Roots 205
Cubes
factoring the sum or difference of two cubes 147

D

Decimal Equations 54
Degree 28
of a polynomial 28
of a term
of a polynomial 28
Denominators
rationalization 212
Determining if a point is on a line whose equation is given 156
Discriminant
of a quadratic equation 268
Disjunction 255
Distance
between two points on a Line 175
Between two Points on a Line in a Pl 176
from a point to a line 178
Distance Formula 176
Distance-Time-Speed Problems
Travel by Water or by Air 195
Distance-Time-Speed-Problems 66
Distributive Property 28

Division
long division 36
of Polynomials 35
of Radicals 212
of Signed Numbers 6

E

Equality Method
for solving systems of linear equations 188
Equation of a line
horizontal line 167
vertical line 168
Equations
checking solutions 49
cross-multiplication 51
decimal equations 54
fractional equations 285
graphical method 189
of Parallel lines 171
of Straight Lines 161
radical equations 289
systems of linear equations 184
Equilateral triangle 301
Evaluation of Algebraic Expressions 14, 17
Exponents 201
rules 21

F

Factorability
of a quadratic trinomial 261
Factorability of a quadratic trinomial 138
Factoring
by Grouping 131
Difference between (of) Two Cubes 147
Difference Between Two Squares 131
greatest common factor (GCF) or divisor 128
Perfect Square Trinomials 304
Polynomials 129
quadratic trinomials (monic trinomials) 133
quadratic trinomials (non-monic trinomials) 138
quadratic trinomials by substitution 139
quadratic trinomials by the ac-method 140
Sum of Two Cubes 147
trinomials quadratic in form 142
Factoring Polynomials 128
Finding the Rate % 82
First Degree Equations 45
Fractional Equations
solving 285
Fractional Exponents 216

Fractions
algebraic fractions 237
Function
definition 271, 272
Function and relation
comparison 273
Functional Notation 275

G

GCD 128
GCF, GCD, HCF 128
Geometry
basic review 292
Graph
of a Horizontal Line 167
Graph of a Straight Line
Slope-Intercept method 154
x-and y-intercepts method 153
Graphical Method
of Solving Systems of Equations 189
Graphing
in the Rectangular Coordinate System 149
ordered pairs 150
Graphs
of inequalities 250
parabolas 278
vertical lines 168
Greatest Common Factor (GCF) 128
Grouping Symbols 30

H

HCF 128
Horizontal line
equation 167

I

Incomplete Factorization 129
Indices
reduction 216
Inequalities 249
intercepts 281
intercepts. 278
Intersection
of lines 189
of two sets 250
Interval Notation 250
Isosceles triangle 301

L

Like Fractions 241
Like Radicals 208
Like terms 28
line contains a given point. 156
line passes through a given point. 156
Linear Equations 45
solutions 50
systems of equations 184
Linear inequalities
solving 252
Lines
equations 161
Literal Equations 55
solutions 55
lowest terms
of fractions 237

M

Measurements, 307
Midpoint
of a Line 174
Minimum Value 284
Mixture and Mixture-Like Problems 85
Monomial 27
Motion Problems 66
Multiplication
of Polynomials 33
of Radicals 209
of Two Signed Numbers 6
Multiplication and Division
of Fractions 238

N

Naming of angles 293

O

Obtuse angle 296
Obtuse triangle 300
Operations Involving Zero 9
Order of Operations 14
Ordered Pair 271

P

Parabola
graphs 281
Parallel Lines
Slopes and Equations 171

Percent Problems 80
Perfect Square Trinomials 304
Perimeter
of a circle 229
Perimeters 224
of a Rectangle 63
Problems 63
Perpendicular Bisector of a Line 178, 181
Perpendicular Lines 298
Slopes and Equations 171
point is on a line. 156
Polynomial 27
Polynomials
multiplication 33
Powers
of Signed Numbers 9
Principal root 204
Proportion
compound proportion 124
direct proportion 99
inverse proportion 100
types of proportion 98, 101
Proportion Problems
methods for solving direct proportion problems 103
methods for solving inverse problems 119
Pythagorean Theorem 220

Q

Quadratic equations
applications 144, 269
discriminant 268
solving by Completing the Square 263
solving by factoring 143, 259
solving by the quadratic formula 266
solving by the square root method. 262
Quadratic Formula 266
Quadratic function., graphs 279
Quadratic Trinomial
factorability 261

R

Radical Equations
solving 289
Radicals 201, 204
addition 208
division 212
multiplication 209
simplifying 206
Ratio 92
comparing quantities 94

dividing a quantity into parts 96
reducing to lowest terms 92
Rational Equations 285
Rationalization of Denominators 212
Ray 293
Real Number Line 1
Real Number Operations 1
Real numbers 1
Reciprocals 7
Rectangles, 224
Rectangular Coordinate System of Axes 149
Reduction of Indices 216
Reflex angle 296
Relation 271
Removing grouping symbols 30
Right angle 296
Right triangle 300
Right Triangle, 220
Roots 204
of Signed Numbers 10
Roster method 249
Rules
of Exponents 21

S

Scientific Notation 25
Sector
of a Circle 230
Sense of an inequality 249
Set Notation 249
Set Notation, 250
Set Operations
Involving Inequalities 250
Set-builder notation 249
Sets 271
intersection 250
union 250
Sets, 249
Signed Numbers 1
signs of the slopes of lines 164
Similar Radicals 208
Simultaneous Equations 184
Sketching
Parabolas 278
Slope
of a horizontal Line 157
of a Vertical Line 158
Slope of a line
given the equation of the line 159

given two points on the line 157
horizontal line 164
sign of 164
vertical line 164
Slopes
parallel lines 171
perpendicular lines 171
Solving
by Quadratic Formula 259
by Factoring 259
by the Square Root Method 259
Linear Inequalities 252
Solving Systems of Linear Equations 184
Solving First Degree Equations 47
Solving Quadratic Equations by Factoring 143
Square Root Method 259
square root method) 144
Square Roots 205
Straight angle 296
Substitution Method
for solving systems of linear equations 187
Subtraction
of Polynomials 31
of Signed Numbers 3
Supplementary angles 297
Systems of Linear Equations
solving by graphing 189

T

Term 28
Trapezoid
area and perimeter' 226
Trapezoids 224
Triangles 300
Triangles, 224
Trinomial 28

U

Union of two sets 250
Unitary Method 108
Units
of Measuring angles 295
Units Label Method 104
Unlike Fractions 241

V

Variable 27
Variables 17
Variation

compound variation 124
direct variation 98
Variation Method 105
Vertex 284
parabola 281
Vertical angles 298
Vertical line
equation 168

W

WORD PROBLEMS 59
age problems 80
aircraft flight problems 196
average problems 85
averages 88
base finding 80, 83
coin problems 190
coin-like problems 192
consecutive even integers 78
consecutive integer problems 78
consecutive odd integer problems 79
inequalities 256
integer problems 78
mixture problems 85
motion problems 66
number problems 61
percent finding 82
perimeter problems 63
phrases and words for translation 59
rate finding 82
solving 61
translating 59
travel by water problems 195
using a single variable 191
using two variables 190
work-time problems 64
Work and Time Problems 64

X

x- and y-intercepts of a line 158
x-intercept 158
of a parabola 281

Y

y-intercept 158
given the equation of the line 159
of a parabola 281

Z

Zero
as an Exponent 9, 21

Some useful conversion factors

Units of length

1 cm = .3937 in = .0328ft = .01094 yd
1 m = 100 cm = 39.3701 in = 3.2808 ft.= 1.0936 yd
1 in. = 2.54 cm = .0833 ft = .0254 m
1 ft = 12 in. = 30.48 cm = .3048 m = .3333 yd
1 mile = 1760 yd = 5280 ft = 1.6093 km
1 yd = 3 ft = 36 in.= .9144 m = 91.44 cm
1 km = .62137 mile = 1000 m = 100,000 cm

Units of mass

1 kg = 1000 g = 2.2046 lb =35.274 oz
1 lb (avdp) = 453.592 g = 16 oz
1 metric ton = 1000 kg = 10^6 g=1.1023 ton
1 ton = 907.1847 kg = 2000 lb = .9072 metric ton
1 gm = .03527 oz = 1000 mg =.0022046 lb
1 oz = 28.3495 g = .0625 lb = 16 drams
1 long ton (British) = 2240 lb

Units of volume

1 liter = 1000 cm^3= 61.0237 $in.^3$ = .26417 gal = 1.0567 qt = .03531 ft^3 = 2.113 pt.
1 gal (U.S.) = 4 qt = 3.7854 liter = 8 pt. = 231 $in.^3$ = .13368 ft^3
1 qt = 2 pt.= .946353 liter = 946.353 cm^3 = 57.75 $in.^3$ = .25 gal = .034201 ft^3
1 cord = 128 ft^3
1 pt = .473 liter
1 ft^3 = 1728 $in.^3$
1 yd^3 = 27 ft^3 = 46656 $in.^3$

Units of area

1 ft^2 (sq. ft.) = 144 $in.^2$ (sq. in.)
1 yd^2 (sq. yd.) = 9 ft^2 (sq. ft.) = 1296 sq. in.
1 $mile^2$ (sq. mile) = 640 acres
1 m^2 = 10^4 cm^2
1 acre = 4840 yd^2

Symbols for units

cm = centimeter	gal = gallon	g = gram
m = meter	qt = quart	kg = kilogram
in. = inch	pt = pint	lb = pound
ft = foot	oz = ounce	mg = milligram
yd = yard		
km = kilometer		
mi = mile		

Some prefixes (International System)

Prefix	Power
tera	10^{12}
giga	10^9
mega	10^6
kilo	10^3
hecto	10^2
deka	10^1
deci	10^{-1}
centi	10^{-2}
milli	10^{-3}
micro	10^{-6}
nano	10^{-9}
pico	10^{-12}

Conversion Factors for Measurements

American System (British System)   Interconversion (Factors)   Metric System

Some " **bridge**s" for converting from one system to the other

1 kilometer (km)	$= 10^3$ m $= 1000$ m
1 hectometer (hm)	$= 10^2$ m $= 100$ m
1 dekameter (dam)	$= 10^1$ m $= 10$ m
1 meter (m)	$= 10^0$ m $= 1$m
1 decimeter (dm)	$= 10^{-1}$ m $= 0.1$m
1 centimeter (cm)	$= 10^{-2}$ m $= 0.01$m
1 millimeter (mm)	$= 10^{-3}$ m $= 0.001$m

Length

12 inches (in) = 1 foot (ft.)
3 feet (ft.)   = 1 yard (yd)
5280 feet   = 1 mile (mi)
1760 yards = 1 mile

1 in. = 2.54 cm
1 yd = 0.9144 m
1 mi = 1.61 km
1 km = 0.62 mi

Some " **bridge**s" for converting from one system to the other

1 kilogram (kg)	$= 10^3$ g $= 1000$ g
1 hectogram (hg)	$= 10^2$ g $= 100$ g
1 dekagram (dag)	$= 10^1$ g $= 10$ g
1 gram (g)	$= 10^0$ g $= 1$ g
1 decigram (dg)	$= 10^{-1}$ g $= 0.1$g
1 centigram (cg))	$= 10^{-2}$ g $= 0.01$g
1 milligram (mg)	$= 10^{-3}$ g $= 0.001$g

Mass

1 *l*b   = 16 oz
1 ton = 2000 *l*b
1 long ton = 2240 *l*b

1 kg = 2.2 *l*b
1 *l*b = 454 g
1 oz = 28.4 g = 16 drams
1 ton = 0.9072 metric ton

Some " **bridge**s" for converting from one system to the other

1 kiloliter (k*l*)	$= 10^3$ *l* $= 1000$ *l*
1 hectoliter (h*l*)	$= 10^2$ *l* $= 100$ *l*
1 dekaliter (da*l*)	$= 10^1$ *l* $= 10$ *l*
1 liter (*l*)	$= 10^0$ *l* $= 1$ *l*
1 deciliter (d*l*)	$= 10^{-1}$ *l* $= 0.1$*l*
1 centiliter (c*l*)	$= 10^{-2}$ *l* $= 0.01$*l*
1 milliliter (m*l*)	$= 10^{-3}$ *l* $= 0.001$ *l*

Volume

16 fluid oz (fl-oz)= 1 pint (pt)
2 pints (pt) = 1 quart (qt)
4 quarts   = 1 gallon (gal)

1 liter (*l*) = 1.057 qt
1 gal = 3.785 *l*
1 liter = 2.1 pt
1 pt = .473 *l*

Must remember the following (metric system:)

100 cm   = 1 m
1000 m = 1 km

1000 mg = 1 g
1000 g   = 1 kg

1000 m*l* = 1 *l*
1 m*l*   = 1 cc = 1cm^3
1000 cc = 1 *l*

Mnemonic device (metric system)

k – ilo – 10^3
h – ecto – 10^2
d – eka – 10^1
d – ec*i* – 10^{-1}
c – enti – 10^{-2}
m – illi– 10^{-3}

Say the following aloud:
Step 1: First go down vertically as kei-eitch-dii-dii-see-em, then Step 2
Step 2: Kilo-hecto-deka-deci-centi-milli, and then note how the powers
       decrease vertically downwards.
    Examples: 1 Kilometer = 10^3meter; 1 milligram =10^{-3}gram;
    1 centimeter =10^{-2} meter =$\frac{1}{100}$ meter --->100 centimeters = 1 meter.

Mathematical Modeling
Some Reciprocal Relationships

1. Arithmetic If A working alone can do a piece of work in time t_A; B working alone can do the same work in time t_B; C working alone can do the same work in time t_C, and if A, B, and C working together, can do the same work in time t_{ABC}, then

$$\frac{1}{t_{ABC}} = \frac{1}{t_A} + \frac{1}{t_B} + \frac{1}{t_C}$$

That is, the reciprocal of the working-together time equals the sum of the reciprocals of working-alone times (individual times).

2. Geometry: For any triangle, the reciprocal of the inradius (R) equals the sum of the reciprocals of the exradii $(r_1, r_2,$ and $r_3)$.

Thus $$\frac{1}{R} = \frac{1}{r_1} + \frac{1}{r_2} + \frac{1}{r_3}$$

3. Physics (Electricity) For electrical resistances in parallel (in an electric circuit), the reciprocal of the combined resistance, R, equals the sum of the reciprocals of the separate resistances, $r_1, r_2,$ and r_3.

Thus $$\frac{1}{R} = \frac{1}{r_1} + \frac{1}{r_2} + \frac{1}{r_3}$$

4. Physics (Optics)
For two thin lenses in contact, the reciprocal of the combined focal length, F, equals the sum of the reciprocals of the separate focal lengths, f_1 and f_2, .

Thus $$\frac{1}{F} = \frac{1}{f_1} + \frac{1}{f_2}$$

5. Physics (Optics) For spherical mirrors and thin lenses, the reciprocal of the focal length F equals the sum of the reciprocals of the object distance, d_o and the image distance d_i.

Thus $$\frac{1}{F} = \frac{1}{d_o} + \frac{1}{d_i}$$

6. Physics (Mechanics). If two bubbles of radii $r_1, r_2,$ coalesce into a double bubble, the radius, R, of the partition is given by

$$\frac{1}{R} = \frac{1}{r_1} - \frac{1}{r_2}$$

www.ingramcontent.com/pod-product-compliance
Lightning Source LLC
Chambersburg PA
CBHW082135210326
41599CB00031B/5994